PROGRESS IN

Nucleic Acid Research and Molecular Biology

Volume 37

PROGRESS IN
Nucleic Acid Research and Molecular Biology

edited by

WALDO E. COHN

Biology Division
Oak Ridge National Laboratory
Oak Ridge, Tennessee

KIVIE MOLDAVE

University of California
Santa Cruz, California

Volume 37

ACADEMIC PRESS, INC.

Harcourt Brace Jovanovich, Publishers

San Diego New York Berkeley Boston
London Sydney Tokyo Toronto

ACADEMIC PRESS, INC.
San Diego, California 92101

United Kingdom Edition published by
ACADEMIC PRESS LIMITED
24-28 Oval Road, London NW1 7DX

LIBRARY OF CONGRESS CATALOG CARD NUMBER: 63-15847

ISBN 0-12-540037-3 (alk. paper)

PRINTED IN THE UNITED STATES OF AMERICA
89 90 91 92 9 8 7 6 5 4 3 2 1

Contents

Eukaryotic DNA Polymerases α and δ: Conserved Properties and Interactions, from Yeast to Mammalian Cells

Peter M. J. Burgers

Structure and Regulation of the Multigene Family Controlling Maltose Fermentation in Budding Yeast

Marco Vanoni, Paul Sollitti, Michael Goldenthal, and Julius Marmur

Abbreviations and Symbols

All contributors to this Series are asked to use the terminology (abbreviations and symbols) recommended by the IUPAC-IUB Commission on Biochemical Nomenclature (CBN) and approved by IUPAC and IUB, and the Editors endeavor to assure conformity. These Recommendations have been published in many journals (*1, 2*) and compendia (*3*) and are available in reprint form from the Office of Biochemical Nomenclature (OBN); they are therefore considered to be generally known. Those used in nucleic acid work, originally set out in section 5 of the first Recommendations (*1*) and subsequently revised and expanded (*2, 3*), are given in condensed form in the frontmatter of Volumes 9–33 of this series. A recent expansion of the one-letter system (*5*) follows.

SINGLE-LETTER CODE RECOMMENDATIONS[a] (*5*)

Symbol	Meaning	Origin of symbol
G	G	Guanosine
A	A	Adenosine
T(U)	T(U)	(ribo)Thymidine (Uridine)
C	C	Cytidine
R	G or A	puRine
Y	T(U) or C	pYrimidine
M	A or C	aMino
K	G or T(U)	Keto
S	G or C	Strong interaction (3 H-bonds)
W[b]	A or T(U)	Weak interaction (2 H-bonds)
H	A or C or T(U)	not G; H follows G in the alphabet
B	G or T(U) or C	not A; B follows A
V	G or C or A	not T (not U); V follows U
D[c]	G or A or T(U)	not C; D follows C
N	G or A or T(U) or C	aNy nucleoside (i.e., unspecified)
Q	Q	Queuosine (nucleoside of queuine)

[a]Modified from *Proc. Natl. Acad. Sci. U.S.A.* **83**, 4 (1986).
[b]W has been used for wyosine, the nucleoside of "base Y" (wye).
[c]D has been used for dihydrouridine (hU or H_2 Urd).

Enzymes

In naming enzymes, the 1984 recommendations of the IUB Commission on Biochemical Nomenclature (*4*) are followed as far as possible. At first mention, each enzyme is described *either* by its systematic name *or* by the equation for the reaction catalyzed *or* by the recommended trivial name, followed by its EC number in parentheses. Thereafter, a trivial name may be used. Enzyme names are not to be abbreviated except when the substrate has an approved abbreviation (e.g., ATPase, but not LDH, is acceptable).

REFERENCES

1. *JBC* **241**, 527 (1966); *Bchem* **5**, 1445 (1966); *BJ* **101**, 1 (1966); *ABB* **115**, 1 (1966), **129**, 1 (1969); and elsewhere.† General.
2. *EJB* **15**, 203 (1970); *JBC* **245**, 5171 (1970); *JMB* **55**, 299 (1971); and elsewhere.†
3. "Handbook of Biochemistry" (G. Fasman, ed.), 3rd ed. Chemical Rubber Co., Cleveland, Ohio, 1970, 1975, Nucleic Acids, Vols. I and II, pp. 3–59. Nucleic acids.
4. "Enzyme Nomenclature" [Recommendations (1984) of the Nomenclature Committee of the IUB]. Academic Press, New York, 1984.
5. *EJB* **150**, 1 (1985). Nucleic Acids (One-letter system).†

Abbreviations of Journal Titles

Journals	*Abbreviations used*
Annu. Rev. Biochem.	ARB
Annu. Rev. Genet.	ARGen
Arch. Biochem. Biophys.	ABB
Biochem. Biophys. Res. Commun.	BBRC
Biochemistry	Bchem
Biochem. J.	BJ
Biochim. Biophys. Acta	BBA
Cold Spring Harbor	CSH
Cold Spring Harbor Lab	CSHLab
Cold Spring Harbor Symp. Quant. Biol.	CSHSQB
Eur. J. Biochem.	EJB
Fed. Proc.	FP
Hoppe-Seyler's Z. Physiol. Chem.	ZpChem
J. Amer. Chem. Soc.	JACS
J. Bacteriol.	J. Bact.
J. Biol. Chem.	JBC
J. Chem. Soc.	JCS
J. Mol. Biol.	JMB
J. Nat. Cancer Inst.	JNCI
Mol. Cell. Biol.	MCBiol
Mol. Cell. Biochem.	MCBchem
Mol. Gen. Genet.	MGG
Nature, New Biology	Nature NB
Nucleic Acid Research	NARes
Proc. Natl. Acad. Sci. U.S.A.	PNAS
Proc. Soc. Exp. Biol. Med.	PSEBM
Progr. Nucl. Acid. Res. Mol. Biol.	This Series

†Reprints available from the Office of Biochemical Nomenclature (W. E. Cohn, Director).

Some Articles Planned for Future Volumes

Retroviral-Mediated Gene Transfer
JEANNE MCLACHLIN, KENNETH CORNETTA, MARTIN A. EGLITIS AND W. FRENCH ANDERSON

Molecular Genetics of Na,K-ATPase
JERRY B. LINGREL, JOHN ORLOWSKI, MARCIA SHULL AND ELMER M. PRICE

Synthesis and Modification of Double-Stranded DNA by Chemical Ligation Methods and Design of Probes
ZOE SHABAROVA

The Chloroplast Ribosome: Unusual Features of the Gene for Its Protein Components
A. R. SUBRAMANIAN

Control of Prokaryotic Translational Initiation by mRNA Secondary Structure
M.-H. DE SMIT AND J. VAN DUIN

Protamine Genes and the Histone/Protamine Replacement Reaction
GORDON H. DIXON

RNA Polymerase Sigma Factor: Promoter Recognition and Control of Transcription
ALEX GOLDFARB, DAVID L. FOX AND SOHAIL MALIK

Transcriptional and Translational Regulation of Gene Expression in the General Control of Amino Acid Biosynthesis in *Saccharomyces cerevisiae*
ALAN G. HINNEBUSCH

Damage to Chromatin DNA Structure from Ionizing Radiations and the Radiation Sensitivities of Mammalian Cells
JOHN T. LETT

Enzymes of DNA Repair
STUART LINN

Structure–Function Relationships in *E. coli* Promoter DNA
MARSHALL HORWITZ AND LAWRENCE LOEB

Gene Expression in Seed Development and Germination
J. DEREK BEWLEY AND ABRAHAM MARCUS

VA RNA and Translational Control in Adenovirus-Infected Cells
MICHAEL B. MATHEWS

Mechanisms Regulating Transient Expression of Mammalian Cytokine and Proto-oncogenes
RAYMOND REEVES AND NANCY S. MAGNUSON

Eukaryotic RNA Processing
G. TOCHINI-VALENTINI AND JOHN ABELSON

Mitochondrial Aminoacyl–tRNA Synthetases
ALEXANDER TZAGOLOFF

The Structure and Expressions of the Insulin-like Growth Factor II Gene
LYDIA VILLA-KOMAROFF AND KENNETH M. ROSEN

Processing of Precursor tRNAs from *Bacillus subtilis* by Catalytic RNA
BARBARA S. VOLD AND CHRISTOPHER J. GREEN

Polynucleotide–Protein Cross-Links Induced by Ultraviolet Light and Their Use for Structural Investigation of Nucleoproteins

EDWARD I. BUDOWSKY AND GULNARA G. ABDURASHIDOVA

N. D. Zelinsky Institute of Organic Chemistry
USSR Academy of Sciences
Leninsky Prospect 47, Moscow B-334, USSR

The functions of nucleic acids are realized when they are complexed with proteins. Therefore, one of the fundamental problems of molecular biology and bioorganic chemistry is investigation of nucleoprotein structure and function. In a first approximation nucleoproteins may be classified into two groups. The first involves complexes required for structuring and transferring nucleic acids—metaphase chromosomes, "informosomes," extracellular viruses. Formation of these nucleoproteins is independent of the nucleotide sequence; the corresponding proteins recognize only the type of polynucleotide—DNA or RNA, single- or double-stranded. In this case, interaction between nucleic bases and amino-acid residues plays no significant role. The second group involves functional

Progress in Nucleic Acid Research and Molecular Biology, Vol. 37

nucleoproteins, in particular, those performing replication, transcription, and translation. The definite primary and higher structures of the corresponding fragments of macromolecules are necessary for the formation and functioning of such complexes. Self-assembly and stabilization of functional nucleoproteins are determined by the cooperativity of multiple noncovalent interactions between components of macromolecules, that is, between nucleic bases and amino-acid residues. Such interactions can occur only between spatially adjacent groups of the macromolecules with suitable electronic structure and proper mutual orientation. In other words, the interaction of the components implies physical contact. The higher structure of functional nucleoproteins is determined by a multiplicity of inter- and intramolecular contacts. Alteration of the functional state of a nucleoprotein is associated with changes in its higher structure that are accompanied and/or are caused by alterations of many intra- and intermolecular contacts. The determination of intermolecular contacts in various functional states makes possible the study of functional dynamics of the complex. On the other hand, the localization of interacting fragments (up to monomer residues) within the primary structure produces information about the functional topography of the respective macromolecules.

It should be stressed that removal or modification of some macromolecules could result in significant alterations in the number and types of their contacts due to cooperativity of interactions that stabilize the complex structure. Therefore, the results obtained with incomplete or modified complexes can differ from those with the complete intact complex, and always require additional confirmation by independent methods. Thus, reliable and strictly interpreted results can be obtained only with complete and unmodified starting complexes.

X-Ray analysis cannot be used for estimating intermolecular contacts in multimacromolecular complexes for a number of functional states. Therefore, the formation and analysis of covalent cross-links between interacting components of macromolecules within intact complexes (specific cross-links) are the only method for solving this problem.

This review deals with theoretical and experimental aspects of the generation of specific polynucleotide–protein cross-links. The use of direct ultraviolet light (UV)-induced cross-links for investigating the structure of nucleoproteins of different types is described in several hundred papers. For the sake of brevity, we consider here only those results obtained from specific UV-induced cross-links in the study of those nucleoproteins involved in protein biosynthesis.

I. Specificity of Cross-Links

Specific cross-links are defined as the covalent bonds arising between those components of macromolecules that were in direct contact (noncovalently interacting) within intact (with regard to higher structure) nucleoprotein. The formation of specific polynucleotide–protein cross-links in nucleoproteins allows us to identify the interacting proteins and polynucleotides, and to isolate and identify the fragments of macromolecules involved in these interactions.

The induced formation of cross-links is widely used for the structural investigation of macromolecular complexes, particularly nucleoproteins (*1–4, 4a*). Heavy metals, bifunctional agents of various types, as well as chemical or photochemical activation of components of macromolecules, are used for this purpose. However it is very difficult, if at all possible, to prove the specificity of the induced cross-links. Only cross-links that occur spontaneously in natural complexes (*4b, 4c*) are evidently specific. Therefore, the specificity of the induced cross-links should be inherent in the method itself. Let us consider from this viewpoint the known methods for the induced formation of intermolecular cross-links, and the conditions required for their specificity.

A. Cross-Linking by Bifunctional Agents

After reaction of one functional group of a bifunctional reagent with some residue of a macromolecule (before or after the formation of a complex), the reaction of the second functional group—that is, the formation of a cross-link—occurs on an already modified nucleoprotein. The cross-links are formed due to the presence of the sufficiently reactive residue in the same or adjacent macromolecule and the distance between cross-linked components cannot exceed the spacer length (the distance between the two functional groups of the reagent). Thus, the formation of cross-links by means of bifunctional reagents is determined by the reactivity of corresponding components of the macromolecules and by the distance between them, but not by the existence of direct contact between these components in the initial complex. Therefore, bifunctional reagents cannot be applied for detection and study of direct intra- and intermolecular contacts (interactions) in nucleoproteins.

B. Cross-Linking by Activation of Interacting Components of Macromolecules

The reactivity of macromolecule components can be enhanced by means of chemical, radiational, or photochemical activation. This may result in formation of covalent bonds (cross-links) between activated and nonactivated components.

1. The Effect of the Lifetime of Activated State on Cross-Link Specificity

Noncovalent interactions, and thus mutual orientation of interacting components of macromolecules and their complexes, is determined by chemical and, hence, the electronic structures of intact components. Any type of activation leads to alteration of electronic structure of the component. Generally, this results in distortion of noncovalent interaction, which can lead to changes in mutual orientation of the components, for example, to disappearance of the contacts existing prior to activation (native) and to occurrence of new (artifactual) contacts (Fig. 1). Therefore, upon activation of a component of the macromolecule, the cross-link may arise between partners in either native or artifactual contact depending on whether the reaction proceeds before or after the appearance of artifactual contact. The principal factor determining the specificity of cross-link formation is the lifetime of the activated state. It is evident that cross-links will arise only between components in contact before activation, if the reaction proceeds via an activated state with a much shorter lifetime than the appearance time of the new contact.

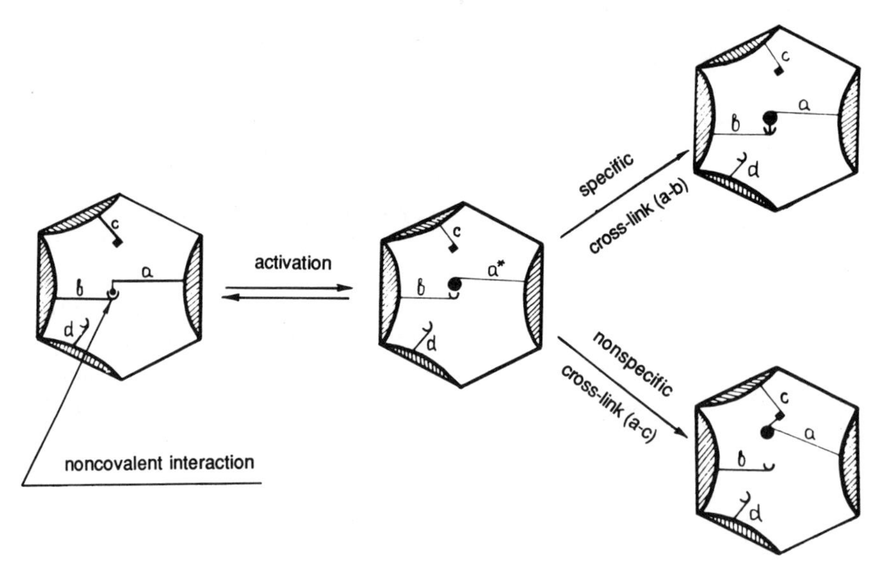

Fig. 1. The effect of the lifetime of the activated state of a macromolecular component on the specificity of cross-link formation.

It follows from this in particular that formation of specific cross-links should kinetically be single-step ones, that is, at least the initial part of the dose–response curve of their formation must be exponential.

The chemical activation of nucleic acid components can be accomplished by generation of aldehyde groups—periodate oxidation of 3′-terminal ribose residues of RNA (*5, 5a*), N-glycosyl bond cleavage in RNA or DNA (*6, 7*)—and by modification of nucleic bases, for example, by saturation of the cytosine double bond (*8–12*). Such activation may result in formation of polynucleotide–protein cross-links (Fig. 2). However, the lifetimes of such activated components are markedly longer than is needed for appearance of artifactual contacts. Therefore, known chemical activation methods cannot lead to the formation of specific cross-links. This approach is widely used for other purposes, e.g. for determination of the mutual position of macromolecules in nucleoproteins (*13, 14*). In contrast to the cross-links generated by the action of bifunctional reagents, chemical activation results in cross-links of zero length, that is, directly between protein amino-acid residues and the sugar–phosphate backbone or the nucleic base of a polynucleotide.

Among the known methods, only radiational (*15, 16*) and photochemical (see below) ones may lead to activated states of macromolecule components with lifetimes sufficiently short to permit formation of specific cross-links.

2. The Effect of Activation Period on Cross-Link Specificity

The higher structure of macromolecules and their complexes is stabilized by multiple cooperative inter- and intramolecular interactions. Both chemical and photochemical activations afford modified components of macromolecules that can distort cooperativity and induce conformational rearrangements of macromolecules and of the whole complex in general. This is coincident with a disappearance of native contacts and the appearance of new artifactual ones that were not present in the starting nucleoprotein (Fig. 3). The formation of nonspecific cross-links evidently results from sequential reactions that begin after formation of artifactual contacts—that is, after a lag period. Hence, only the specific contacts accumulate during the lag period. Therefore, the second factor determining the specificity of the cross-links formed and thus the possible proper interpretation of the results is the duration of activation.

The duration of the lag period is inversely proportional to the rate

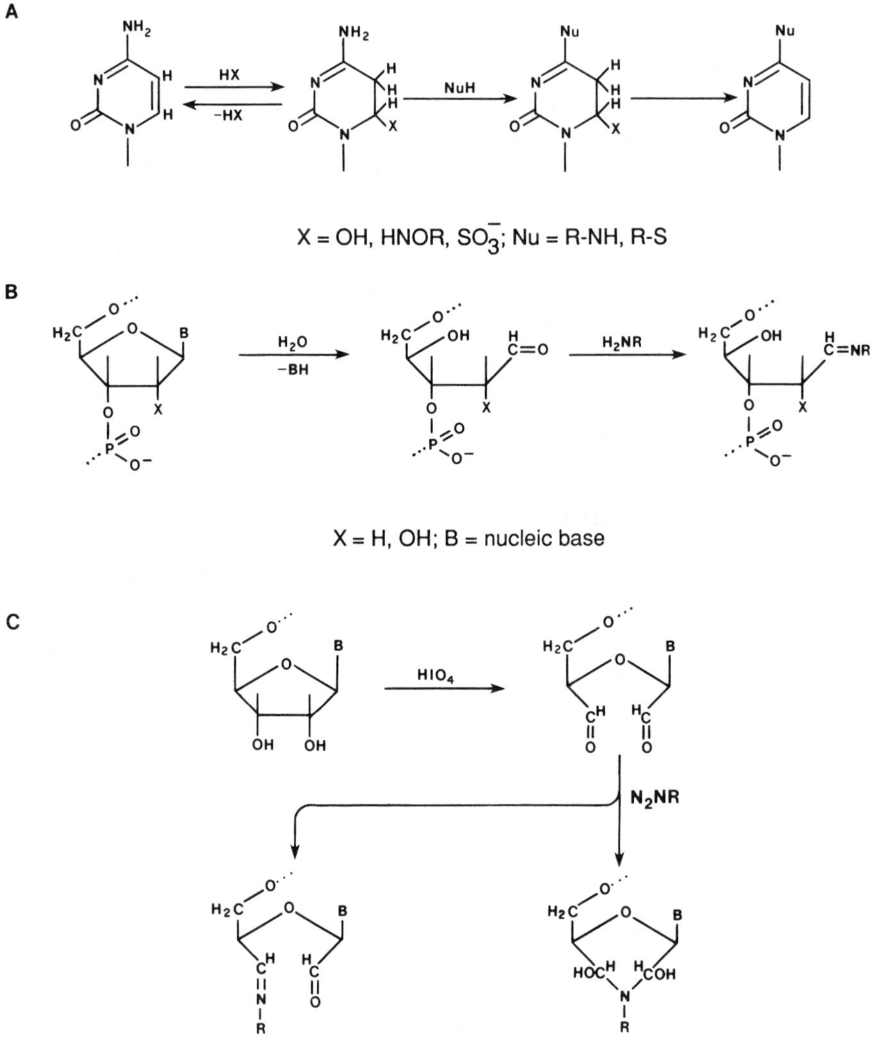

A

X = OH, HNOR, SO_3^-; Nu = R-NH, R-S

B

X = H, OH; B = nucleic base

C

FIG. 2. Polynucleotide–protein cross-links arising from chemical activation of nucleoside unit. (A) saturation of C5–C6 double bond of cytosine residues favors nucleophilic substitution of exocyclic amino groups (*8–13*). (B) cleavage of *N*-glycosyl bond in a polynucleotide yields an aldehyde group in a sugar residue, which results in a Schiff base with a free amino group of an amino-acid residue of a protein (*6, 7*). (C) periodate oxidation of a glycol group at the 3′-terminal nucleoside unit of an RNA leads to formation of a dialdehyde, which forms a Schiff base or morpholine derivative with a free amino group of an amino-acid residue of the protein (*5, 5a*).

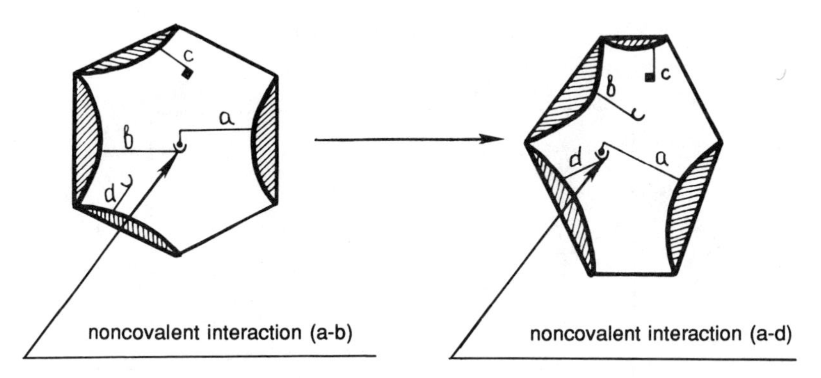

FIG. 3. The effect of change of nucleoprotein conformation induced by modification of macromolecular components on the specificity of cross-link formation.

of conformational rearrangement, which is different for various nucleoproteins and depends on the activation conditions. Therefore, the duration of the lag period should be estimated experimentally in every case. The modification of macromolecular components results, as a rule, in a loosening of the higher structure of natural nucleoproteins, that is, in a reduction in the total number of noncovalent interactions between components of macromolecules. Therefore, the modification-induced conformational rearrangements of nucleoproteins are accompanied by a reduction of the total rate of cross-link formation. Due to this effect, the deviation from the exponential of the dose–response curve for net cross-link formation indicates the end of the lag period and the possible accumulation of nonspecific cross-links (Fig. 4) (*17, 18*).

Other explanations for a deviation of the dose–response curve from the exponential are also possible. The association of the macromolecules A and B (formation of an A·B complex) can be expressed by the equation

$$A + B \underset{K_{-1}}{\overset{K_1}{\rightleftharpoons}} A \cdot B$$

Provided the dissociation rate constant is high and the modification of the free macromolecules prevents association, the concentration of the complex in the course of activation will decrease and the dose–response curve will deviate from the exponential even if modification does not lead to conformational rearrangement of the complex (*19*).

Both conditions for the formation of specific cross-links may be fulfilled in radiational and photochemical activations of nucleoprotein

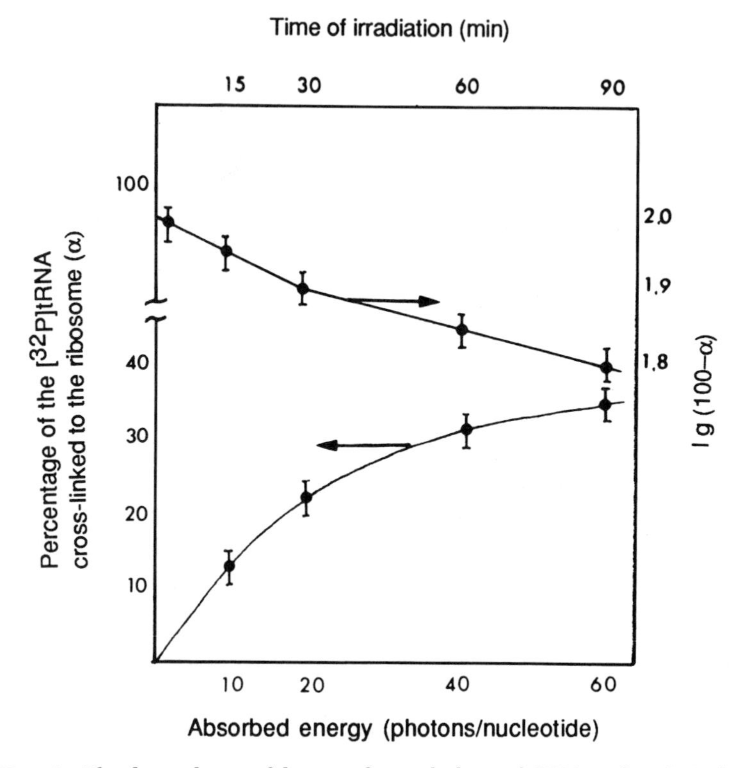

FIG. 4. The dependence of degree of cross-linking of tRNA with a 70-S ribosome on duration of irradiation (absorbed dose) of the complex Phe-tRNA^{Phe} · 70S · poly(U). Lower curve, percentage of cross-linking; upper curve, semilogarithmic anamorphose of curve 1 (*18*).

components. The activation of the macromolecules by ionizing radiation in dilute aqueous solutions generally occurs indirectly by water radiolysis products (*20–23*). But in this case, modification of both polynucleotide and protein components takes place. Photochemical activation of nucleoprotein components can be realized with much higher specificity. Due to the difference in the UV-absorption spectra of polynucleotides and proteins, it is possible to induce selective activation of nucleic bases in nucleoproteins by choosing suitable UV wavelength. The selectivity of photoexcitation can be varied by changing the wavelength of irradiation, and by addition of sensitizers and quenchers of photoexcited states. Taking into account these facts and the availability of UV-light sources; photoactivation has significant advantages over ionizing radiation for the formation of specific polynucleotide–protein cross-links within nucleoproteins.

However, the application of photochemical activation, as with any other method, can be successful only if all limitations and feasibilities of it are taken into account. In order to formulate the feasibilities and limitations of the method of photo-induced formation of polynucleotide–protein cross-links in nucleoproteins, we consider below theoretical and methodological aspects of this approach.

II. UV-Induced Formation of Polynucleotide–Protein Cross-Links

The formation of covalent bonds between DNA and proteins upon UV-irradiation of mammalian cells was discovered in 1962 (*24, 25*). Many papers dealing with the formation of polynucleotide–protein cross-links upon UV-irradiation of cells, extracellular viruses, and various nucleoproteins as well as model mixtures of nucleic-acid components with amino acids were published in the following years. The experimental data given in these papers are summarized in several reviews (*2, 26–28*).

It was proposed in 1974 that UV-irradiation of nucleoproteins can result in formation of specific polynucleotide–protein cross-links (*29*). However, the unique possibilities for investigation of nucleoproteins offered by this approach have been used only to a small extent. This seems accounted for by the lack of a theoretical basis for the method and distinctly formulated requirements necessary for formation of specific cross-links.

A. Theoretical Aspects

1. Light Absorption by Components of Nucleoproteins

Each nucleoside unit of a polynucleotide contains a heterocyclic base. The maximum of the first absorption band of the nucleic bases (neutral forms) is near 260 nm, and the molar extinction value is in the range of $8–15 \times 10^3$ (*30, 30a*). The maximum of protein absorption is at 280 nm; it is determined by the aromatic amino acids (*31*), the content of which in proteins does not exceed a few percent (*32*). Therefore, upon UV-irradiation of nucleoproteins in the range of 250–270 nm, the light is absorbed mainly by the nucleic bases.

2. Photophysics of Nucleic Bases

The absorption of a photon at the first band induces transfer of the base from the ground state (S_0) to the excited singlet state (S_1) (Fig. 5).

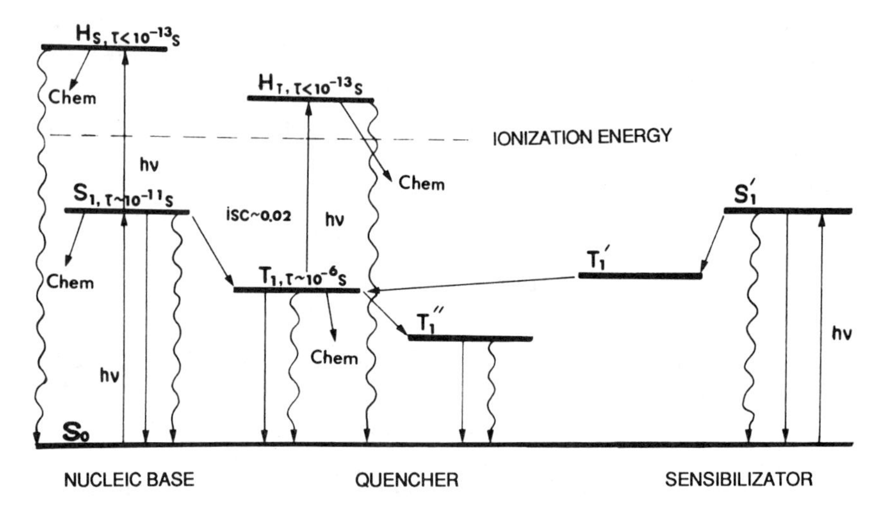

FIG. 5. Simplified Jablonsky diagram, reflecting excited states and their transitions for nucleic bases, quenchers, and sensitizers.

The lifetime of this state for nucleic bases in liquid aqueous solution does not exceed 10^{-11} sec (*33, 34*). During this period, the excited molecule undergoes either chemical (photochemical) transformation or a transition (radiative or nonradiative) into the S_0 state or into the first-triplet state (T_1). During its lifetime [$\approx 10^{-6}$ sec (*34*)] in the triplet state, the base undergoes either chemical transformation or transfers (radiatively or nonradiatively) into the ground state. The efficiency of $S_1 \rightarrow T_1$ transitions for nucleic bases does not exceed several percent (*33, 34*). The T_1 state may be populated not only due to the $S_1 \rightarrow T_1$ transition, but also by energy transfer using a suitable donor [sensitizer (*35*)]. On the other hand, the population in T_1 from S_1 can be decreased by a suitable quencher (*35*).

Under the normal (low) intensity of light (less than 10^{18} photons cm^{-2} sec^{-1}, 10^8 W m^{-2} at 260 nm) only the levels S_1 and T_1 of nucleic bases are populated. Further increase of the intensity, possibly with lasers, strongly enhances the probability of absorption of a second photon by a molecule already in the T_1 state ($\geq 10^9$ W m^{-2}) or in the S_1 state ($\geq 10^{12}$ W m^{-2}) (*36–38*). This results in a population at higher excited states (H_T and H_S, respectively), the lifetimes of which do not exceed 10^{-13} sec (*39*). During this period, the bases either undergo chemical transformations or return to the ground state, apparently bypassing lower excited states.

3. Photochemistry of Nucleic Acids and Their Components

The electronic structures (bond orders, charge distribution, etc.) of nucleic bases in the states S_0, S_1, T_1, H_T, and H_S are markedly different (*30*). Therefore, the bases in these states have different chemical properties; they are varied in transformation mechanisms and in the structures of the final reaction products.

Thus, photohydration and formation of noncyclobutane dimers of pyrimidines proceed via S_1 (*40, 41*), the formation of cyclobutane dimers of pyrimidines in aqueous solution proceeds via T_1 (*42*), while the transition of bases into higher excited states leads to products that are not formed from lower excited states (*43–45*). In other words, each of the nucleic bases, under UV-irradiation of normal intensity, can transfer into two, and under highly intensive irradiation into four, excited states, which are quite different in their chemical properties.

Because of the additional energy that nucleic bases acquire upon absorption of a photon, the rates of chemical transformations of excited states are many orders of magnitude higher than those of the ground state. Thus, although the components of nucleic acids in the ground state are quite stable, the rate constants of their transformations via excited states may attain 10^7, 10^{11}, and 10^{13} sec^{-1} for T_1, S_1, and H_T states, respectively, as estimated from the lifetimes of these states and the quantum yields of the corresponding reactions (*39, 41–44*).

Since the lifetimes of the excited states of nucleic bases do not exceed 10^{-6} sec (see above), the reactions proceeding directly via excited states can be considered, with sufficient fit, as single-step ones. But the final products of photochemical transformations can arise not only directly from the excited states but also via relatively long-lived intermediates arising from kinetically non-single-step reactions. For example, the substitution of the exocyclic amino group of a cytosine residue may proceed via the intermediate formation of the corresponding photohydrate (9) (Fig. 2). The energy received by a nucleic base after absorption of two photons at $\leq$270 nm exceeds the ionization energy of the bases in aqueous solutions (*44*). Therefore, the products of biphotonic photolysis of nucleic bases may arise, in particular, via corresponding radicals or ion-radicals (*44, 45*), with markedly longer lifetimes than in the case of bases in higher excited states (*23*).

4. Quantum Yields of Cross-Link Formation

UV-induced polynucleotide–protein cross-links are formed as a result of bimolecular photoreactions between an excited base of a polynucleotide and an amino-acid residue of a protein. The quantum yield of each cross-link formation is affected by a series of factors: (i) the reactivity of the base in the corresponding excited state; (ii) the reactivity of the amino-acid residue adjacent to this base (iii) the excitation efficiency of this base; (iv) the lifetime of the corresponding excited state; (v) the mutual positions and relative mobilities of the components to be cross-linked. It follows from this that the quantum yields of cross-links are independent of the energy of the corresponding noncovalent interaction and of their roles in the stabilization of structure and the function of the nucleoprotein. However, the last three factors depend on the microsurroundings of the base involved in the cross-link formation. Therefore, the quantum yield of cross-links is rather sensitive to alterations in nucleoprotein conformation induced by changing the conditions as well as by the composition and functional state of the complex.

a. Efficiency of excitation. The efficiency of photoexcitation of a nucleic base in a nucleoprotein depends on its light absorption cross-section, shielding by other macromolecule components, and on the possibility of indirect excitation due to energy transfer to or from other components or the nucleoprotein.

The interplanar (stacking) interactions of bases in a polynucleotide chain reduce the light absorption cross-section (hypochromic effect) (*46*). The amount of this effect depends on both the nature of the adjacent bases and the conformation of the polynucleotide (*46*). Thus, the efficiency of excitation depends not only on the type of base but on its position in the polynucleotide chain and the local conformation of the relevant polynucleotide fragment.

The degree of shielding of a nucleic base in nucleoprotein is affected not only by the nature and number of other chromophores absorbing light at the corresponding wavelength, but also by the spatial distribution of chromophores in the complex, i.e., by the conformation of the nucleoprotein. For UV-radiation, such chromophores in nucleoproteins, apart from nucleic bases, are aromatic groups of amino-acid residues. Thus, the degree of shielding of each nucleic base depends not only on the composition but also on the conformational state of nucleoprotein.

The excitation of nucleic bases in nucleoproteins may also occur by means of electronic excitation transfer from other chromophores within the complex, excited by photon absorption. The efficiency of excitation transfer depends on the energy levels of donor and acceptor as well as on the distance between them and their mutual orientation. The latter two factors are determined by nucleoprotein conformation.

Thus, all factors affecting the efficiency of excitation of bases in nucleoproteins depend on the conformation of each macromolecule and of the complex as a whole. Because of this, the conformational changes in nucleoproteins should result in alterations in excitation efficiency of nucleic bases, and thus in changes of the quantum yield of cross-links.

b. Lifetime of the excited state. Due to the mutual mobility of the interacting components of macromolecules, the quantum yield of the bimolecular reaction (cross-link formation) is proportional to the lifetime of the corresponding excited state and to the reaction rate constant. The enhancement of excessive energy of the excited state leads to an increase of the rate constants of chemical reactions, but to a decrease of lifetime. Therefore, the transfer into an excited state with higher energy is not always accompanied by an increase of quantum yield of starting compound transformation.

The lifetime of the excited state is inversely proportional to the sum of the rates of all relaxation processes (Fig. 5), some of which are chemical reactions. Therefore, the appearance of a new efficient relaxation path leads to a reduction of quantum yield of phototransformations via the corresponding excited state. One relaxation pathway is the transfer of electronic energy by migration of excitation along the polynucleotide chain (*47*) and Förster energy transfer (*48*) on other acceptors in the nucleoprotein positioned several nanometers away from the excited residue. As mentioned above, the efficiency of transfer of electronic energy in a nucleoprotein depends on its conformation. Thus, the changes in conformation of macromolecules and in the complex as a whole may result in alterations of the quantum yield of cross-link formation due to changes in the lifetime of the excited state.

c. Mutual orientation and relative mobility of cross-linked components. The dependence of the quantum yield of UV-induced polynucleotide–protein cross-links in nucleoproteins on the mutual orientation and the relative mobilities of the contacting components of macromolecules has not been studied. However, these problems have been studied in detail for a similar reaction: the UV-induced cross-linking (dimerization) of pyrimidines.

To form a pyrimidine dimer, a certain mutual orientation of the two bases is necessary (Fig. 6). After photo-induced cleavage of

cis-syn A (meso) | trans-syn B(d,l) | cis-anti C (d,l) | trans-anti D (meso)

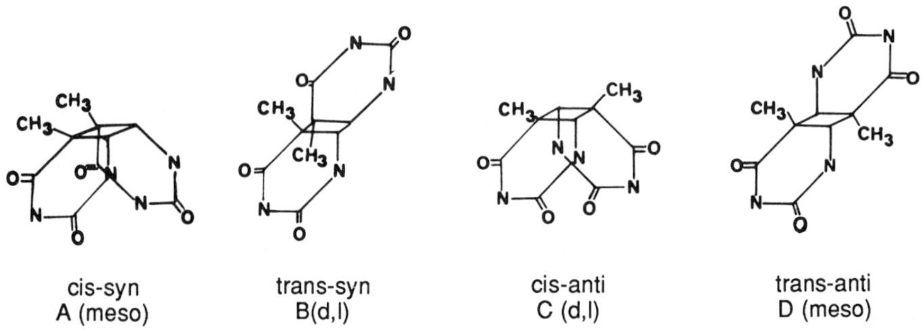

FIG. 6. Cyclobutane dimers of thymine.

dimers frozen in a glass, the same mutual orientation of bases as in the initial dimer is retained (the base planes are parallel, the distance between them is 0.28 nm, and double bonds are parallel and drawn together). Upon subsequent irradiation, the quantum yield of redimerization is close to unity (*49*). In crystalline thymine monohydrate, the mutual orientation of bases is the same, but the distance between planes is 0.36 nm. The quantum yield of dimerization is also lower (0.3) due to this effect (*50*). In the case of homoassociates (aggregates) arising in concentrated liquid aqueous solutions of thymine and after freezing aqueous solutions, the quantum yield of dimerization is even lower (0.04) (*41*).

This seems to be caused by less suitable (for dimerization) and/or less rigid mutual orientation of bases in aggregates as compared with glasses or crystals. It is worth noting that after dehydration of crystalline thymine monohydrate, the mutual orientation of bases becomes random, and in spite of retention of volume concentration of bases, the quantum yield of dimerization decreases several orders of magnitude (*50*).

In dilute ($\leq 10^{-4}$ *M*) liquid aqueous solutions of thymine, not only mutual orientation but mutual disposition of bases is not fixed, and the formation of dimers is a true bimolecular photoreaction. In this case, the quantum yield of dimerization is determined by the probability of appearance of a productive mutual orientation of excited and nonexcited bases during the lifetime of the excited state. Therefore, the quantum yield of dimerization in aqueous solutions is proportional to the thymine concentration with regard to its aggregation (*41*).

In liquid aqueous solutions of poly(dT) and of denatured DNA, the mobility of bases is limited due to the presence of internucleotide bonds and weak stacking interactions, which are optimal for ensuring parallel planes of the bases. The quantum yield of thymine dimerization in polynucleotides is independent of its concentration, as the reaction proceeds entirely inside the polynucleotide. At each moment, only a part of the adjacent thymines in a polynucleotide has a mutual orientation suitable for dimer formation. Therefore, the quantum yield of dimerization in polynucleotides is lower than in crystals but higher than in monomer solution at the same thymine concentration. Raising the temperature of a solution weakens stacking interactions (*51*), which decreases the proportion of properly oriented adjacent bases and reduces the quantum yield of dimerization (Fig. 7).

In native DNA, the mobility of the bases is even more limited than in denatured DNA due to cooperative stabilization of secondary structure by complementary pairing of bases. But the energetically optimal mutual orientation of the bases in native DNA differs from what is optimal for dimer formation. Thus, although the base planes are parallel and the mean distances between them are even less than

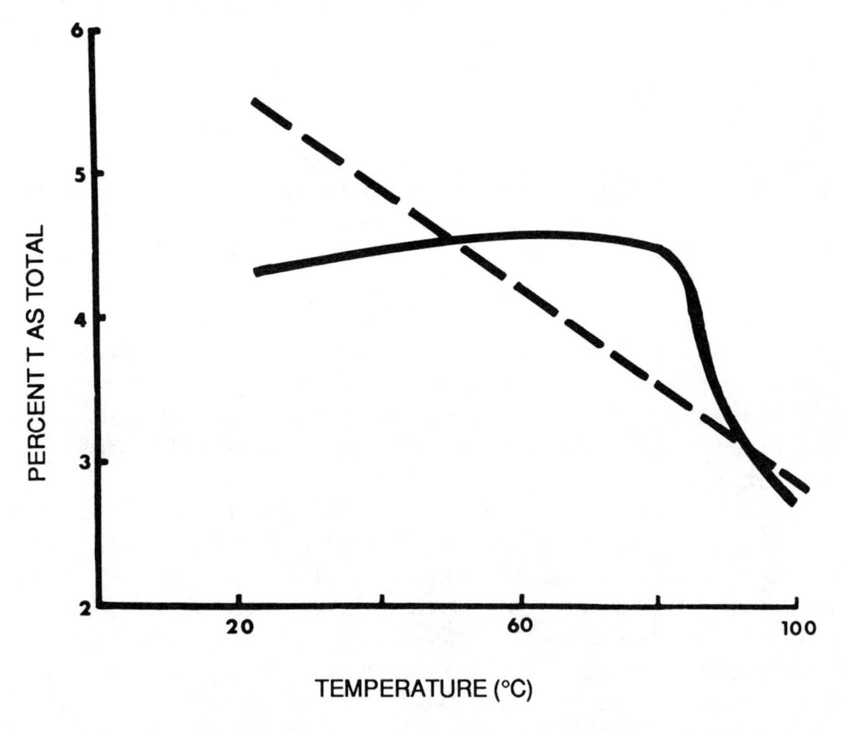

FIG. 7. The effect of temperature on the quantum yield of dimerization in single- and double-stranded DNA (broken and solid lines, respectively) (*52*).

in crystalline thymine monohydrate (0.34 nm), the angle between the C5–C6 double bond of ajacent pyrimidines for the B-form of DNA is 36°. Therefore, dimerization proceeds only because of the mobility of the bases in DNA, and only those adjacent pyrimidines that can get into suitable mutual orientation during the lifetime of the excited state undergo dimerization. Because of this, the quantum yield of thymine dimerization in DNA is an order of magnitude lower than it is in crystalline thymine monohydrate (0.02 and 0.3, respectively), and it slowly increases with increase in temperature—that is, with enhanced mobility of bases (Fig. 7).

The equilibrium distribution of the mutual orientation of the bases in liquid aqueous solution of native DNA is restored during irradiation, and the reaction proceeds up to dimerization of 7–9% thymine in DNA. But in a frozen (140° K) solution of DNA, the reaction reaches a plateau at 1% thymine dimerization. However, if an irradiated sample is thawed and refrozen, reirradiation leads to further dimerization, suggesting that freezing of DNA locks into place a small percentage of suitably oriented thymines; thawing and refreezing then introduce new populations of potentially dimerizable pairs. Therefore, less than 10% of adjacent thymines are in a mutual orientation suitable for formation of dimers in the initial equilibrium distribution.

The quantum yields of thymine dimerization in a crystal and redimerization in a glass are substantially higher than the quantum yield of the triplet state of thymine (0.014–0.02) (*34, 53*). Therefore, one can propose that in these cases (i.e., at rigid fixation of a suitable mutual orientation of bases), the dimerization proceeds via the short-lived excited state S_1; but in the absence of rigid fixation, the suitable orientation is one among many others that exist in dynamic equilibrium. It is evident that the longer the lifetime of the excited state, the higher the probability of the appearance of a suitable mutual orientation of bases during this time in a dynamic system, and occurrence of dimerization.

The lifetime of the state T_1 is almost four orders of magnitude longer than of state S_1. Therefore, it is not surprising that in dynamic systems (liquid solutions of monomers and polynucleotides), dimerization of pyrimidines proceeds mainly via state T_1 (*41, 54*). The transfer into a higher excited state sharply increases the efficiency of some phototransformations of thymine (*42–44, 55*). The lack of the effect of this transfer on efficiency of dimerization of thymine in oligo- and polynucleotides (*55*) seems to be caused by a shorter lifetime of the higher excited state.

5. Mutual Orientation of Components and Structure of Cross-Links

There are several phototransformation pathways for every nucleic base through even one certain excited state. The ratio of final products of bimolecular phototransformations depends not only on the concentration of starting compounds but on their mutual orientation at the moment of reaction.

The rigidly fixed mutual orientation of the cross-linked residues results only in those types of cross-links possible at this orientation. Thus, the redimerization of thymine in a glass affords the same dimer (*49*), and crystalline thymine monohydrate yields only the *cis–syn* cyclobutane dimer of thymine (*50*) (Fig. 6). Any mutual orientation of bases is possible in liquid aqueous solution of thymine, and all four types of productive (suitable for formation of cyclobutane dimers) mutual orientation of adjacent bases are possible (Fig. 6). Therefore, all four types of cyclobutane dimers of thymine are formed in comparable amounts upon irradiation in liquid aqueous solutions (*41*). The same set of dimers is obtained in the case of thymine aggregates formed in aqueous solution at high concentration or after freezing the solution (*41*). These facts demonstrate that all four productive orientations for the formation of cyclobutane dimers are present in the same ratio both in aggregates and in solution. Anti-orientation of adjacent thymines in polynucleotides (Fig. 6) is prohibited due to limitations caused by the presence of internucleotide linkages. Therefore, the irradiation of polynucleotides affords only *syn–cis* and *syn–trans* cyclobutane dimers of thymine (*52*).

The other type of pyrimidine dimer formed upon UV-irradiation of DNA is the noncyclobutane dimer Thy(6-4)Pyo[1] (Fig. 8), formed from adjacent cytosine and thymine bases (*56–59*). This reaction proceeds probably via S_1 (*58*) and it requires the close proximity of the C4–NH_2 bond of cytosine and the C5–C6 double bond of thymine. The probability of such mutual orientation of bases during the lifetime period of the state S_1 is higher in the sequence C-T than T-C (Fig. 8) (*59*). Therefore, Thy(6-4)Pyo is the main product from C-T (*57*). Most probably the difference in the mutual disposition of adjacent adenine nuclei is also responsible for UV-induced formation of adenine dimers only in oligo- or polydeoxyribonucleotides (*59a*).

The data given above demonstrate that the ratio of quantum yields

[1]Nomenclature proposed by W. E. Cohn, N. J. Leonard and S. Y. Wang, *Photochem. Photobiol.* **19**, 89 (1974). Pyo = 2-pyrimidinone.

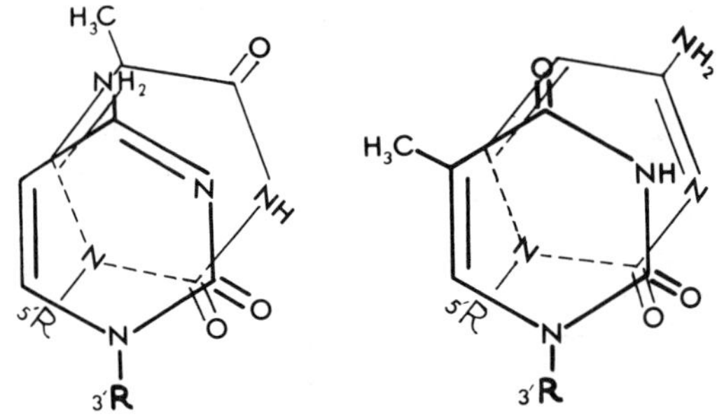

T–C C–T

FIG. 8. The scheme of formation of noncyclobutane dimers of T-C and C-T in DNA upon UV irradiation (59).

of various products of biomolecular photoreactions, in particular, the cross-links of different types, depends on the positions and relative mobilities of the respective residues. Because of this, the products of bimolecular photoreactions obtained from model mixtures of monomeric nucleic acid components and amino acids (*3, 60–62*) may not be formed at all between the same components within nucleoprotein due to the fixed mutual orientation and limited relative mobility of the latter. On the same basis, the photoreactions yielding efficient formation of plynucleotide–protein cross-links in nucleoproteins may not be observed upon UV-irradiation of mixtures of the corresponding monomers in solution.

6. The Yields of Cross-Links and Interpretation of the Results

It is necessary for proper interpretation of the results that radiation should induce only specific cross-links, which indicate the noncovalent interactions of the components of macromolecules in the initial nucleoprotein. The yield of specific cross-links formed under UV-irradiation of normal intensity is proportional to the absorbed dose and to the quantum yields of corresponding reactions only during the lag period of conformational rearrangement of the nucleoprotein (Section I,B,2). For this reason, the maximum yield of specific cross-links is limited; it usually does not exceed several percent.

The duration of the lag period is more dependent on the duration of irradiation than on the degree of modification of the macromolecular components, which is proportional to the absorbed dose. Therefore, an increase of radiation intensity allows one to increase the dose absorbed during the lag period; i.e., to enhance the yield of specific cross-links. However, the increase of total yield of photo-induced cross-links is accompanied by enhancement of probability of multiple cross-link formation; i.e., the covalent bonding of one macromolecule with others (*63, 64*). This makes analysis so complex that the increase of total cross-links above several percent becomes purposeless.

The advantages should be especially noted for the formation of specific cross-links by means of laser-pluse UV-irradiation with intensity above 10^{22} photons cm^{-2} sec^{-1}. At such an intensity, the cross-links can be formed via higher excited states between those components that are not cross-linked at lower excited states. The possibilities of the photochemical method of cross-link induction are markedly extended on the basis of this effect (*64*). The quantum efficiency of photoreactions of nucleic acid components via higher excited states is higher than that via lower ones (*42–44*) (see, however, refs. *38*, *65*, and *66*), which permits a decrease in dose of an order of magnitude to obtain the required quantity of cross-links. A dose sufficient to yield an amount of cross-links sufficient for analysis may be obtained in one pulse [10^{-8} sec (*64*)]. It is evident that a conformational rearrangement of a nucleoprotein cannot take place during such a short period, so that only specific cross-links can arise. Besides, the use of short pulses allows us to study the labile nucleoproteins and the intermediate functional states with short lifetimes.

Since the allowable yield of UV-induced polynucleotide–protein cross-links in nucleoprotein does not exceed several percent, the validity of this approach of elucidation of the structure of these complexes should be discussed. The main doubts that the results obtained reflect the structure of native complexes are based on the possible contribution of cross-links of aberrant complexes present in the starting population. Such complexes may differ from the native ones in composition and conformation and, hence, in the set of polynucleotide–protein interactions. The aberrant complexes are the nightmare of immunoelectron microscopists, whose authority is raised by the progress in the structural investigation of complicated complexes, including ribosomes. The choice (from a great number) of those images that correlate with a predicted structure may, and often does, lead to a wrong interpretation, even if aberrant complexes represent only several percent of the total population.

Unlike electron microscopy, chemical modification, including formation of cross-links, affords an unprejudiced mean result for the entire population of molecules. Therefore, UV-induced polynucleotide–protein cross-links, provided they are obtained under the aforementioned conditions, reflect the structure of the predominant part of a population, even if the yield of cross-links does not exceed a few percent. The relative yield of each cross-link is proportional to the quantum yield of its formation in a certain complex and its content in the entire population. The quantum yields of tRNA–protein cross-links can be as high as 10^{-2} to 10^{-3} (*64*); i.e., they are comparable with the quantum yield of thymine dimerization in DNA—the most efficient cross-link between adjacent bases in polynucleotides (*52, 57*).

Today, even such multicomponent nucleoproteins as ribosomal complexes can be obtained with a homogeneity (by composition and functional tests) of up to 90% (*67, 68*). But even at 80% homogeneity (*69*), the cross-links in aberrrant complexes may contribute noticeably and lead to distorted results only if the major part of such complexes has a similar structure, and if the yield of cross-links in these complexes is much higher than for the most efficient cross-links in the native complex. Both the structural homogeneity of the aberrant part of the complex population and the exclusively high yields of cross-links are of low probability. Therefore, UV-induced polynucleotide–protein cross-links, provided they are obtained in suitable conditions and properly interpreted, afford reliable information about direct polynucleotide–protein interactions in native functional nucleoproteins.

One should keep in mind in the course of interpretation that the quantum yields of photochemical reactions, including cross-link formation, may differ by several orders of magnitude. Besides, not all cross-links are sufficiently stable under irradiation conditions and in the course of the following analysis (see, e.g., *70*). Because of this, and due to the limited sensitivity of detection methods, not all cross-links can be determined experimentally. Therefore, if the cross-links are not detected, this alone does not disprove the existence of direct interaction between macromolecules. But the reliable determination of cross-link formation is the proof of such interactions.

As mentioned above (Section II,A,4), the quantum yields of cross-links are not correlated with the importance of corresponding interactions for functioning and stability of nucleoprotein, but they are dependent on some other, in particular, conformational, factors. Therefore, even local conformational changes induced by alterations of conditions, as well as of the composition and functional state of nucleoprotein, may be accompanied by changes of quantum yields of separate cross-links. In other words, the ratio of yields of specific cross-links reflects the conformation of nucleoprotein, while the change of this ratio upon changing of functional state reflects functional dynamics of nucleoprotein structure.

B. Practical Aspects

As mentioned above (Section I,B,2), the absorption of an excessive dose of UV-irradiation by nucleoprotein can lead to a significant amount of nonspecific cross-links, whereas the absorption of an insufficient dose, although it affords practically only specific cross-links, may yield less than what is necessary for reliable results in the subsequent analysis. Therefore, in order to obtain reasonably interpreted and reliable data, the selection of a suitable dose is of crucial importance. The absorbed dose and thus the duration of irradiation depend on the geometry of the irradiated sample, the optical density of the irradiated layer, the intensity and spectral characteristics of radiation, etc. The effects of these factors on the dose absorbed by nucleoprotein molecules as well as the calculation of dose and duration of irradiation in various experiments are considered below.

1. Geometry of the Sample and Calculation of the Absorbed Dose

The geometry of the irradiated sample may be either simple—with a uniform thickness of the layer under irradiation along the whole volume (for example, in rectangular quartz cells or irradiation from above the solution in a Petri dish)—or intricate—with a nonuniform thickness of the layer (irradiation of the drop placed on hydrophobic support (e.g., *71*) or in a cylindrical vessel with the light beam perpendicular to the cylinder axis).

Provided one can neglect the mutual internal shielding of the molecules (self-absorption)—the optical density does not exceed 0.1 at the wavelength of the incident light even in the thickest part of the solution—then all molecules in the solution absorb an equal dose independently from the geometry of the irradiated sample and from their location in the solution. In this case, the duration of irradiation (t, in seconds) needed for absorption of a given dose (D, photons per molecule) can be calculated according to the equation

$$t = \frac{D}{I_0 \sigma} \tag{1}$$

where I_0 is the intensity of the incident light, in photons per square centimeter per second, σ is the mean cross-section of absorption at the wavelength of incident light for the molecule, in cm^2 per molecule.

The absorbed dose can be expressed as the number of photons per nucleoprotein molecule (29). But since polynucleotide–protein cross-links are formed as a result of excitation of nucleic bases, it is more convenient to calculate the dose as the number of photons per nucleotide residue (*64, 69*). The mean value of cross-section of absorption (σ) for a polynucleotide is about 4×10^{-18} cm^2 per nucleotide (*30*).

If the optical density of the irradiated layer is higher than 0.1, the absorbed dose for each molecule depends on its position in the solution with respect ot the source of irradiation. Therefore, it is necessary to stir the solution during irradiation so that all solute molecules absorb an equal dose.

In this case, the duration of irradiation for absorption of a given dose can be calculated from the following equation

$$t = \frac{2.3(\mathrm{AD})}{I_0\sigma(1 - 10^{-A})} \qquad (2)$$

where A is the optical density of the layer under irradiation.

It is noteworthy that this equation is applicable only for samples with simple geometry, as the optical density of the irradiated layer is nonuniform in the case of complex geometry.

2. Intensity of the Light Flux

It follows from Eq. (1) and (2) that it is necessary to determine the intensity of the incident light for the calculation of the duration of irradiation. One can use for this purpose physical photometers (*72*), for sources of irradiation of normal intensity; however, their accuracy and reproducibility is not always satisfactory. Chemical actinometry is much more precise and reliable (*72, 73*). Uridine is a rather convenient compound for chemical actinometry in the range of 250–290 nm (*74*). In this case, it is not necessary to darken the room during actinometry.

The irradiation of a dilute ($\leq 10^{-3}$ *M*) solution of uridine affords practically a single photoproduct, the photohydrate of uridine, which does not absorb light essentially at 260 nm (*39*) and is rather stable ($\tau \approx 150$ hours at pH 7.0 and 20°C) (*39*). The change of the optical density of a neutral uridine solution at 260 nm during irradiation allows a determination of the intensity of incident light (in photons per square centimeter per second) from the equation

$$I_0 = \frac{6.02 \times 10^{20}}{\varepsilon\psi t} \log \frac{10^{A_0} - 1}{10^{A_t} - 1} \qquad (3)$$

where A_0 and A_t are optical densities of the solution of 260 nm before and after t seconds of irradiation, ϵ is the molar extinction coefficient of uridine at 254 nm [8900 $cm^2\ mol^{-1}$ (*30*)], and ψ is the quantum yield of photohydration of uridine [2.16×10^{-2} (*75*)]. This formula is applicable at any density of the irradiated layer and without stirring the actinometric solution during irradiation.

It should be stressed that uridine actinometry cannot be applied for intensities of light above 10^{20} photons $cm^{-2}\ sec^{-1}$, which is produced by the focused beam of pulse-UV lasers. At such intensities the biphotonic transformations of uridine make significant contribution to changes of the optical density of the solution (unpublished data).

3. Sources of UV Radiation

With nonmonochromatic sources of radiation, the calculation of the absorbed dose is very difficult, if not impossible, as the distribution of the light intensities along the wavelengths and the absorption spectra of the irradiated substances must be taken into account. On the other hand, the enhancement of intensity of radiation is associated with a reduction of the duration of accumulation of the required dose, which favors the increase of the yield of the specific cross-links (see *76, 77*). Therefore, a rather intense source of monochromatic UV-irradiation should be used to induce polynucleotide–protein cross-links. Low-pressure mercury lamps are widely used as such a source. Almost all radiation from these lamps in the range of 200–300 nm (which comprises about 85% of the total energy) is concentrated at 254 nm (*78*). In the case of nucleoproteins, this irradiation is mainly absorbed by nucleic bases (Section II,A,1). When irradiation with other wavelenghts is required (in the case of cross-links involving 4-thiouridine (*79, 80*) or in photosensitized cross-link formation (*81*)), one can use high-pressure lamps with monochromators or filters (*78*) as well as various lasers. It should be noted that, at incident light intensities $\geq 10^{14}\ W\ m^{-1}$, there is photolysis of water (*82*). This results in the same chemical transformations of the solute as in the case of radiolysis (see, e.g., *83*).

4. Determination of the Total Degree of Cross-Links

The main step in determination of the degree of polynucleotide–protein cross-links is the complete dissociation of the non-cross-linked nucleoproteins followed by the quantitative determination of one of the components, either in free state or within the cross-linked nucleoprotein.

a. According to the amount of free or cross-linked polynucleotide. The separation of free polynucleotide from cross-linked nucleoprotein can be performed by various methods. The first method applied was extraction of DNA from homogenates of irradiated cells (*24, 25, 84, 85*). In this case, DNA containing a sufficient amount of cross-linked protein is transferred to an organic phase or interphase that is the measure of the cross-link formation.

A simpler and more convenient method for separation of the cross-linked polynucleotide is sorption on nitrocellulose filters (*86, 87*). In this method, the portion of polynucleotide sorbed on the filter is the measure of cross-link formation. Both extractivity and sorption on nitrocellulose filters depend on the ratio of the masses of protein and polynucleotide in the cross-linked nucleoprotein, as well as on the experimental conditions. Thus, both methods allow one to control the increase of cross-linking only in a limited interval of dose, but not the real extent of cross-linking. Due to the different buoyant densities of polynucleotides and proteins (*88*), it is possible to separate free polynucleotides from cross-linked nucleoproteins by equilibrium centrifugation in a salt density gradient (*89, 90*). This method has the same limitations as the two preceding ones.

Free polynucleotides can be separated from the cross-linked nucleoproteins (independently from the relative amounts of the components) by sorption on an affinity support containing, e.g. immobilized antibodies against the corresponding proteins (*91*). This method is rather sensitive and its only limitation is availability of the respective antibodies or other ligands that can be immobilized and that specifically bind the cross-linked proteins.

b. According to the amount of free or cross-linked protein. Provided the M_r of polynucleotide is markedly greater than that of the cross-linked protein, the latter can be separated from free protein in a denaturating sucrose gradient (*92, 92*). If the M_r of the protein is comparable to or above that of the polynucleotide, the cross-linked nucleoprotein can usually be separated from free protein by electrophoresis under denaturing conditions (*94*).

However, the simpler and more quantitative method of separation of free protein from cross-linked nucleoprotein is deproteinization of the irradiated sample sorbed on PEI- or DEAE-cellulose; only proteins that are cross-linked to polynucleotide are retained even after washing the sorbent with phenol (*95*).

5. The Choice of the Optimal Dose

For each nucleoprotein, there is a certain range of UV-irradiation dose that can lead to reliable and reasonably interpreted results, that is, sufficient to yield enough specific cross-links for analysis. The minimal dose is determined by the sensitivity of the analytical method chosen, and by the availability of starting material, but the maximal one—leading to the accumulation of nonspecific cross-links due to conformational changes of nucleoprotein induced by phototransformations of its components (Section I,B,2), or to multiple cross-links—makes difficult or even impossible the subsequent analysis (*64*). Thus, the important step in using UV-induced cross-links for the investigation of nucleoprotein structure is the determination of the maximum acceptable dose.

In the case of pulse-laser sources with a high density of UV irradiation power (above 10^{22} photons cm^{-2} sec^{-1}), the maximum dose is defined practically only by accumulation of multiple cross-links (*64*). Since the duration of the pulse is usually fixed, a change of dose may be realized by alteration of either the intensity of the light flux (*64, 65*) or by the number of pulses (*66*). In the first case, if the cross-links result from biphotonic excitation, the yield of cross-links can be proportional to the square of dose in a certain interval of intensity (*64*). But in the second case, the yield of cross-links is always directly proportional to the dose independently of the appearance of cross-links via mono- or biphotonic pathways (unpublished).

The maximum acceptable dose from UV sources of normal intensity is determined mainly by the conformational changes of the nucleoprotein. The rate of these changes is increased after the lag period, and affects the kinetics of accumulation of cross-links (Section I,B,2). During the lag period, the formation of photoinduced cross-links follows single-step kinetics, and the dose-dependence of their accumulation is exponential. The deviation of this dependence from an exponential indicates, as a rule, the end of the lag period and the possible accumulation of nonspecific cross-links (see, however, Section I,B,2). Thus, the dose at which the cross-link accumulation curve deviates from the exponential is the maximum acceptable dose. The duration of the lag period is more dependent on irradiation time than on absorbed dose (Section II,A,6). [It should be stressed that we refer to absorbed dose, which depends not only on the time and intensity of radiation, but also on the optical density and geometry of the irradiated layer, on the light wavelenght, etc.] The duration of the lag period and, thus, the maximal acceptable absorbed dose at a given light intensity depend on the nature of nucleoprotein.

6. Analysis of the Proteins Cross-Linked with Polynucleotides

In most cases, two problems should be solved by such analysis: (a) identification of the proteins cross-linked to a certain polynucleotide, and (b) determination of the amount and/or the ratio of the cross-linked proteins. In the case of nucleoprotein containing single molecules of both protein and polynucleotide (*71*), these two problems are solved simultaneously by determination of the total degree of cross-linking (Section II,B,4). We consider below only more complex cases—the analysis of the cross-links arising from UV-irradiation of multicomponent nucleoproteins. But sometimes only the first problem is solved, although the solution of both tasks yields much more important information.

a. Identification of the cross-linked proteins. The most reliable identification of the cross-linked proteins can be achieved by immunochemical methods (*91, 96–99*), provided the corresponding antibodies of sufficient purity are available. Even a slight admixture of antibodies specific to other proteins of this nucleoprotein may lead to false identification. [It should be stressed that it is desirable to use total non-monoclonal antibodies against every protein for identification of the cross-linked proteins, since some antigenic determinants of the protein may be shielded or distorted by the cross-linked polynucleotide or its fragment.]

The preparation of pure antibodies against all proteins involved in a given nucleoprotein is not always possible. Therefore, other approaches are widely used, based either on the separation and quantitative determination of each of the free (non-cross-linked) proteins or on the separation of the cross-linked proteins after chemical or enzymatic digestion of the corresponding polynucleotide. In the first case, the cross-linked protein is identified by a difference in the amount of free protein before and after UV-irradiation of nucleoprotein (*101*). But quantitative determination enables the reliable identification of the protein only at a significant ($\geq$10%) degree of cross-linking.

Under the conditions of preferential formation of specific UV-induced cross-links, the degree of cross-linking of every protein does not exceed several percent (Section II,A,6). In this case, the cross-linked proteins are identified according to their mobilities in polyacrylamide gel electrophoresis after cleavage of the polynucleotide (*102–107*). Under conditions of sufficiently great chemical degradation of polynucleotides, many proteins also undergo destruction (unpublished), which makes difficult or even impossible the identification of the cross-linked proteins. Therefore, enzymatic cleavage of polynucleotides is preferable, although oligonucleotides covalently bound to proteins and containing several nucleotides remain. The presence of such oligonucleotides changes slightly and specifically the electrophoretic mobility of the protein but, as a rule, does not prevent its identification (*106*). In this approach, the decisive condition for strict identification of the proteins cross-linked with polynucleotide is complete removal of free (non-cross-linked) proteins.

Not all deproteinization methods remove the free proteins completely. Moreover, the degree of removal may be rather different for various proteins within the nucleoprotein (*108*). Since the degree of cross-linking is low (Section II,A,6), the proteins not covalently bound to polynucleotide but removed incompletely during deproteinization may be mistakenly identified as cross-linked ones.

b. Determination of the amount and/or the ratio of the cross-linked proteins. Since individual nucleoproteins are, as a rule, available only in limited amounts, and the yield of the cross-links is low, a radioactive label is usually introduced into the polynucleotide or the protein to aid in the analysis. It is also essential for identification of the cross-linked macromolecules, but it is absolutely necessary for their quantitative estimation.

The label in the components of the initial complex can be introduced *in vivo*—by growing cells in the presence of radioactive phosphate or labeled amino acids, nucleic bases or nucleosides (*109, 110*). One can obtain uniformly labeled preparations with rather high specific activities. However, the high radioactivity of the starting material makes more difficult the isolation and purification of the complexes and their components, as well as the functional characterization and further experimentation with the nucleoproteins. Besides, one must take into account the possible autoradiolysis of the initial products before, during, and after purification.

A second and more convenient approach is the introduction of the label in prepurified components of nucleoprotein *in vitro*, followed by their incorporation into the complex. This method involves the introduction of ^{125}I into proteins (*111*), radioactive phosphate in the 5′ terminus of polynucleotide (*114*), labeled nucleotide residues (^{32}P, ^{14}C, ^{3}H) at the 3′ end of polynucleotides (*109, 110*), etc.

The most convenient is the third appraoch, which allows the introduction of the label after irradiation of the complex. It is possible to employ for this purpose either iodination of the proteins under conditions where iodination of polynucleotides in negligible (*112*), or introduction of the label at the 5′ or 3′ end of oligonucleotides already covalently bound to the protein by means of polynucleotide kinase or the respective ligase (*113, 114*).

It is noteworthy that the efficiencies of iodination of every protein (*114a*) and of labeling of oligonucleotides cross-linked to the protein depend on the composition and the structure of the protein, on the length and shielding of oligonucleotide ends by the protein, and on the experimental conditions. Therefore, the results obtained with this approach prevent the determinations, without additional information, of the yield of cross-links or the molar ratio of the cross-linked proteins. However, the relative radioactivity of the cross-linked proteins is fairly reproducible and reflects the relative yield of the cross-links of corresponding proteins. Since the relative yield of the cross-links depends on the set of polynucleotide–protein interactions within nucleoprotein (i.e., on its conformation), the relative radioactivity of the cross-linked proteins is a rather sensitive characteristic of the state (conformation) of the nucleoprotein.

7. Isolation and Analysis of the Cross-Linked Fragments

The identification of the proteins cross-linked with a given polynucleotide after UV-irradiation of the nucleoprotein only permits the determination of the existence of direct interaction between these macromolecules within the complex. Usually the primary and often the secondary structures of the polynucleotides and proteins that are a part of the nucleoproteins under study are known. Therefore, the identification of the cross-linked fragments makes possible the localization of the sites of interaction of macromolecules on their primary and higher structures and thus a determination of their mutual orientation within the nucleoprotein as well as the functional topography of the corresponding macromolecules. It should be stressed that the unequivocal localization of the fragment within the protein or within a polynucleotide containing several thousands of nucleotide residues usually requires the determination of a sequence comprising less than ten monomer units.

The solution of this problem involves the isolation of the cross-linked fragments, the elucidation of their primary structure and, desirable, the determination of the single residue participating in the cross-link formation. The problem becomes substantially more simple provided each macromolecule has only one cross-link. This is one of the reasons for limiting the total yield of cross-links upon UV-irradiation of the nucleoprotein (Section II,B,5).

a. Identification of the cross-linked fragments of polynucleotide. For short polynucleotides, e.g. tRNA, the identification of the cross-linked fragment can be performed by comparison of fingerprints of the initial and cross-linked polynucleotides after exhaustive digestion by some specific endonuclease(s) (*19*). For long nucleotides (several hundreds or more nucleotide units), such an approach is complicated, if at all possible. In this case, the first step is isolation of a sufficiently short fragment cross-linked to a definite protein. To this end, the irradiated nucleoprotein is digested by a specific nuclease, and the respective protein containing cross-linked oligonucleotide can be isolated by means of electrophoresis (*115*) or affinity sorption (*91*).

The primary structure of the isolated oligonucleotides (after or without previous proteinase digestion of the protein) can be determined by standard methods. Thus, RNA fragments may be subjected to complete or partial cleavage by endonucleases. In the first case, the determination is finished by identification of the oligonucleotides obtained (*19*), whereas in the second one it is by electrophoretic separation of the set of oligonucleotides according to their length (*91, 116, 117*). Sequencing is also possible with the help of oligonucleotide maps after statistical chemical cleavage of internucleotide bonds in the fragment (*63*). If uniformly labeled polynucleotide is used for preparation of the initial complex, conventional methods for identification of oligonucleotides can be used after the action of specific endonucleases on the cross-linked fragment (*118*) (cf., however, Section II,B,6a). Fragments labeled at one end are necessary for sequencing after statistical cleavage of internucleotide bonds or partial digestion by endonucleases.

Provided the polynucleotide–protein cross-links are retained (fully or partially) during sample preparation and analytical procedures, it is possible to localize the nucleotide residue participating in their formation. Even short peptides remaining after proteolysis of the cross-linked protein change the position of the corresponding oligonucleotides. The subsequent determination of the oligonucleotide composition permits localization of the nucleotide residue participating in the formation of polynucleotide–protein cross-links (*114, 115, 119*). Upon sequencing by separation of the oligonucleotide mixture arising from statistical cleavage of internucleotide bonds, the presence of the molecule of cross-linked protein results in a gap in the ladder or nucleotide map. This gap appears after the residue preceding (from the labeled end) that cross-linked with the protein. Thus, it is possible to localize in the polynucleotide not only the fragment taking part in interaction with a given protein, but the nucleotide residue directly involved in this interaction.

The limited digestion of the polynucleotide by endonucleases affords a set of fragments cross-linked with the protein, which vary in the distances from the cross-linking point to 5′ and/or 3′ ends. However, this does not prevent the localization on the polynucleotide chain of the fragment or even the individual nucleotide residue cross-linked with the protein (see, e.g., (*91*). The problem is simplified if these fragments can be separated (usually after complete proteolysis of the fixed protein) (*91*, *112*).

b. Identification of the cross-linked protein fragments. Cross-linked poly- or oligonucleotides change peptide mobility in chromatography and electrophoresis. Therefore, after isolation and identification of the protein cross-linked with polynucleotide, one can detect the cross-linked peptide by comparison of the sets of peptides formed after cleavage of the free and cross-linked proteins by specific proteinases (*89*, *120*, *121*). But the complete removal of free proteins is not always possible, and in the case of large proteins, it is not so easy to separate all peptides after cleavage of the protein. However, the cross-linked peptide can be separated from all other peptides due to the negative charge of poly- or oligonucleotide covalently bonded to it. Analysis of this peptide is possible by the conventional methods of protein chemistry. The amino-acid composition of the peptide fragment is often sufficient for its localization in the known primary structure of the protein. From such data, it is sometimes possible to locate the position in the polypeptide of the amino acid that participates in the cross-link formation, i.e., is involved in direct interaction with polynucleotide within the complex.

Some other approaches for this purpose have recently been described (*121a*, *121b*, *121c*). Iodination leads to introduction of the label mainly into tyrosine-containing peptides (*112*), and reductive hydroxymethylation also appears rather specific; not all residues in a polypeptide chain are modified and, hence, labeled (*122*). Therefore, to identify the cross-linked unlabeled protein fragments, it is necessary to use either a large amount of the initial nucleoprotein (*118*), or to apply proteins uniformly labeled *in vivo*. Thermal activation of tritium is also possible for the labeling of proteins *in vitro* (*123*). However, this approach is not yet sufficiently developed for the structural analysis of proteins. Because of these problems, information about the protein fragments cross-linked with polynucleotides upon UV-irradiation of intact natural complexes is more scarce than about fragments of polynucleotides cross-linked with proteins.

III. The Application of UV-Induced Cross-Links for the Investigation of RNA–Protein Interactions in the Translation Machinery of *E. coli*

The machinery for synthesis of the protein in the cell (translation) is composed of many RNAs and proteins. The nucleus of this machinery is the ribosome, one of the most complex functional nucleoproteins. Although genes of rRNAs and proteins are situated in different regions of the *E. coli* genome, their syntheses are strictly coordinated, and the close-to-equimolar amounts of all components of ribosomes appear in the cytoplasm (*124*). The ribosomal subunits are spontaneously formed (self-assembled) in the cytoplasm—the small one (30 S) containing one RNA molecule (16 S) and 21 molecules of protein, and the large one (50 S) comprising two RNA molecules (23 S and 5 S) and 35 molecules of protein. The first step of translation is the formation of a preinitiation complex, the association of a 30-S subunit with mRNA and initiator tRNA. This complex is formed with the help of three initiation factors and GTP.

Only after hydrolysis of one molecule of GTP into GDP does the association of the preinitiation complex with the 50-S subunit proceed, and the initiation complex (70 S) begins elongation of the peptide chain (*124*, *125*). In the course of elongation, the ribosome selects those tRNAs that correspond to the respective codons independently from the type of amino acid attached to this tRNA (*126*). Therefore, the fidelity of translation is determined to a great extent by the precision of aminoacylation, which is performed by aminoacyl-tRNA synthetases (*127*). In recent years, there has been much progress in the study of the translating machinery reviewed in (*125*), but many problems concerning the structure, dynamics, and functional topography of translational complexes and their macromolecular components remain unsolved.

The important data for solution of these problems can be obtained upon investigation of intermolecular, in particular, RNA–protein contacts in translational complexes, as well as the changing of these contacts in the course of alteration of their functional state. The UV-induced formation and analysis of the polynucleotide–protein cross-links are widely used for this purpose (see, e.g., *2*, *3*). It should be stressed that the capabilities of this approach are far from being fully used. On the other hand, inadequate understanding of the limitations inherent in the method leads to misuse that unfairly discredits this approach. We consider below some results obtained from the investigation of translating nucleoproteins by the correct use of UV-induced RNA–protein cross-links.

A. Complexes of tRNA with Aminoacyl-tRNA Synthetases

The incorporation of the proper amino acid into the growing peptide chain requires the participation of the corresponding tRNA and aminoacyl-tRNA synthetase. The specificity of aminoacyl-tRNA formation is determined by a specific binding with an enzyme of the corresponding amino acid and tRNA. The cleavage of tRNA at its anticodon loop and even the removal of the anticodon do not prevent enzymatic aminoacylation (*128, 129*). Thus, both specific and nonspecific binding of a tRNA with an aminoacyl-tRNA synthetase is determined by interactions (contacts) of the enzyme with other regions of tRNA apart from the anticodon loop.

Upon UV-irradiation (254 nm) of the complexes of tyrosyl-tRNA synthetase with *E. coli* $tRNA_1^{Tyr}$ and $tRNA_2^{Tyr}$ [the latter differs from the former by the substitution of U-C by C-A in positions 50 and 51 (*71*)],[2] the same three fragments of tRNA are cross-linked to the enzyme in both complexes (one in the dihydrouridine stem, the second in the anticodon loop, and the third in the variable loop). The difference in the primary structure of these tRNA is manifested in the fact that the third fragment of $tRNA_1^{Tyr}$ is cross-linked to synthetase with higher efficiency than is that of $tRNA_2^{Tyr}$ (*71*).

In complexes of *E. coli* $tRNA^{Ile}$ with *E. coli* Ile-tRNA synthetase and yeast Val-tRNA synthetase, which aminoacylates this tRNA by valine, three fragments are cross-linked to the enzyme in both cases. Two of them are common for these complexes, those in the dihydrouridine loop and stem. But the third fragment, in the complexes with its own and the foreign synthetase, is at the opposite ends of the L-shaped structure of tRNA, in the regions of the CCA end and the anticodon loop, respectively (*19*). Since the same tRNA is involved in both complexes, the various localizations of the third cross-linked fragment indicate the difference in direct interactions (contacts) between tRNA and these enzymes.

In similar work, the complexes of yeast $tRNA^{Phe}$ with yeast Phe-tRNA synthetase and *E. coli* Ile- and Val-tRNA synthetases were irradiated by UV light (*130*). Three fragments of tRNA were cross-linked to the enzyme in each complex; two of them (the same for all three complexes) are situated from the 5′ end of tRNA up to the beginning of the dihydrouridine loop. However, the third fragments are in various regions of tRNA: from the 3′ end of the pseudouridine stem up to the middle of the corresponding loop (complex with Phe-tRNA synthetase), at the 3′ end of the acceptor stem (complex with Val-tRNA synthetase), and in the anticodon stem and corresponding loop (complex with Ile-tRNA synthetase) (Fig. 9).

On the basis of these data, the authors of the above papers

[2] Positions 47:2 and 47:3 in the numbering system proposed by Singhal *et al.* in Vol. 23 of this series (p. 289). [Eds.]

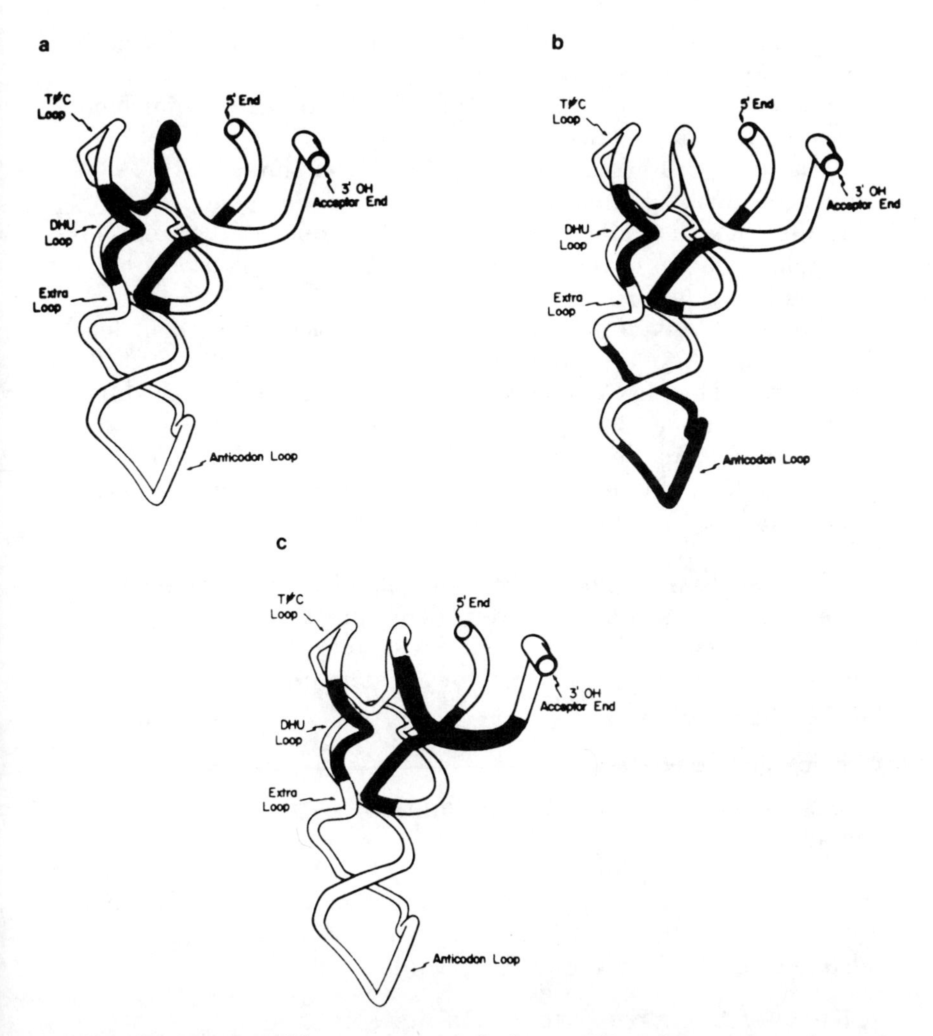

FIG. 9. Location of fragments cross-linked upon UV irradiation of complexes of yeast $tRNA^{Phe}$ with the following aminoacyl-tRNA synthetases: (a) phenylalanine from yeast; (b) isoleucine from *E. coli*; (c) valine from *E. coli*. The fragments cross-linked are indicated by shading.

conclude that, in all of these complexes, the fragments of tRNA in the region of the angle of the L-shaped structure are involved in interaction with synthetases. The additional sites of interaction may be positioned both in the acceptor stem and in the anticodon loop.

UV-irradiation of the complexes of Met-tRNA synthetase (*E. coli*) with $tRNA_f^{Met}$ and $tRNA_m^{Met}$ from *E. coli*, yeast, and wheat germ results in cross-linking of the protein with three tRNA fragments: 3′-terminal hexanucleotide, anticodon loop, and pseudouridine loop (*131*). But at low doses, there was binding solely at the 3′-terminal hexanucleotide for the same system (*113*).

It is noteworthy that irradiation of the mixtures of tRNA with aminoacyl-tRNA synthetases inactivates both components (*19, 70*). This effect does not allow us to obtain high yields of the cross-links, although alteration of the wavelength of UV-irradiation may help to increase this value (*70*).

The results given above permit some conclusions about localization in tRNA of the regions interacting with aminoacyl-tRNA synthetases. The tRNA fragments interacting directly with this enzyme are positioned in the dihydrouridine and pseudouridine stems and loops, i.e., in the region of the angle of the tRNA L-shaped structure. The additional sites of interaction are distributed along the whole molecule, from the acceptor stem up to the anticodon loop of tRNA.

In all of these papers, only fragments of tRNA interacting with aminoacyl-tRNA synthetases were determined. Certainly the more precise localization of interactions (cross-links) to a specific nucleotide would permit another step toward determining the functional topography of tRNA.

B. Ribosomal Complexes

The self-assembly of subunits, formation of ribosomal complexes, and alterations of their functional states are conjugated with and/or caused by conformational changes of both separate macromolecules and the complex as a whole. It is shown below that this also affects RNA–protein interactions that are reflected in alterations of the relative amount of UV-induced cross-links.

1. The 30-S Subunit of the *E. coli* Ribosome

The 30-S subunit, being a constituent of the 70-S ribosome, functions as a component of the latter during the whole translation process. However, at the first stage of translation—initiation—it behaves as a separate nucleoprotein particle (Fig. 10).

a. The changes in RNA–protein contacts upon self-assembly of the 30-S subunit. One of the self-assembly stages of the 30-S subunit of the *E. coli* ribosome *in vitro* is the transformation of the so-called ribosomal intermediate particles (RI-particles) into RI*-particles (*135, 136*). The addition of 16-S RNA to a full mixture of 30-S subunit proteins at 0–10°C results in formation of RI-particles—complexes of

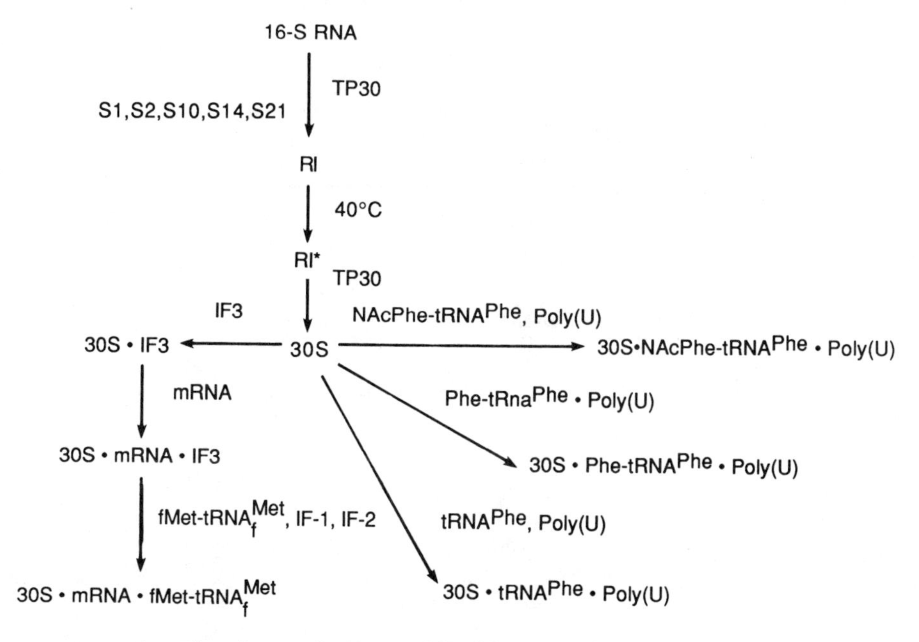

FIG. 10. The scheme of self-assembly of the 30-S subunit of *E. coli* ribosomes and of the formation of the 30-S preinitiation complex and its analogues. TP30 = total protein of the 30-S subunit; RI, RI*= the ribosomal intermediate nucleoproteins before and after termal activation, respectively.

16-S RNA with only part of the proteins (proteins S1, S2, S10, S14, and S21 are absent). Incubation of RI-particles at 40°C is required for completion of the self-assembly, the transformation of RI-particles into RI*-particles. The last can bind all the lacking proteins and form functionally active 30-S subunits. The RI → RI* transition is accompanied by an alteration of the physicochemical characteristics of this nucleoprotein: the sedimentation constant is changed and the sensitivity to salt deproteinization is decreased (*137, 138*). Also, this transition is accompanied by changes of 16-S RNA conformation (*139*). But a substantial change of RNA–protein contacts is also observed in the course of temperature activation of the RI particles (Table I). Thus, within RI-particles, only proteins S4 and S7 are cross-linked to 16-S RNA in the ratio ≈4:1; within RI*, there are five proteins, and the ratio of the cross-linked proteins S4 and S7 is increased to ≈1:2. The data obtained demonstrate that the change of nucleoprotein conformation is manifested in alterations of the set and ratio of proteins cross-linked with polynucleotide upon UV-irradiation of the complex.

TABLE I
RIBOSOMAL PROTEINS CROSS-LINKED IN 16-S RNA UPON UV-IRRADIATION (254 nm) OF RIBOSOMAL INTERMEDIATE NUCLEOPROTEIN BEFORE (RI) AND AFTER (RI*) THERMAL ACTIVATION *(140)*[a]

Ribosomal proteins	% of total ^{125}I-labeled protein RI	RI*
S4	80	25
S7	20	45
S9/11	—	15
S12	—	4
S15/16/17	—	11

[a] The values are the percent of total radioactivity of gel zones containing ^{125}I-labeled proteins.

b. The application of pulse-laser UV-irradiation for induction of RNA–protein cross-links in the 30-S subunit. This review deals with the data obtained upon application of UV-irradiation of ordinary intensity (up to 10^{17} photons cm^{-2} sec^{-1}). However, laser-pulse sources giving monochromatic irradiation with intensity up to 10^{30} photons cm^{-2} sec^{-1} and pulse durations of 10^{-8} to 10^{-13} sec have recently become available. This technique presents new possibilities for the investigation of polynucleotide–protein interactions (*64–66*). But the advantages of laser irradiation for these purposes are so far not widely known. Therefore, it is worthwhile to compare the results obtained with the 30-S subunits of *E. coli* ribosomes with ordinary and pulse-laser UV light sources.

Upon pulse-laser irradiation (fourth harmonic Nd^{+3}-YAG laser, $\lambda = 266$ nm, duration of the pulse 10 nsec, intensity of the incident light up to 10^{25} photons cm^{-2} sec^{-1}) of the 30-S subunit, the efficiency of cross-linking (percentage of cross-linked protein per absorbed photon) is proportional to the intensity of the light flux (Fig. 11), i.e., the cross-links are formed due to biphotonic excitation via higher excited states. The efficiency of the cross-linking at the intensity of the light flux 2×10^{24} photons cm^{-2} sec^{-1} is an order of magnitude higher than upon irradiation at normal intensity (Table II), when the cross-links are formed via lower excited states. In the case of laser

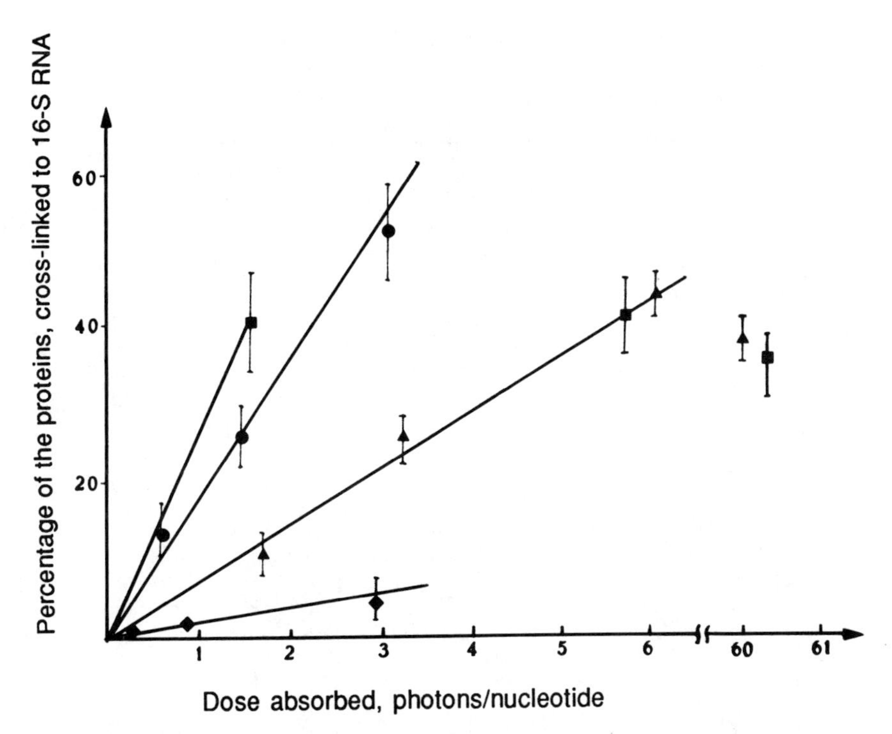

FIG. 11. The dependence of percent of proteins cross-linked to 16-S RNA on absorbed dose after laser UV irradiation of the 30-S subunit of *E. coli* ribosomes (in percent in relation to ^{125}I-radioactivity of all proteins of the subunit). Incident light intensity (photons cm^{-2} sec^{-1} × 10^{24}): 1.6 (◆), 3.4 (▲), 5.8 (●), 6.8 (■). Irradiation was in the presence of 6 m*M* β-mercaptoethanol (*64*).

irradiation, in addition to proteins S4, S7, and S9/S11 (cross-linked to 16-S RNA in normal irradiation), proteins S3, S18, and S20 are also cross-linked (Table II), and all of these proteins are cross-linked via higher excited states of the macromolecular components.

It was shown earlier (*118*) that upon UV-irradiation of common intensity, protein S7 in the 30-S subunit is cross-linked to 16-S RNA at U-1239, whereas upon pulse-laser irradiation, this protein is cross-linked at both U-1239 and G-1240 (*64*). The efficiency of cross-linking to both bases is comparable; i.e., both cross-links are formed via higher excited states.

Thus, the application of high-intensity pulse-irradiation not only increases the efficiency of cross-linking but allows us to detect polynucleotide–protein interactions not shown by cross-links formed by UV-irradiation at the usual intensity. It should be added that the

TABLE II
THE CROSS-LINKING OF PROTEINS TO 16-S RNA INDUCED BY NORMAL AND PULSE-LASER UV-IRRADIATION OF THE 30-S SUBUNIT OF *E. coli* RIBOSOMES (*64*)[a]

Proteins of 30-S subunit	Molar ratio of cross-linked proteins		Cross-section of cross-link $\alpha_0 \times 10^{21}$ cm^2 per photon	
	a	b	a	b
S3	—	2	—	520
S4	1	1	2.5	260
S7	8	2	20.0	520
S9/11	1	1	2.5	260
S18	—	2	—	520
S20	—	4	—	1000

[a] a: normal (low-intensity) irradiation, 10^{15} photons cm^{-2} sec^{-1} at 254 nm; absorbed dose, 20 photons per nucleotide; duration of irradiation, 1.2×10^3 sec; total cross-linking of the protein, $\approx$ 2.5%

b: laser irradiation, 2×10^{24} photons cm^{-2} sec^{-1} at 266 nm; absorbed dose, 0.5 photons per nucleotide; duration of irradiation, one pulse (10^{-8} sec); total cross-linking of the protein, $\approx$ 6%.

required dose can be absorbed by nucleoprotein during a single pulse, i.e., in 10^{-8} to 10^{-13} sec. Since the lifetime of the higher excited states does not exceed 10^{-13} sec, it is evident that all cross-links arise from a single pulse and, if formed directly via higher excited states, are specific.

c. The change of RNA–protein contacts upon formation of the complex 30-S · IF3. The formation of the complex of the *E. coli* 30-S subunit with initiation factor 3 (IF3) is necessary for the initiation of protein synthesis in natural mRNA (*141*), and it markedly increases the formation rate and stability of the complex of the 30-S subunit with poly(U) and aminoacyl-tRNA (*142*). The results (*143–145*) permit us to propose that the formation of the 30-S · IF3 complex is accompanied by an alteration of the 30-S subunit conformation.

This is proved also by the difference in the sets of proteins cross-linked to 16-S RNA upon UV-irradiation of free 30-S subunit and the complex 30-S · IF3. The IF3 in the latter complex is also cross-linked to 16-S RNA (*146, 147*). We do not refer here to the real data, since they were obtained at high doses that can lead to accumulation of significant amounts of nonspecific cross-links (Section II,B,5). Nevertheless, the results obtained correlate with the concept that at least one function of IF3 is induction of conformational changes of the 30-S subunit that favor binding of mRNA (*148–150*).

2. Preinitiation 30-S Complexes

a. mRNA–protein contacts in complexes of 30-S mRNA. The first step in formation of a preinitiation complex is binding of a 30-S subunit to mRNA (*151*). In productive binding of natural mRNA, the initiation codon of the corresponding cistron should be correctly positioned in the decoding region of the 30-S subunit. Shine–Dalgarno interactions betwen mRNA and 16-S RNA play a certain role in the proper binding of mRNA (*152*). Also, one of the initiation factors (IF3) (*153*) and ribosomal protein S1 (*154*) are important in productive binding of mRNA.

The degree of association of 30-S subunits with natural mRNA (phage MS2 RNA) at 0°C is low and remains practically unchanged after addition of IF3 (*155*). However, in the presence of IF3, the set of proteins cross-linked to mRNA upon UV-irradiation is sharply increased (Fig. 12). After incubation at 37°C in the absence of IF3, almost the same set of proteins as at 0°C with IF3 is cross-linked to mRNA. The total yield of cross-links is increased by almost an order of magnitude when the temperature is raised from 0° to 37°C, which correlates with an enhancement of the degree of association of mRNA with the 30-S subunit. The addition of IF3 at 37°C results both in an increase in the total yield of cross-links and in changes of the set of cross-linked proteins (Fig. 12). IF3 is efficiently cross-linked to mRNA both at 0°C and at 37°C.

The comparison of these results with the effect of IF3 on the structure of the 30-S subunit (Section III,B,3) implies that a definite conformational state of the subunit is required for productive binding of mRNA. The transition of the 30-S subunit to a conformational state suitable for binding with mRNA is promoted by its association with IF3. However, the conformational mobility of nucleoproteins at 0°C is low, and during the incubation period only a small part of the molecules is transferred into a new conformational state induced by binding with IF3.

That the close-to-optimal conformational state of the 30-S subunit may arise at 37°C even in the absence of IF3 follows from the degree of association of mRNA with the 30-S subunit and from the set of the proteins cross-linked to it. But it also follows from these data that the optimal conformation appears or is stabilized only after binding of the subunit to IF3. Since IF3 interacts directly with mRNA, the effect of this interaction on mRNA conformation cannot be excluded.

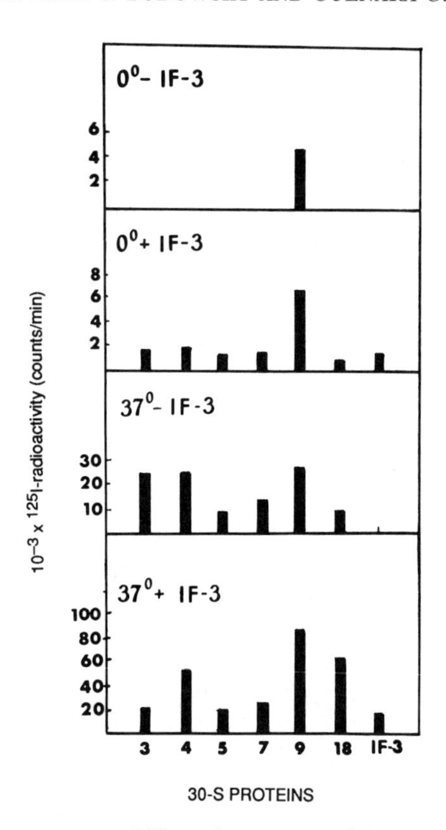

FIG. 12. Absolute amount of ^{125}I in the proteins of the 30-S subunit cross-linked to tRNA upon UV irradiation (254 nm, 15 photons per nucleotide) of a complex of 30-S subunit and phage MS2 RNA at + 4° and 37° C, with and without factor IF-3 (*155*).

Protein S1 plays an important role in the formation of the complex of the 30-S subunit with both natural and artificial mRNA (*154*). This protein belongs to the group of so-called split-proteins, i.e., it is weakly bound to the 30-S subunit (*156*). Protein S1 does not cross-link with 16-S RNA upon UV irradiation of the 30-S subunit (*157*), and it is bound to it by protein–protein rather than RNA–protein interactions (*158*). In the paper considered above (*155*), the cross-linked proteins are identified by means of two-dimensional electrophoresis where the position of protein S1 coincides with the position of protein aggregates. This prevents any conclusion about the interaction of S1 protein with 16-S RNA or mRNA.

However, immobilized anti-S1 antibodies (*91*) permit detection of a cross-link of protein S1 with mRNA upon irradiation of the complex of 30-S subunit with phage MS2 RNA, as well as the isolation and determination of the structure and localization of the mRNA fragment interacting with (cross-linked to) protein S1. *Escherichia coli* ribosomes translate *in vitro* mainly the coat-protein cistron of phage MS2 (*159*); the native higher structure of the phage RNA is necessary for the correct initiation of translation (*160*). However, protein S1 within the complex of the 30-S subunit with phage MS2 RNA interacts with the RNA fragment located about 700 nucleotides downstream from the initiator region of the coat-protein cistron, situated in the replicase cistron (Fig. 13) (*91*).

In the RNA of phage f_r, which belongs to the same serological group as phage MS2 (*161*), as well as in the RNA of phage Q_β, which belongs to the other group (*161*), the regions cross-linked with protein S1 upon irradiation of the corresponding 30-S mRNA complexes are also located several hundred nucleotides away, but upstream from the initiator sites of the predominantly translated coat-protein cistrons (*162*). Nevertheless, it is evident that the mRNA regions of translation initiation and binding of protein S1 are close to each other in space because of the peculiarities of the higher structure of these mRNAs.

The presence of such a region in mRNA may be an important factor in reducing the duration of idle "running" of ribosome—from termination of the previous polypeptide synthesis to initiation of synthesis of the next one. Thus, it may be suggested that the primary interaction in cytoplasm proceeds between free protein S1, the amount of which is rather high (*154*), and the corresponding region of mRNA. Because of the affinity of protein S1 for the 30-S subunit, the local concentration of the complex of S1 with mRNA will be markedly higher in the vicinity of the 30-S subunit than in the whole volume, resulting in an increase in the formation rate of the 30-S · mRNA complex.

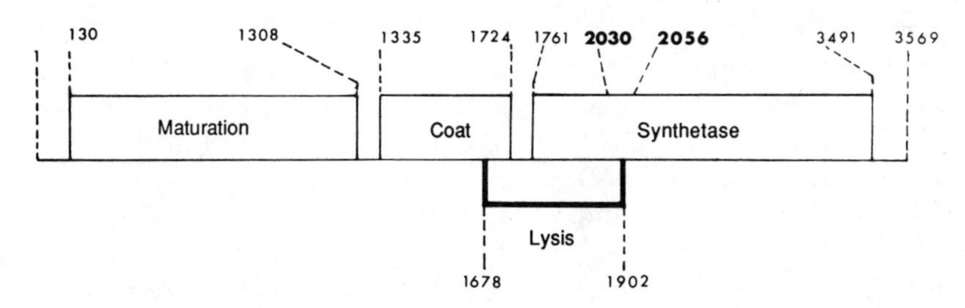

FIG. 13. Genetic map of phage MS2 RNA (*159*). The fragment cross-linked with protein S1 upon irradiation of a complex of S1 with phage MS2 RNA is indicated by boldface numbers.

The presence of regulatory regions of mRNA positioned apart from the beginning of the corresponding cistron in any direction resembles the situation with enhancers in DNA, affecting transcription efficiency (*163*). From this point of view, these regions in mRNA may be considered as enhancers of translation (cf. *164*).

b. Complexes of the 30-S subunits with mRNA and tRNA. The formation of the prokaryotic preinitiation complex is terminated by association of the complex 30-S · mRNA with initiator fMet-$tRNA_f^{Met}$ (*165*). The translation of poly(U) in a cell-free system is widely used in investigations of various aspects of protein biosynthesis (*166*). In this case, the 30-S preinitiation complex is formed by the association of a 30-S subunit with AcPhe-$tRNA^{Phe}$ or Phe-$tRNA^{Phe}$ and any region of poly(U) (Fig. 10). However, it is unknown what degree of structural similarity exists between a real 30-S preinitiation complex containing fMet-$tRNA_f^{Met}$ and a native mRNA, and artificial analogues containing AcPhe-$tRNA^{Phe}$ or Phe-$tRNA^{Phe}$ and poly(U). The correctness in using complexes of 30-S subunits with poly(U) and deacylated $tRNA^{Phe}$ for studies of the initiation of translation is even less evident, since aminoacylation of tRNA and N-acylation of aminoacyl-tRNA may result in changes of tRNA conformation and/or its position in the complex.

In order to clarify the structural similarity of a natural preinitiation complex and its analogues containing poly(U) and AcPhe-$tRNA^{Phe}$, Phe-$tRNA^{Phe}$, and deacetylated $tRNA^{Phe}$ (complexes I, II, and III, respectively), UV-induced cross-linkages of 16-S RNA and tRNA with ribosomal proteins within these complexes have been examined (*93*, *167*).

i. *tRNA–protein cross-links.* Of the artificial complexes I, II, and III, the one closest to a natural preinitiation complex is I, containing N-acetylated aminoacyl-tRNA. The proteins predominantly cross-linked to tRNA upon irradiation are common to complex I and the natural preinitiation complex are S4 and S5 (cf. *167* and Fig. 14), which are components of the decoding region of the 30-S subunit (*168*), and S9/S11. This fact indicates the similar positions of fMet-$tRNA_f^{Met}$ and AcPhe-$tRNA^{Phe}$ in the 30-S subunit within these complexes. A difference is observed only for weakly cross-linked proteins: S7, S17, and S18 in complex I, and S14 and S15 in the natural complex. Since the efficiency of formation of each cross-link depends on the nature of the components of the macromolecules in contact, such differences may be caused by differences in the primary structures of $tRNA_f^{Met}$ and $tRNA^{Phe}$.

Complexes I, II, and III contain tRNA with the same primary

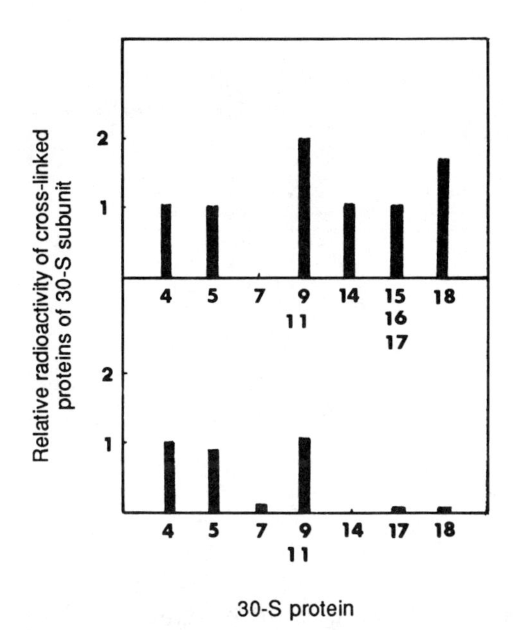

FIG. 14. The proteins cross-linked with fMet-tRNA$_f^{Met}$ (*167*) and AcPhe-tRNA^{Phe} (*93*) upon UV-irradiation (254 nm, 15 photons per nucleotide) of the complexes fMet-tRN$_f^{Met}$ · 30S · RNA of phage MS2 (upper panel) and AcPhe-tRNA^{Phe} · 30S · poly(U) (lower panel). The ratios of the cross-linked proteins are given (according to ^{125}I radioactivity).

structure. Therefore, the differences in the sets of proteins cross-linked with tRNA within these complexes may be caused solely by differences in the tRNA–protein interactions. The set of proteins cross-linked to tRNA in complexes I and II is essentially the same (Table III), which demonstrates the similarity of the structures of these complexes and the position of tRNA in them. But the set of proteins cross-linked to tRNA in complex III is markedly different from that in complexes I and II (Table III). This may be caused by a change in tRNA conformation upon deacylation (*169*) and/or another conformation of the 30-S subunit in this complex. The first proposition correlates with data about the alteration of tRNA conformation (or ratio of conformers) upon aminoacylation (*170*).

ii. *16-S RNA–protein cross-links.* Compared with the free 30-S subunit, many more proteins are cross-linked to 16-S RNA in preinitiation complexes (Table IV). The set of cross-linked proteins and their ratio are essentially the same for complexes I and II. But a comparison of these complexes with complex III, containing deacylated tRNA,

TABLE III

THE PROTEINS CROSS-LINKED TO $tRNA^{Phe}$ BY UV-IRRADIATION (254 nm, 15 PHOTONS PER NUCLEOTIDE) OF THE COMPLEXES 30S · POLY(U) WITH AcPhe-$tRNA^{Phe}$ (I), Phe-$tRNA^{Phe}$ (II), AND DEACYLATED $tRNA^{Phe}$ (III) (*93*)[a]

Proteins	Complex I	Complex II	Complex II
S4	34	31	21
S5	27	24	18
S6	—	8	16
S7	4	5	—
S8	—	6	13
S9/11	31	26	17
S15	—	—	—
S17	2	—	—
S18	2	—	—

[a] The radioactivity of separate ^{125}I proteins is given in percent of total ^{125}I.

shows that the cross-linking of proteins S8 and S9/S11 is markedly decreased, the portion of protein S7 is increased, and a considerable amount of protein S13/S14 appears.

Thus, the binding of poly(U) and tRNA results in a substantial alteration of the conformation of the 30-S subunit. This alteration is

TABLE IV

THE PROTEINS CROSS-LINKED IN 16-S RNA UPON UV-IRRADIATION (254 nm, 15 PHOTONS PER NUCLEOTIDE) OF FREE 30-S SUBUNITS (*93*) AND COMPLEXES OF 30S · POLY(U) WITH AcPhe-$tRNA^{Phe}$ (I), Phe-$tRNA^{Phe}$ (II), AND DEACYLATED $tRNA^{Phe}$ (III) (*93*)

Proteins	Free 30-S Subunit	Complex I	Complex II	Complex III
S4	20	16	17	14
S5	—	7	9	8
S7	64	39	35	45
S8	—	16	14	5
S9/11	16	8	6	3
S10	—	3	5	5
S13/14	—	—	—	12
S15	—	4	5	2
S16	—	3	5	3
S17	—	4	4	3

[a] The relative radioactivity of some ^{125}I proteins is presented in percent of total gel radioactivity.

the same with Phe-tRNA^{Phe} and AcPhe-tRNA^{Phe}, but differs from that induced by the binding of deacylated tRNA^{Phe}.

3. 70-S Ribosomal Complexes

Association of a preinitiation 30-S complex with a 50-S ribosome subunit results in the formation of a 70-S initiation complex. This event terminates the initiation of translation, which precedes the elongation step, i.e., the synthesis of polypeptide. Elongation is a cyclic process, repeating with the incorporation of each amino-acid residue. Each cycle of elongation may be arbitrarily divided into several steps (Fig. 15). This division is not only convenient didactically, but indicates that one may stop the elongation process at one of

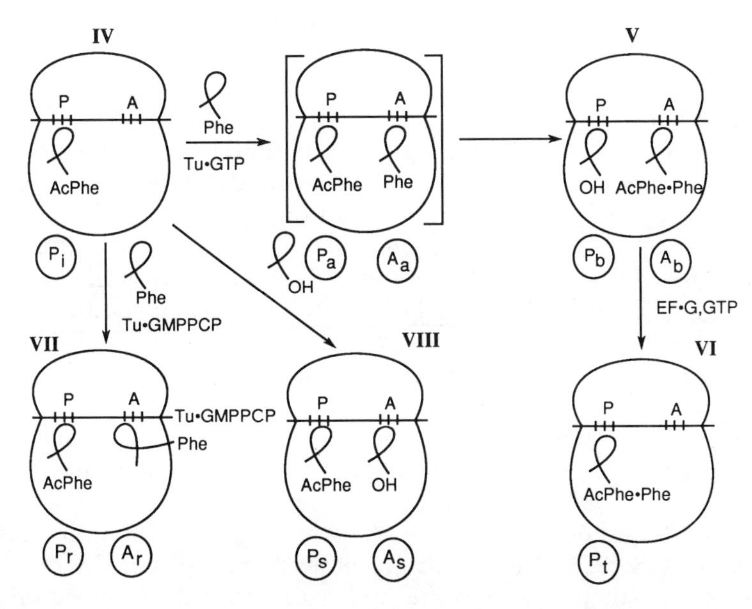

Fig. 15. The scheme of the elongation process. *Complex IV:* AcPhe-tRNA^{Phe} · 70-S · poly(U) obtained by direct binding of AcPhe-tRNA^{Phe} with 70-S ribosomes in the presence of poly(U). *Complex Ve:* tRNA^{Phe} · 70-S · poly(U) · AcPhePhe-tRNA^{Phe}, obtained by addition of the ternary complex, Phe-tRNA^{Phe} · Tu · GTP, to complex IV (pretranslocated state). *Complex Vn:* tRNA^{Phe} · 70-S · poly(U) · AcPhePhe-tRNA^{Phe}, obtained by addition of Phe-tRNA^{Phe} to complex IV (pretranslocated state). *Complex VI:* AcPhePhe-tRNA^{Phe} · 70-S · poly(U), obtained by addition of elongation factor G and GTP to complex V (posttranslocated state). *Complex VII:* [AcPhe-tRNA^{Phe} · 70-S · poly(U)] [Phe-tRNA^{Phe} · Tu · GMPPCP], obtained by addition of the ternary complex Phe-tRNA^{Phe} · Tu · GMPPCP to complex IV. *Complex VIII.* [AcPhe-tRNA^{Phe} · 70-S · poly(U) · tRNA^{Phe}], obtained by addition of deacylated tRNA^{Phe} to complex IV.

these stages artificially, which permits one to study separately the respective states of the complex. The transition from one step to the other, i.e., the change of the composition and the functional state of the elongating complex, can be coupled with the alteration not only of the structure of the ribosomal subunits but also of their interactions. Hence, the change of the functional state of the elongating complex can be conjugated with the alteration of the intra- and intersubunit RNA–protein contacts. This could affect the sets and the ratios of UV-induced cross-links.

Among the 70-S elongating complexes presented in Fig. 15, only complexes IV, V, and VI were obtained in a sufficiently homogeneous state for analysis of cross-links between ribosomal RNAs and proteins. Therefore, they were used for the investigation of alterations of inter- and intrasubunit RNA–protein contacts with changes in the functional state of 70-S ribosome. For comparison, such contacts were also determined in an artificial complex—the free 70-S ribosome (tight couple) (*171*).

Complex $tRNA^{Phe}$·70 S·poly(U)·AcPhePhe-$tRNA^{Phe}$ results from transpeptidation after addition of Phe-$tRNA^{Phe}$·Tu·GTP to AcPhe-$tRNA^{Phe}$·70 S·poly(U) (*172*). After addition of elongation factor EF-G and GTP to complex V, the conjugated migration of mRNA and AcPhePhe-$tRNA^{Phe}$ in the ribosome (translocation) proceeds, and deacylated tRNA leaves the ribosome (*173*). Thus, complexes V and VI are the pre- and posttranslocated ribosomes.

a. Intrasubunit RNA–protein contracts in the 30-S subunit. As mentioned above (Section III,B,2,b), the binding of the 30-S subunit to poly(U) and AcPhe-$tRNA^{Phe}$ (the formation of the 30-S analogue of preinitiation complex I) is accompanied by substantial changes in the contacts of 16-S RNA with protein (Tables IV and V). The association of free 30-S and 50-S subunits leading to the formation of the tightly coupled 70-S ribosomes results in other changes in the contacts of 16-S RNA with proteins of the 30-S subunit (Table V). Significant changes of these contacts are observed upon association of 30-S preinitiation complex I with a 50-S subunit, i.e., upon formation of the analogue of the 70-S initiation complex IV. Complex IV is similar in composition to posttranslocated complex VI (Fig. 15). Both the set and the ratio of intrasubunit cross-links (between 16-S RNA and the proteins of the 30-S subunit) are practically the same in these two complexes. But the intrasubunit RNA–protein cross-links in the 30-S subunit within post- and pretranslocated complexes (VI and V, respectively) are markedly different (Table V).

Thus, the change of the functional state of the 30-S subunit (apart

TABLE V

THE PROTEINS OF THE 30-S SUBUNIT CROSS-LINKED TO 16-S RNA UPON UV IRRADIATION (254 nm, 20–30 PHOTONS PER NUCLEOTIDE) OF FREE 30-S SUBUNIT, 30-S PREINITATING COMPLEX (I) (*93*), FREE 70-S RIBOSOME (TIGHT COUPLE), 70-S RIBOSOME BOUND TO POLY(U) AND AcPhe-tRNA^{Phe} (IV), PRETRANSLOCATED (V) AND POSTTRANSLOCATED (VI) RIBOSOMAL COMPLEXES (*174*)[a]

Proteins	Free 30-S Subunit	30-S · poly(U) · AcPhe-tRNA^{Phe} (I)	Free 70-S Subunit (Tight Couple)	AcPhe-tRNA^{Phe} · 70S · Poly(U) (IV)	AcPhePhe-tRNA^{Phe} · 70S · poly(U) · tRNA^{Phe} (V)	AcPhePhe-tRNA^{Phe} · 70S·poly(U) (VI)
S3	—	—	—	8	—	6
S4	20	16	17	18	10	17
S5	—	7	—	—	—	—
S6	—	—	—	—	14	—
S7	64	39	34	36	36	35
S8	—	16	—	7	—	7
S9/11	16	8	12	9	9	11
S10	—	3	—	—	—	—
S13/14	—	—	9	5	4	7
S15	—	4	5	8	4	7
S16	—	3	—	—	—	—
S17	—	4	—	—	—	—
S18	—	—	8	—	5	—
S19	—	—	—	—	4	—
S20/21, L20	—	—	15	9	14	10

[a] The relative radioactivity of some ^{125}I proteins is presented in percent of the total radioactivity of 30-S subunit proteins.

from complexes IV and VI) in all complexes studied is correlated with the change in its structure manifested in the set and ratio of proteins of this subunit cross-linked with 16-S RNA.

The comparison of results obtained with free 70-S ribosomes and pre- and posttranslocated complexes V and VI (Table V) shows that the changes of the functional state of the 50-S subunit are also correlated with the changes in its structure manifested in changes of the contacts of 23-S RNA with the proteins of this subunit. The minimum alterations are observed by transfer from the free 70-S ribosome to pretranslocated 70-S complex V (Table VI). Compared with these complexes, the contacts of 23-S RNA with proteins L6, L9, and L27 disappear in the posttranslocated complex, but the contact with L3 appears.

b. Intersubunit RNA–protein contacts. It is evident that the association of subunits and the stabilization of 70-S ribosomal complexes at all steps of protein biosynthesis are determined by specific interactions between the components of two subunits. Three types of such interactions are possible: protein–protein, RNA–RNA, and RNA–protein. The question as to which one of these types of interactions plays a decisive role seems to be incorrect in principle, although RNA–RNA interactions are considered by some to be preferable (see, e.g., (*175, 176*)).

TABLE VI

THE PROTEINS OF THE 50-S SUBUNIT CROSS-LINKED TO 23-S RNA BY IRRADIATION (254 nm, 20–30 PHOTONS PER NUCLEOTIDE) OF FREE 70-S RIBOSOME (TIGHT COUPLE), 70-S RIBOSOME BOUND TO POLY(U) AND AcPhe-tRNA^{Phe} (IV), PRETRANSLOCATED (V) AND POSTTRANSLOCATED (VI) RIBOSOMAL COMPLEXES (*174*)[a]

Protein	70-S Ribosome (Tight Couple)	AcPhe-tRNA^{Phe} 70S · poly(U) (IV)	tRNA^{Phe} · 70S · poly(U) · AcPhePhe-tRNA^{Phe} (V)	AcPhePhe-tRN^{Phe} · 70S · poly(U) (VI)
L1	19	20	20	17
L2	25	34	24	31
L3	—	12	—	14
L4	15	22	17	23
L5	5	12	9	14
L6	7	—	—	—
L9	20	—	15	—
L27	9	—	15	1

[a] The relative radioactivity of separate ^{125}I proteins is given as a percent of the total radioactivity of the 50-S subunit protein.

For identification of the components participating in intersubunit interactions, the methods of chemical modification of the proteins (*177*) and RNA (*178*) have been applied. Alteration of the degree of modification of separate macromolecules upon association of subunits can be caused by both intersubunit and intrasubunit shielding. In the first case, the accessibility of the reaction center of one subunit is decreased by the drawing together with the macromolecule of another subunit; in the second one, it is because of changes in macromolecular conformation or its approach to other macromolecule(s) of the same subunit. Since the association of subunits is conjugated with changes of their conformation (Section III,B,2,a), the results obtained by means of chemical modification cannot be interpreted unambiguously. The results obtained with the help of bifunctional agents (*179, 180*) fail strict interpretation for the same reason.

UV-induced cross-links allow one to obtain reliable information about intersubunit interactions, but of only one type, RNA–protein.

Upon irradiation of free 70-S ribosomes in either pre- or posttranslocated complexes, the ribosomal RNAs may be cross-linked to the proteins of not only the same subunit (Section III,B,2) but also to those of the other subunit (Fig. 16). The latter effect demonstrates intersubunit RNA–protein interactions in these complexes.

The proteins interacting with a ribosomal RNA of another subunit can be divided into two groups. The first group involves the proteins that participate in intersubunit RNA–protein interactions in one or two of the three complexes studied, i.e., they take part in variable RNA–protein contacts (Fig. 16). The number of such contacts is minimal in the posttranslocated complex. When converted into a pretranslocated complex, the set of contacts is dramatically changed and increased. The greatest number of contacts is found in the "tight-couple" 70 S, while three of these proteins also participate in intersubunit RNA–protein interactions in pretranslocated complexes and one is involved in the posttranslocated ones. Due to the total number of intersubunit RNA–protein contacts, "tight-couples" are also actually the tightest; the posttranslocated state is less tight.

The alteration of the set of intersubunit contacts upon transition from the pre- to the posttranslocated state is correlated with the proposition that translocation unlocks the subunits within the 70-S complex while locking occurs upon transition to the pretranslocated state (*181*). On the basis of this, the proteins of the second group (S6, S7, S9/11, S15, S20/S21/L27, L1, and L9), which are involved in intersubunit RNA–protein interactions in all three complexes studied, may be called "hinge" proteins. Those hinge proteins whose

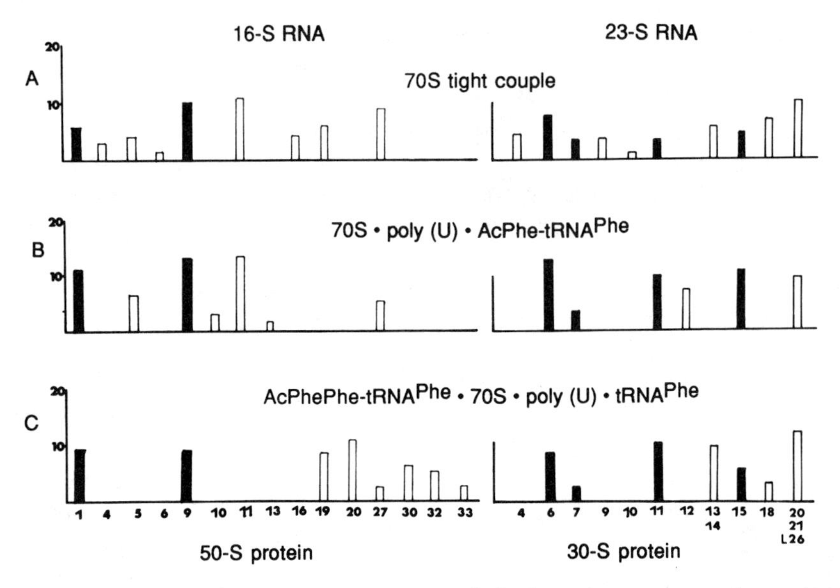

FIG. 16. Intersubunit RNA–protein cross-links formed upon UV-irradiation (254 nm, 20–30 photons per nucleotide) of free 70-S ribosomes (tight couple), pretranslocated (V) and postranslocated (IV) ribosomal complexes (A, B, and C, respectively). Radioactivity of ^{125}I-proteins is given as a percent of the total radioactivity of the proteins of the 30-S and 50-S subunits (left and right panels, respectively). Solid columns: intersubunit RNA–protein contacts common for these complexes (hinge contacts). Empty columns: variable intersubunit RNA–protein contacts.

positions in the ribosome are known (*182–184*) are positioned on both subunits along one edge of intersubunit surface (Fig. 17).

Thus, the cyclic changes in the course of elongation involve not only the conformation of ribosomal subunits (Section III,B,2) but also the interactions between them, i.e., the conformation of the whole

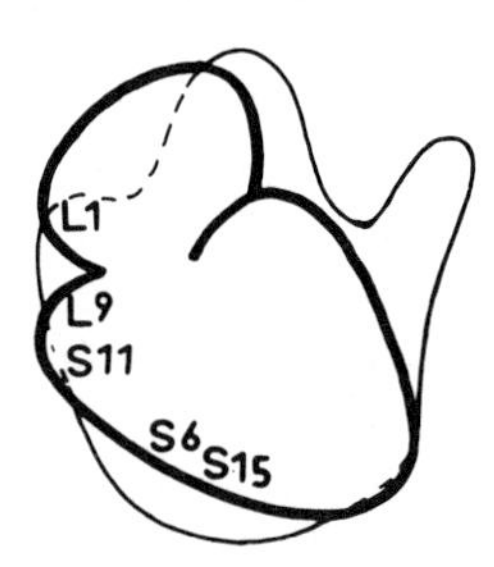

FIG. 17. The location of the intersubunit "hinge" proteins involved in intersubunit RNA–protein interactions on the model of *E. coli* ribosome (*171*).

ribosomal nucleoprotein. Changes in the number of variable contacts may be caused by locking/unlocking of the subunits upon transition of the elongating complex from post- to pretranslocated state and back, and/or by alteration of the conformation of each subunit.

c. The tRNA-binding sites in the 70-S ribosome. The problem of tRNA binding sites of ribosomes is one of the oldest in "ribosomology" (*185–187*) and it is probably more entangled than complicated. This is caused by the absence of a generally accepted definition of these sites. Since two molecules of tRNA must be present simultaneously in a ribosome for its role in transpeptidation, Watson proposed the presence, in elongating ribosomes, of at least two sites of codon-dependent binding prior to transpeptidation, one for pep-tRNA and the second for aminoacyl-tRNA (P and A sites, respectively) (*188*). This definition was further expanded, without any justification, to other ribosomal complexes even if they were unrelated to any of intermediate steps of protein biosynthesis.

This interpretation takes into account the binding of tRNA with only one of the codons (P or A, respectively) participating simultaneously in the elongation cycle. Codon-dependent binding implies only a definite position of the anticodon loop of tRNA in the ribosome, which then migrates along the ribosome due to translocation. But the other tRNA fragments migrate along the ribosome even in the case of a fixed position of the anticodon loop. Thus, the CCA-terminus of aminoacyl-tRNA reaches the ribosome later than does the anticodon loop—only after the release of Tu-GDP. Therefore, the characterization of a tRNA binding site only by codon interaction with tRNA is obviously insufficient and ambiguous.

The codon-dependent binding of tRNA with an elongating ribosome is determined not only by codon–anticodon interaction, which is specific for each tRNA, but also by interaction of other fragments of tRNA with components of the ribosome. Such interactions are necessary for correct positioning of any tRNA and its migration along the ribosome in the course of translation. It is evident that a change of location of a tRNA in a ribosome conjugated with an alteration of the functional state of the translation complex should result in a change of the set of these interactions and thus the set of UV-induced cross-links of tRNA with ribosomal proteins. Naturally, the set of the cross-links characterizes tRNA binding sites of ribosome more fully than only codon–anticodon interactions. This allows us to classify in more detail the tRNA-binding sites of ribosome.

Although 70-S ribosomal complexes containing three tRNA molecules are known (*189–191*), we consider below only those that contain one or two codon-dependent bound molecules of tRNA (Fig. 15) (69). The tRNA binding sites in these complexes may be divided into two groups. The first involves those sites in which a tRNA anticodon interacts with the codon responsible for the binding of pep-tRNA in Watson's pretranslocated complex. The second group includes those sites in which a tRNA anticodon interacts with the codon responsible for the binding of aminoacyl-tRNA in this complex. These groups may be named the P and A groups of tRNA binding sites.

A more precise definition of the sites should take into account the functional state of the complexes. The following nomenclature for specific tRNA binding sites of ribosome is proposed: the capital letters (A and P) denote the group of the sites, and the subscripts indicate the functional state of ribosomal complex (the translation step): i, initiation of translation; t, posttranslocated step; r, initial selection (recognition) of aminoacyl-tRNA; a, pretranspeptidation step, b, posttranspeptidation step; s, stringent control induction step (Table VII). It should be noted that this nomenclature cannot be applied to complexes containing only one ribosomal subunit or to artifical complexes not formed during translation.

UV irradiation of all 70-S ribosomal complexes studied (Fig. 15) results in cross-linking of the proteins of both 30-S and 50-S subunits to tRNA (Figs. 18 and 19). Thus, both ribosome subunits are involved in the formation of these tRNA binding sites. Therefore, the change of not only the conformation of the subunits (Sections III,B,2,a and III,B,3,a) but also of their mutual disposition (Section III,B,3,b) may lead to an alteration of the set of tRNA–protein interactions, i.e., the structure of the tRNA binding sites.

TABLE VII
tRNA BINDING SITES IN RIBOSOMAL COMPLEXES

Steps of translation	Site		Type of tRNA bound
Initiation	P_i		Initiator tRNA
Primary selection of tRNA	P_r		pep-tRNA
		A_r	aa-tRNA · Tu · GTP
Binding of aa-tRNA			
Before transpeptidation	P_a		pep-tRNA
		A_a	aa-tRNA
After transpeptidation	P_b		Deacylated tRNA
		A_b	pep′-tRNA
Translocation	P_t		pep′-tRNA
Switching on of stringent control	P_s		pep-tRNA
		A_s	Deacylated tRNA

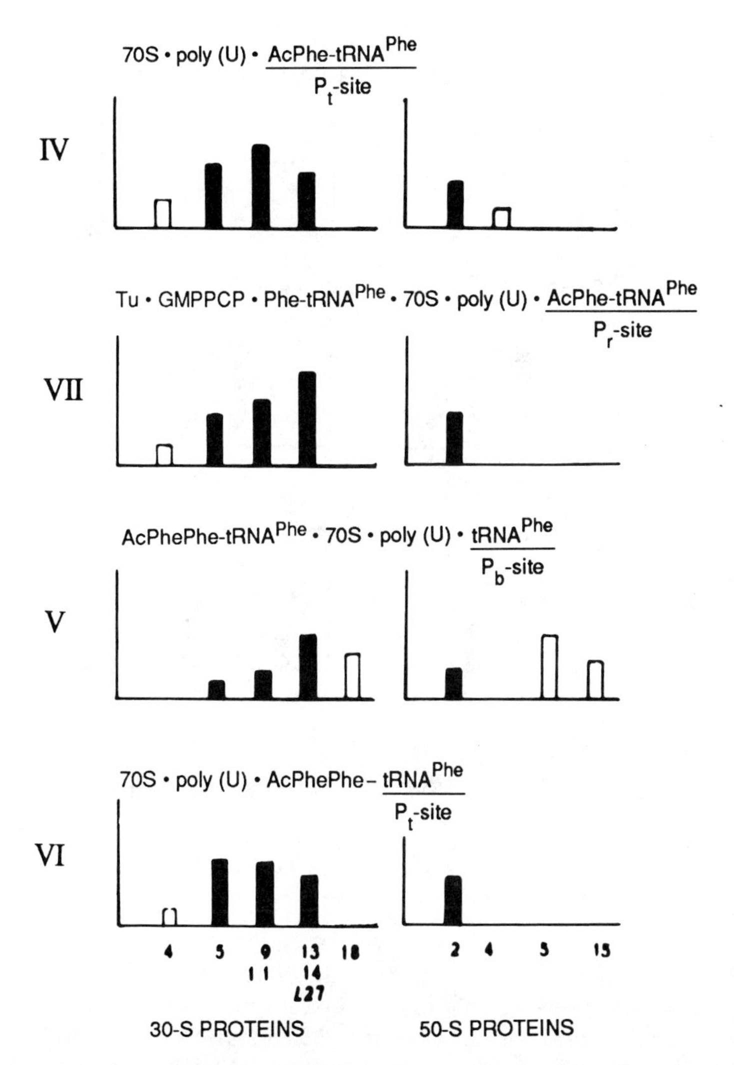

FIG. 18. The proteins cross-linked to tRNA molecules located in the P-group sites. The respective tRNA molecules are underlined in the complexes; below the lines are shown the locations of these tRNAs. The figure shows relative radioactivity of the cross-linked proteins (total radioactivity of all the cross-linked proteins is taken as 100%).

Thus, AcPhe-$tRNA^{Phe}$ in complexes IV and VII, and deacetylated $tRNA^{Phe}$ in complex V, as well as AcPhePhe-$tRNA^{Phe}$ in complex VI, are located by the classic definition in the P-site (*185–187*) or, by the aforementioned definition, in sites of the P-group.

Complexes IV and VI are the closest in composition (Fig. 15). But

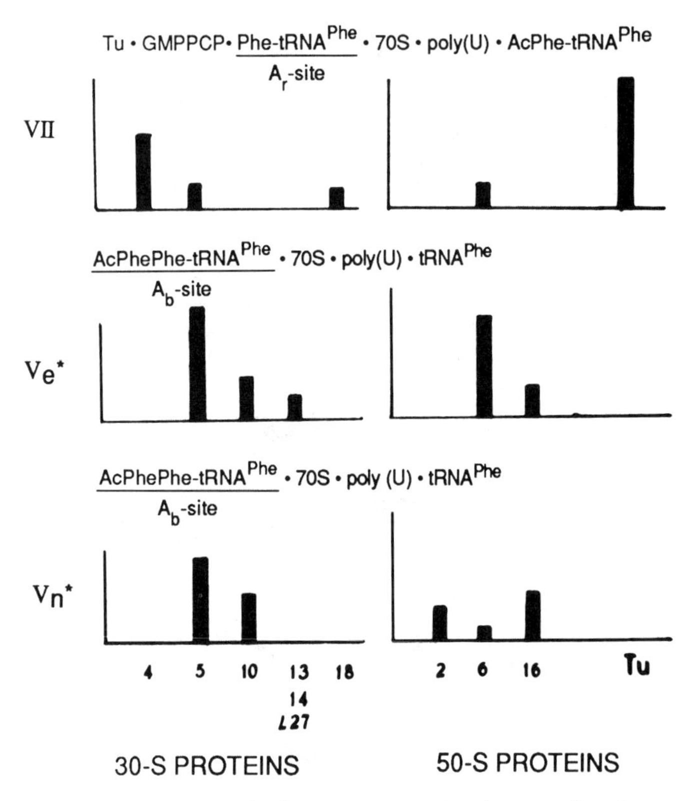

FIG. 19. Proteins cross-linked to tRNA molecules in the A-group sites. Presentation of the data is the same as in Fig. 18. *, Obtained by binding of aa-tRNA in the presence (e) and obsence (n) of Tu · GTP.

the first is obtained by direct binding of AcPhe-tRNA to the 70-S ribosome in the presence of poly(U), i.e., it imitates the 70-S initiation complex; the second complex is formed upon EF-G-promoted translocation of complex V. Therefore, AcPhe-tRNA^{Phe} and AcPhePhe-tRNA^{Phe} in complexes IV and VI are located in the P_i and P_t-sites, respectively. The difference between these sites is revealed by the presence of an additional crosslink of tRNA with protein L4 in the P_i-site (Fig. 18). Only a slight difference in ratio of cross-linked proteins (Fig. 18) is observed for tRNA in the P_r- and P_t-sites (complexes VII and VI, Fig. 15). These results demonstrate the similarity (but not identity) of P_i-, P_t-, and P_r-sites and, hence, of the conformation of subunits as well as of their mutual disposition in complexes IV, V, and VII.

The substitution of pep-tRNA for deacylated tRNA (upon transition from pretranslocated complex V to posttranslocated complex VI) is accompanied by a significant alteration of the set of tRNA–protein cross-links (Fig. 18). Thus, the transition from a pre- to a posttranslocated complex is accompanied by an alteration of not only the conformation of subunits (Sections III,B,2,a and III,B,3,a) and their mutual disposition (Sections III,B,3,b) but also of the structure of the tRNA binding site of the P-group.

Thus, at least four sites in the P-group can be distinguished on the basis of tRNA–protein interactions among tRNA binding sites within the 70-S ribosome. The greatest difference is observed between the tRNA binding P-sites in pre- and posttranslocated complexes.

If a ribosome contains a tRNA molecule in one of the sites of the P-group, the codon-dependent binding of the second tRNA molecule proceeds at sites of the A-group. Upon binding of the ternary complex Phe-$tRNA^{Phe}$ · Tu · GMPPCP with the posttranslocated ribosome, the anticodon of Phe-$tRNA^{Phe}$ interacts with the next codon of mRNA. But the significant part of the tRNA is bound to Tu (*192*); therefore, in complex VII, this part of the tRNA is located outside the ribosome. This complex reflects the recognition step by the ribosome of the incoming aminoacyl-tRNA and, according to the above nomenclature, Phe-$tRNA^{Phe}$ is located in this complex in the A_r-site [earlier called R- (*193*) and entry (*194*) site]. Upon UV-irradiation of complex VII, two proteins from each subunit are cross-linked with Phe-$tRNA^{Phe}$ apart from factor EF-Tu (Fig. 19).

Association of the ternary complex containing GTP with complex IV or VI results in cleavage of a γ,β-pyrophosphate bond, and Tu · GDP is released from the ribosome (*192*). After this, the corresponding aminoacyl-tRNA migrates to an A_a-site accompanied by transpeptidation (*195–197*) and transition of the ribosome into the pretranslocated state. The AcPhePhe-$tRNA^{Phe}$ formed is located, in the above nomenclature, in an A_b-site of a pretranslocated ribosome. Upon passage from an A_r- to an A_b-site, the set of proteins cross-linked with tRNA is markedly changed, although the proteins S5 and L6 are common for these two sites. The nonenzymatic (in the absence of factor Tu) binding of aminoacyl-tRNA with complex IV is also possible, followed by transpeptidation with formation of complex V. But in the case of enzymatic and nonenzymatic binding, i.e., in complexes Ve and Vn, a slightly different set of proteins cross-linked with AcPhePhe-$tRNA^{Phe}$ is formed (Fig. 19). Thus, after enzymatic and nonenzymatic binding, the position of Phe-$tRNA^{Phe}$ in the ribosome is not the same, and this difference is retained also after transpeptidation and transition of the ribosome from a post- to a pre-translocated state.

In amino-acid starvation, the concentration of deacylated tRNAs in the cell is increased and the codon-specific binding of deacylated tRNA with a postranslocated complex takes place (*198, 199*). This binding proceeds at an A_s-site [previously pre-A (*200*), A-like- (*201*), and S-sites (*202*)], since it imitates the first step of switching of so-called "stringent control" in prokaryotes (*203*). The set of proteins cross-linked to tRNA in the A_s-site is quite different from all other sites of A-group studied—S7, L5, and L11 (*202*).

Thus, all things considered, the tRNA-binding sites of the A-group are markedly varied in the set of tRNA–protein contacts, and the difference is observed even in the cases of factor-promoted and nonfactor binding of Phe-$tRNA^{Phe}$ with the posttranslocated complex.

d. Migration of tRNA inside the ribosome upon elongation. Despite approximate and incomplete localization of proteins in the ribosome (*182–184*), the position of the proteins cross-linked with tRNA molecules permits some conclusions about the arrangement of tRNA in ribosomal complexes at different translation steps. One should keep in mind that the accuracy of localization of the protein antigenic determinants by means of immunoelectron microscopy is usually about several nanometers. On the other hand, the distance between an antigenic determinant and the part of the protein molecule that interacts with tRNA can also attain the size of the protein molecule, i.e., also several nanometers. Finally, the localization of antigenic determinants was performed on free ribosomal subunits, whereas the transfer from them to 70-S ribosomal complexes may change the mutual disposition of determinants, since such transition is accompanied by significant alterations of the higher structure of the subunits (Section III,B,3a). Neutron scattering determines another parameter, the distance between centers of mass of proteins (*204*). But almost all remarks considered above for immunoelectron microscopy are valid for this method. Therefore, the correctness of the aforementioned arrangement is limited by the accuracy of protein localization. This seems to be one of the reasons why tRNA interacts simultaneously with the proteins, the antigenic determinants of which are separated from each other by distances exceeding the size of the L-shaped tRNA. The other reason could be an increase in the angle between the branches of the tRNA molecules in ribosomal complexes.

The most probable arrangement of tRNA bound to an A_r-site of the ribosome (Fig. 20) can be deduced from the localization of proteins cross-linked to tRNA and elongation factor EF-Tu (*205, 205a*). The

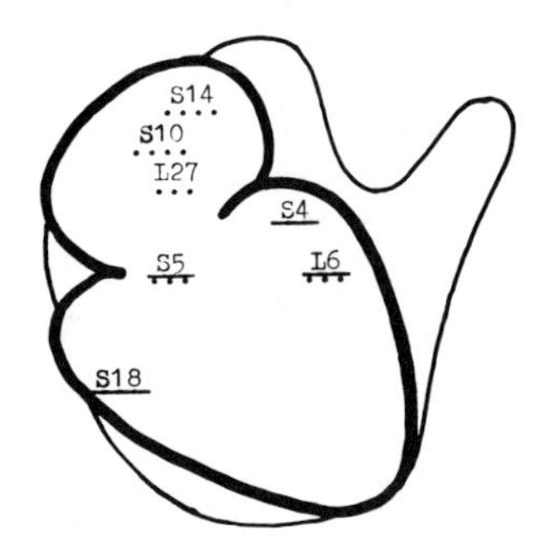

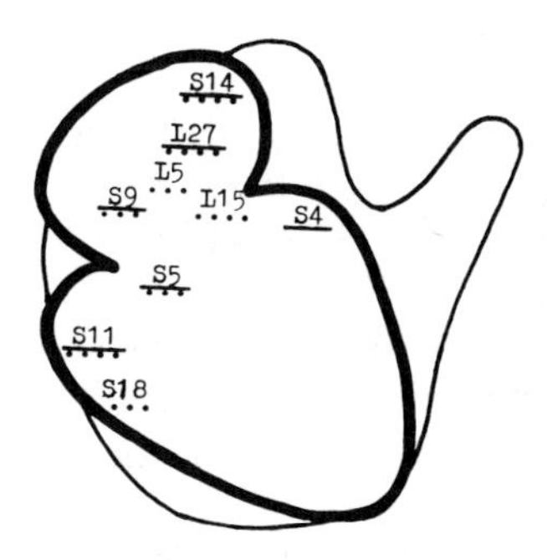

FIG. 20. Location of proteins directly interacting with tRNA molecules in A_r and A_b sites (to the left) and in P_r and P_b sites (to the right). Symbols of proteins interacting with tRNA in the P_r and A_r sites are underlined with solid lines, those in the P_b and A_b sites are shown with broken lines.

ACC terminus of a tRNA in the A_r-site is fixed outside the ribosome because of interactions with EF-Tu. Therefore, proteins S5 and S18, which are the most remote from the Tu-binding site, belong to or are adjacent to the decoding center of the ribosome. Passage of tRNA from the A_r- to the A_b-site is accompanied by liberation of the ACC-terminus and cross-linking of tRNA to proteins S10 and S14/L27 belonging to, or located near, the peptidyltransferase center (*206*, *207*).

Despite the structural differences between complex VII and complex V and VI, a similar arrangement of tRNA molecules bound to the P_r-, P_t-, and P_b-sites of the ribosome is revealed by the similarity of the sets of cross-linked proteins. tRNA molecules bound to these P-sites as well as to both A-group sites contact protein S5 (decoding site of the ribosome). Other proteins cross-linked to tRNA in P-group sites (L2 and S14/L27) belong or are adjacent to the peptidyltransferase center (*56*, *57*). These observations indicate the route of tRNA migration inside the ribosome upon elongation: upon transposition from the A_r to the P_t-site, the anticodon part of tRNA remains attached to the ribosomal decoding site, and the ACC end is shifted counterclockwise (Fig. 20).

Thus, the identification of proteins cross-linked to tRNA and the determination of their ratio demonstrate that the structural characteristics of tRNA binding sites depend on the functional and, hence, on the structural state of a ribosomal complex. This supports the validity of the proposed classification of tRNA binding sites, which takes into account the functional state of complexes, and although this classification is also based on the functional state of the tRNA, it reflects also the structural characteristics of tRNA binding sites.

IV. Conclusion

The general principles, as well as the theoretical and experimental limitations and possibilities of application, of the UV-induced formation of specific polynucleotide–protein cross-links for the investigation of functional nucleoproteins are formulated in this review. The observations presented show that this approach can be used for the solution of a wide variety of problems, from identification of proteins interacting with a certain polynucleotide within nucleoprotein, to investigation of interacting fragments of proteins and polynucleotides. Only some data obtained for well-defined nucleoproteins of the translation system are discussed. Not included are many papers dealing with the application of this method for the investigation of other nucleoproteins [complexes of RNA and DNA polymerases with synthetic polynucleotides and DNA (*86, 87, 209*), complexes of eukaryotic mRNA with proteins of the cytoplasm (*210, 211*), splicing complexes (*212*), etc.]

The appearance of new methods of isolation and analysis of cross-links evidently expands the applicability of the method. It is of great importance the recently introduced method of hybridization selection (*213*), which allows the isolation of definite polynucleotides and the proteins cross-linked to them directly from homogenates of irradiated cells, i.e., to study nucleoproteins existing in intact cells.

UV-induced cross-links allow us to identify the peptides directly involved in the binding of nucleotide coenzymes (e.g., ATP) with corresponding proteins (*214*). Nucleotide-coenzymes are necessary components in the biosynthesis of polynucleotides (ribo- and deoxyribonucleoside triphosphates), transformation of carbohydrates (nucleoside diphosphate sugars), protein biosynthesis (ATP and GTP), redox reactions (NAD, FAD), etc. Since any nucleic base may participate in the formation of UV-induced cross-links with proteins (*64, 121, 215, 216*), this approach can be applied to reveal nucleotide binding sites in the corresponding proteins, and to stimulate studies in protein engineering.

Acknowledgments

We are much indebted to G. A. Titova for valuable help in the preparation of the manuscript, and to M. A. Chlenov for translation of this article.

References

1. W. B. Jacoby and M. Wilchek (eds.), *Methods Enzymol.* **46** (1982).
2. E. I. Budowsky, *in* "Trends in Photobiology" (C. Helene, M. Charlier, T. Montenay-Garestier and G. Laustriat, eds.), p. 93. Plenum, New York, 1982.

3. M. D. Shetlar, *Photochem. Photobiol. Rev.* **5,** 105 (1980).
4. D. G. Knorre (ed.), "Affinity Modification of Biopolymers." Nauka Press, Novosibirsk, USSR, 1983 (in Russian).
4a. A. Wedrychowsky, W. N. Schmidt and L. S. Hnilica, *JBC* **261,** 3370 (1986).
4b. R. M. Starzuk, S. W. Koontz and P. R. Schimmel, Nature **298,** 136 (1982).
4c. F. Tejedor and J. P. G. Ballesta, *Bchem* **25,** 7725 (1986).
5. F. Hansske and R. Cramer, *Methods Enzymol.* **59,** 172 (1979).
5a. P. Senepathy, M. A. Ali and M. T. Jacob, *FEBS Lett.* **19,** 337 (1985).
6. W. Wintermeyer and H. G. Zachau, *FEBS Lett.* **11,** 160 (1970).
7. A. D. Mirzabekov, V. V. Shick, A. V. Belyavsky and S. G. Bavykin, *PNAS* **75,** 4184 (1978).
8. T. I. Tikhonenko, N. P. Kisseleva, A. I. Zintchenko, B. P. Ulanov and E. I. Budowsky, *JMB* **73,** 109 (1973).
9. I. V. Boni and E. I. Budowsky, *J. Biochem. (Tokyo)* **73,** 827 (1973).
10. M. F. Turchinsky, K. S. Kussova and E. I. Budowsky, *FEBS Lett.* **38,** 304 (1974).
11. E. I. Budowsky, N. A. Simukova, M. F. Turchinsky, I. V. Boni and Y. M. Skoblov, *NARes* **3,** 261 (1976).
12. R. Shapiro, *in* "Aging, Carcinogenesis and Radiation Biology" (K. C. Smith, ed.), p. 225. Plenum, New York, 1976.
13. R. Hinnen, R. Schafer and R. M. Franklin, *EJB* **50,** 1 (1974).
14. A. D. Mirzabekov, *Itogi Nauki Tekh.: Gen. Problems Phys.-Chem. Biol.*, 9 (1985) (in Russian).
15. N. Giocanti and B. Ehert, *Int. J. Radiat. Biol.* **38,** 69 (1980).
16. N. Giocanti and B. Ehert, *Int. J. Radiat. Biol.* **40,** 507 (1981).
17. L. Gorelic, *Bchem* **15,** 3579 (1976).
18. G. G. Abdurashidova, M. F. Turchinsky, K. A. Aslanov and E. I. Budowsky, *NARes* **6,** 12 (1979).
19. G. P. Budzik, S. S. M. Lam, H. J. P. Schoemaker and P. R. Schimmel, *JBC* **250,** 4433 (1975).
20. A. K. Pikaev and S. A. Kabakchi, "Reactivity of Primary Products of Water, Radiolysis." Energoizdat, Moscow, 1982 (in Russian).
21. R. Teoule and J. Cadet, *in* "Effects of Ionizing Radiation on DNA" (G. Huttermann, W. Kohlein and R. Teoule, eds.), p. 171. Springer-Verlag, Berlin, 1978.
22. M. Simic and M. Dizdaroglu, *Bchem* **24,** 233 (1985).
23. G. Huttermann, W. Kohnlien, R. Teoule and A. Bertigchamps (eds.), "Effect of Ionizing Radiation on DNA. Physical, Chemical and Biological Aspects." Springer-Verlag, Berlin, 1978.
24. P. Alexander and H. Moroson, *Nature* **194,** 882 (1962).
25. K. C. Smith, *BBRC* **8,** 157 (1962).
26. K. C. Smith, *in* "Photochemistry and Photobiology of Nucleic Acids" (S. Y. Wang, ed.), Vol. 2, p. 187. Academic Press, New York, 1976.
27. M. P. Gordon, *in* "Aging, Carcinogenesis and Radiation Biology" (K. C. Smith, ed.), p. 105. Plenum, New York, 1976.
28. I. Saito, H. Sugiyama and T. Matsuura, *Photochem. Photobiol.* **38,** 735 (1983).
29. N. A. Simukova and E. I. Budowsky, *FEBS Lett.* **38,** 299 (1974).
30. A. V. Borodawkin, E. I. Budowsky, Y. V. Morosov, F. A. Savin and N. A. Simukova, *Itogi Nauki Tekh.: Mol. Biol.* **14,** 123 (1977) (in Russian).
30a. D. Voet, R. A. Cox and P. Doty, *Biopolymers* **1,** 193 (1963).
31. H. Neurath and K. Bailey (eds.), "The Proteins: Chemistry, Biological Activity and Methods," Vol. 1, Parts A and B. Academic Press, New York, 1953.

32. M. O. Dayhoff (ed.), "Atlas of Protein Sequence and Structure." Natl. Biomed. Res. Found., Silver Spring, Maryland, 1972–1976.
33. M. Daniels, *in* "Photochemistry and Photobiology of Nucleic Acids" (S. Y. Wang, ed.), Vol. 1, p. 23. Academic Press, New York, 1976.
34. W. W. Hauswirth and M. Daniels, *in* "Photochemistry and Photobiology of Nucleic Acids" (S. Y. Wang, ed.), Vol. 1, p. 110. Academic Press, New York, 1976.
35. N. J. Turro, "Molecular Photochemistry." Benjamin Press, Reading, Massachusetts, 1965.
36. K. S. Bagdasaryan, "Two-Quantum Photochemistry." Nauka, Moscow, 1976 (in Russian).
37. V. S. Letokhov, "Nonlinear Selective Photoprocesses in Atoms and Molecules." Nauka, Moscow, 1983 (in Russian).
38. O. I. Kovalsky and E. I. Budowsky, *Izv. Acad. Nauk SSSR, Ser. Biol.* in press (1987) (in Russian).
39. G. J. Fisher and H. E. Johns, *in* "Photochemistry and Photobiology of Nucleic Acids" (S. Y. Wang, ed.), Vol. 1, p. 169. Academic Press, New York, 1976.
40. M. E. Umlas, W. A. Franklin, G. L. Chan and W. A. Haseltine, *Photochem. Photobiol.* **42,** 265 (1985).
41. G. J. Fisher and H. E. Johns, *in* "Photochemistry and Photobiology of Nucleic Acids" (S. Y. Wang, ed.), Vol. 1, p. 226. Academic Press, New York, 1976.
42. P. L. Kryukov, V. S. Letochov, D. N. Nikogosyan, A. V. Borodawkin, E. I. Budowsky and N. A. Simukova, *Chem. Phys. Lett.* **61,** 375 (1979).
43. T. N. Menshonkova, N. A. Simukova, E. I. Budowsky and L. B. Rubin, *FEBS Lett.* **112,** 299 (1980).
44. L. B. Rubin, T. N. Menshonkova, N. A. Simukova and E. I. Budowsky, *Photochem. Photobiol.* **34,** 339 (1981).
45. J. Cadet, M. Berger, C. Decarroz, J. R. Wagner, J. E. Van Lier, Y. M. Linot and P. Vigny, *Biochimie* **68,** 813 (1986).
46. C. Cantor and P. Schimmel, "Biophysical Chemistry," Part II. Freeman, San Francisco, California, 1984.
47. D. M. Rayner, A. G. Szabo, R. O. Loutfy and R. W. Yip, *J. Phys. Chem.* **84,** 289 (1980).
48. J. A. Barltrop and J. D. Coyle, "Excited States in Organic Chemistry." Wiley, London, 1975.
49. J. Eisinger and A. A. Lamola, *Mol. Photochem.* **1,** 209 (1971).
50. S. Y. Wang, *Nature* **200,** 879 (1963).
51. M. Leng and G. Felsenfeld, *JMB* **15,** 455 (1966).
52. M. H. Patrick and R. O. Rahn, *in* "Photochemistry and Photobiology of Nucleic Acids" (S. Y. Wang, ed.), Vol. 2, p. 35. Academic Press, New York, 1976.
53. R. V. Bensasson, E. J. Lana and T. Y. Truscott, "Flash Photolysis and Pulse Radiolysis." Pergamon, Oxford, England, 1984.
54. R. O. Rahn and M. H. Patrick, *in* "Photochemistry and Photobiology of Nucleic Acids" (S. Y. Wang, ed.), Vol. 2, p. 97. Academic Press, New York, 1976.
55. E. I. Budowsky, O. I. Kovalsky, N. A. Simukova, V. V. Tsishewsky, D. Y. Yakovlev and L. B. Rubin, *Lasers Life Sci.* **1,** 151 (1986).
56. S. Y. Wang, *in* "Photochemistry and Photobiology of Nucleic Acids" (S. Y. Wang, ed.), Vol. 1, p. 296. Academic Press, New York, 1976.
57. J. A. Lippke, L. K. Gordon, D. E. Brash and W. A. Haseltine, *PNAS* **78,** 3383 (1981).
58. M. E. Umlas, W. A. Franklin, G. L. Chan and W. A. Haseltine, *Photochem. Photobiol.* **42,** 265 (1985).

59. W. A. Franklin, K. M. Lo and W. A. Haseltine, *JMB* **257,** 135 (1982).
59a. S. Humar, N. D. Sharma, R. J. Davies, D. W. Phillipson and J. A. McCloskey, *NARes* **15,** 1199 (1987).
60. K. Smith, *BBRC* **34,** 354 (1969).
61. A. J. Varghese, *in* "Aging, Carcinogenesis and Radiation Biology" (K. C. Smith, ed.), p. 207. Plenum, New York, 1976.
62. M. D. Shetlar, K. Hom, J. Carbon, D. Moy, E. Stlady and M. Watanabe, *Photochem. Photobiol.* **39,** 135 (1984).
63. G. G. Abdurashidova, M. F. Turchinsky, K. A. Aslanov and E. I. Budowsky, *Bioorg. Khim.* **3,** 1570 (1977) (in Russian).
64. E. I. Budowsky, M. S. Axentyeva, G. G. Abdurashidova, N. A. Simukova and L. B. Rubin, *EJB* **159,** 95 (1968).
65. C. A. Harrison, D. H. Turner and D. C. Hinkle, *NARes* **10,** 2399 (1982).
66. J. W. Hockensmith, W. L. Kubazek, W. R. Vorachek and P. H. von Hippel, *JBC* **261,** 3512 (1986).
67. S. V. Kirillov, K. S. Kemhadze, E. M. Makarov, V. I. Makhno, V. B. Odinzov and Y. P. Semenkov, *FEBS Lett.* **120,** 221 (1980).
68. S. V. Kirillov, V. I. Makhno and Y. P. Semenkov, *NARes* **8,** 183 (1980).
69. G. G. Abdurashidova, I. O. Baskaeva, A. A. Chernyi, L. B. Kaminir and E. I. Budowsky, *EJB* **159,** 103 (1986).
70. M. Renaud, H. Bacha, A. Dietrich, P. Remy and J. P. Ebel, *BBA* **653,** 145 (1981).
71. H. J. E. Schoemaker and P. R. Schimmel, *JMB* **84,** 503 (1974).
72. J.-F. Rabek, "Experimental Methods in Photochemistry and Photophysics," Part 2. Wiley, Chichester, England, 1982.
73. J. Calvert and J. Pitts, "Photochemistry." Wiley, New York, 1968.
74. L. H. Shulman, I. Kucan, R. Edelman and R. W. Chambers, *Bchem* **12,** 201 (1973).
75. Z. Kazimirczuk, J. Giziewicz and D. Shugar, *Acta Biochim. Pol.* **20,** 169 (1973).
76. C. S. Park, Z. Hillel and C.-W. Wu, *JBC* **257,** 6944 (1982).
77. C. S. Park, F. Y.-H. Wu and C.-W. Wu, *JBC 257,* 6950 (1982).
78. J.-F. Rabek, "Experimental Methods in Photochemistry and Photophysics," Part 1. Wiley, Chichester, England, 1982.
79. A. Favre, M. Yaniv and A. M. Michelson, *BBRC* **37,** 266 (1969).
80. M. Baltzinger, F. Fasiolo and P. Remy, *EJB* **97,** 481 (1979).
81. A. Havron and I. Sperling, *Bchem* **25,** 5631 (1977).
82. D. N. Nikogosyan, A. A. Oraevsky and V. I. Rupasov, *Chem. Phys.* **41,** 73 (1973).
83. E. I. Budowsky, D. N. Nikogosyan, A. A. Oraevsky, N. A. Simukova and D. Y. Yakovlev, *Photobiochem. Photobiophys.* **4,** 233 (1982).
84. P. Todd and A. Han, *in* "Aging, Carcinogenesis and Radiation Biology" (K. C. Smith, ed.), p. 83. Plenum, New York, 1975.
85. R. Mandel, G. Kolomijtseva and J. G. Brahms, *EJB* **96,** 257 (1979).
86. A. Markovitz, *BBA* **281,** 522 (1972).
87. G. F. Strniste and D. A. Smith, *Bchem* **13,** 485 (1974).
88. G. D. Birnie, *in* "Centrifugal Separation in Molecular and Cell Biology" (G. D. Birnie and D. Rickwood, eds.), p. 169. Butterworth, London, 1978.
89. B. Ehresmann, J. Reinbolt and J. P. Ebel, *FEBS Lett.* **58,** 106 (1975).
90. A. J. H. Wagenmakers, R. J. Reinders and W. J. Van Venrooig, *EJB* **112,** 323 (1980).
91. I. V. Boni, D. M. Isaeva and E. I. Budowsky, *Bioorg. Khim.* **12,** 293 (1986) (in Russian).
92. K. Möller, C. Zwieb and R. Brimacombe, *JMB* **126,** 489 (1978).

93. G. G. Abdurashidova, M. G. Nargizyan, N. V. Rudenko, M. F. Turchinsky and E. I. Budowsky, *Mol. Biol. (Moscow)* **19**, 553 (1985) (in Russian).
94. E. Ulmer, M. Meinke, A. Ross, G. Fink and R. Brimacombe, *MGG* **160**, 183 (1978).
95. G. G. Abdurashidova, V. A. Ovsepyan and E. I. Budowsky, *Mol. Biol. (Moscow)* **19**, 1148 (1985) (in Russian).
96. G. Stoffler and H. G. Wittmann, *PNAS* **68**, 2283 (1971).
97. A. P. Czernilofsky, C. G. Kurland and G. Stoffler, *FEBS Lett.* **58**, 281 (1975).
98. I. Wower, J. Wower, M. Meinke and R. Brimacombe, *NARes* **17**, 4285 (1981).
99. A. Kyriatsoulis, P. Maly, B. Greuer, R. Brimacombe, G. Stoffler, R. Frank and H. Blocker, *NARes* **14**, 1171 (1986).
101. L. Gorelic, *BBA* **390**, 209 (1975).
102. K. Möller and R. Brimacombe, *MGG* **141**, 343 (1975).
103. I. G. Baca and J. W. Bodeley, *BBRC* **70**, 1091 (1976).
104. A. M. Reboud, M. Buisson, M. J. Marion and J. B. Reboud, *EJB* **90**, 421 (1978).
105. T. Uchiumi, K. Terao and K. Ogata, *EJB* **132**, 495 (1983).
106. M. F. Turchinsky, N. E. Broude, K. S. Kussova, G. G. Abdurashidova and E. I. Budowsky, *Bioorg. Khim.* **3**, 1013 (1977) (in Russian).
107. N. Riehl, P. Remy, J. P. Ebel and B. Ehresmann, *EJB* **128**, 427 (1982).
108. H. G. Wittmann, J. A. Littlechild and B. Wittmann-Leibold, *in* "Ribosomes: Structure, Function and Genetics" (G. Chambliss, G. R. Craven, J. Davies, K. Davis, L. Kahan and M. Nomura, eds.), p. 51. University Park Press, Baltimore, Maryland, 1980.
109. L. Grossman and K. Moldave (eds.), *Methods Enzymol.* **65** (1980).
110. L. A. Osterman, "Methods in Protein and Nucleic Acids Research," Vol. 2. Springer-Verlag, Berlin, 1984.
111. M. F. Turchinsky, N. E. Broude, K. S. Kussova, G. G. Abdurashidova, E. V. Muchamedganova, J. N. Schatsky, T. F. Bystrova and E. I. Budowsky, *EJB* **90**, 83 (1978).
112. O. A. Roholt and D. N. Pressman, *Methods Enzymol.* **25**, 438 (1972).
113. J. Ackerman and P. B. Sigler, *in* "Structural Aspects of Recognition and Assembly in Biological Macromolecules" (M. Balaban, L. Weissman, W. Traut and A. Yonath, eds.), p. 645. Balaban ISS, Rehovot, Israel, 1981.
114. G. G. Abdurachidova, E. A. Tsvetkova and E. I. Budowsky, *Bioorg. Khim.* **11**, 1353 (1985) (in Russian).
114a. R. V. Miller and P. S. Sypherd, *JMB* **78**, 527 (1973).
115. R. Brimacombe, J. Atmadja, A. Kuriatsoulis and W. Stiege, *in* "Structure, Function and Genetics of Ribosomes" (B. Hardesty and G. Kramer eds.), p. 184. Springer-Verlag, New York, 1986.
116. C. Ehresmann, B. Ehresmann, R. Millon, J. P. Ebel, K. Nurse and J. Ofengand, *Bchem* **23**, 429 (1984).
117. J. Ciesiolk, K. Nurse, J. Klein and J. Ofengand, *Bchem* **24**, 3233 (1985).
118. C. Zwieb and R. Brimacombe, *NARes* **6**, 1775 (1979).
119. J. P. Ebel, B. Ehresmann, C. Ehresmann, M. Mougel, R. Giege and D. Moras, *in* "Structure, Function and Genetics of Ribosomes" (B. Hardesty and G. Kramer, eds.), p. 271. Springer-Verlag, New York, 1986.
120. P. Maly, J. Rinke, E. Ulmer, C. Zwieb and R. Brimacombe, *Bchem* **19**, 4179 (1984).
121. V. T. Yue and P. R. Schimmel, *Bchem* **16**, 4678 (1977).
121a. P. R. Paradiso and W. Konigsberg, *JBC* **257**, 1462 (1982).
121b. L. Lica and D. S.Ray, *JBC* **261**, 3512 (1986).
121c. L. P. Coorotchkina, A. A. Komissarov and C. Y. Kolomijtseva, *Biokhimia (Moscow)* **52**, 1678 (1987) (in Russian).

122. L. P. Nelles and J. R. Bamburg, *Anal. Biochem.* **255,** 150 (1979).
123. L. A. Neiman, V. S. Smolyakov and A. V. Shishkov, *Itogi Nauki Tekh. Obschie Voprosy Fis. Chem. Biol.* **2,** 78 (1985) (in Russian).
124. B. Alberts, D. Bray, J. Lewis, M. Raff, K. Roberts and J. Watson, "Molecular Biology of the Cell." Garland, New York, 1983.
125. B. Hardesty and G. Kramer (eds.), "Structure, Function and Genetics of Ribosomes." Springer-Verlag, New York, 1986.
126. F. Chapeville, F. Lipmann, G. Von Ehrenstein, B. Weisblum, W. J. Ray, Jr., and S. Benzer, *PNAS* **48,** 1086 (1962).
127. P. R. Schimmel and D. Soll, *ARB* **48,** 1013 (1979).
128. R. Thiebe, K. Harbers and H. G. Zachau, *EJB* **26,** 144 (1972).
129. A. D. Mirzabekov, D. Lastity, E. S. Levina and A. A. Bayev, *Nature NB* **229,** 21 (1971).
130. H. J. P. Schoemaker, G. P. Budzik, R. C. Giege and P. R. Schimmel, *JBC* **250,** 4440 (1975).
131. J. J.Rosa, M. D. Rosa and P. P. Sigler, *Bchem* **18,** 637 (1979).
135. S. Mizushima and M. Nomura, *Nature* **226,** 1214 (1970).
136. M. Nomura and W. A. Held, *in* "Ribosomes" (M. Nomura, A. Tissieres and P. Lengyel, eds.), p. 193. CSH Laboratory, Cold Spring Harbor, New York, 1974.
137. W. A. Held and M. Nomura, *Bchem* **12,** 3273 (1975).
138. A. M. Kopylov, E. S. Shalaeva and A. A. Bogdanov, *Dokl. Akad. Nauk SSSR* **216,** 1178 (1974) (in Russian).
139. J. M. Dunn and K.-P. Wong, *JBC* **16,** 7705 (1979).
140. A. D. Pivazyan, G. G. Abdurashidova, M. F. Turchinsky and E. I. Budowsky, *Bioorg. Khim.* **6,** 461 (1980) (in Russian).
141. H. Weissbach and S. Ochoa, *ARB* **45,** 191 (1976).
142. C. Gualerzi, G. Risuleo and C. L. Pon, *Bchem* **16,** 1684 (1977).
143. H. Paradies, A. Franz, C. Pon and C. Gualerzi, *BBRC* **59,** 600 (1974).
144. C. Pon and C. Gualerzi, *PNAS* **71,** 4950 (1974).
145. C. Gualerzi, M. Grandolfo, H. H. Paradies and C. Pon, *JMB* **95,** 569 (1975).
146. N. E. Broude, K. S. Kussova, N. I. Medvedeva and E. I. Budowsky, *Bioorg. Khim.* **4,** 1687 (1978) (in Russian).
147. E. I. Budowsky, M. F. Turchinsky, N. E. Broude, I. V. Boni, I. V. Zlatkin, K. S. Kussova, N. I. Medvedeva, G. G. Abdurashidova, K. A. Aslanov and T. A. Salikhov, *in* "Frontiers of Bioorganic Chemistry and Molecular Biology" (S. N. Ananchenko, ed.), p. 389. Pergamon, Oxford, England, 1980.
148. R. Ewald, C. Pon and C. Gualerzi, *Bchem* **15,** 4786 (1976).
149. C. J. Michalsky, B. H. Sells and A. J. Wahba, *FEBS Lett.* **71,** 347 (1976).
150. N. E. Broude, K. S. Kussova, N. I. Medvedeva and E. I. Budowsky, *Bioorg. Khim.* **6,** 1303 (1980) (in Russian).
151. M. Takanami and T. Okamoto, *JMB* **7,** 323 (1963).
152. J. Shine and L. Dalgarno, *PNAS* **71,** 1342 (1974).
153. C. Vermeer, W. von Alphen, P. von Knippenberg and L. Bosen, *EJB* **40,** 295 (1973).
154. A. R. Subramanian, *This Series* **28,** 101 (1983).
155. N. E. Broude, K. S. Kussova, N. I. Medvedeva and E. I. Budowsky, *EJB* **132,** 139 (1983).
156. W. A. Held, B. Ballou, S. Mizuchima and M. Nomura, *JBC* **249,** 3103 (1974).
157. I. V. Boni, I. V. Zlatkin and E. I. Budowsky, *Bioorg. Khim.* **5,** 1633 (1979) (in Russian).
158. I. V. Boni, I. V. Zlatkin and E. I. Budowsky, *EJB* **121,** 371 (1982).
159. R. A. Kastelein, E. Remaut, W. Fiers and J. van Duin, *Nature* **295,** 35 (1982).

160. H. F. Lodish, *Nature* **226,** 705 (1970).
161. K. Horiuchi, *in* "RNA Phages" (N. D. Zinder, ed.), Cold Spring Harbor Monogr. Ser., p. 29. CSH Laboratory, Cold Spring Habor, New York, 1975.
162. I. V. Boni and D. M. Isaeva, *Dokl. Akad. Nauk SSSR* **298,** 1015 (1988) (in Russian).
163. Y. Gluzman and T. Shenk (eds.), "Eucaryotic Transcription: The Role of cis- and trans-Acting Elements in Initiation." CSH Laboratory, Cold Spring Harbor, New York, 1985.
164. J. E. McCarty, H. H. Schairer and W. Sebalt, *EMBO J.* **4,** 519 (1985).
165. M. Revel, *in* "Molecular Mechanisms of Protein Biosynthesis" (H. Weissbach and S. Pestka), p. 245. Academic Press, New York, 1977.
166. G. Chambliss, G. R. Craven, J. Davies, K. Davis, L. Kahan and M. Nomura (eds.), "Ribosomes: Structure, Function and Genetics." University Park Press, Baltimore, Mayland, 1980.
167. N. E. Broude, N. I. Medvedeva, K. S. Kussova and E. I. Budowsky, *Mol. Biol. (Moscow)* **19,** 1269 (1985) (in Russian).
168. L. Gorini, *Nature* **234,** 261 (1971).
169. D. C. Fritzinger and M. J. Fournier, *NARes* **10,** 2419 (1982).
170. G. N. Bondarev, V. V. Isaev-Ivanov, L. S. Isaeva-Ivanova, V. B. Odinzov, O. Y. Sidorov and V. N. Fomichev, *Mol. Biol. (Moscow)* **16,** 352 (1982) (in Russian).
171. G. G. Abdurashidova, E. A. Tsvetkova, A. A. Chernyi, L. B. Kaminir and E. I. Budowsky, *FEBS Lett.* **185,** 291 (1985).
172. Y. P. Semenkov, E. M. Makarov, V. I. Makhno and S. V. Kirillov, *FEBS Lett.* **144,** 125 (1982).
173. W. Wintermeyer, R. Lill, H. Paulsen and J. M. Robertson, *in* "Structure, Function and Genetics of Ribosomes" (B. Hardesty and G. Kramer, eds.), p. 286. Springer-Verlag, New York, 1986.
174. G. G. Abdurashidova, E. A. Tsvetkova and E. I. Budowsky, *Bioorg. Khim.* **11,** 417 (1985) (in Russian).
175. D. P. Burma, B. Nag and D. S. Fewori, *PNAS* **80,** 4875 (1983).
176. M. M. Yusupov and A. S. Spirin, *FEBS Lett.* **197,** 229 (1986).
177. C. Gualerzi, H. Ohsawa, G. Risuleo and C. L. Pon, *in* "Structural Aspects of Recognition and Assembly in Biological Macromolecules" (M. Balaban, J. L. Sussman, W. Traut and A. Yonath, eds.), p. 793. Balaban ISS, Rehovot, Israel, 1981.
178. N. Meier and R. Wagner, *EJB* **146,** 83 (1985).
179. S. E. Skold, *Biochimie* **63,** 53 (1981).
180. J. A. Cover, J. M. Lambert, C. M. Norman and R. R. Traut, *Bchem* **20,** 2843 (1981).
181. A. S. Spirin, *Dokl. Akad. Nauk SSSR* **179,** 1467 (1968) (in Russian).
182. M. Oakes, E. Henderson, A. Scheinman, M. Clark and J. A. Lake, *in* "Structure, Function and Genetics of Ribosomes" (B. Hardesty and G. Kramer, eds.), p. 47. Springer-Verlag, New York, 1986.
183. G. Stoffler and M. Stoffler-Meilicke, *in* "Structure, Function and Genetics of Ribosomes" (B. Hardesty and G. Kramer, eds.), p. 28. Springer-Verlag, New York, 1986.
184. R. R. Traut, D. S. Tewari, A. Sommer, G. R. Gavino, H. M. Olson and D. G. Glitz, *in* "Structure, Function and Genetics of Ribosomes" (B. Hardesty and G. Kramer, eds.), p. 286. Springer-Verlag, New York, 1986.
185. I. B. Prince and R. A. Garrett, *TIBS* **7,** 79 (1982).
186. K. H. Nierhaus and H. J. Rheinberger, *TIBS* **9,** 428 (1984).
187. I. B. Prince and R. A. Garrett, *TIBS* **7,** 280 (1982).
188. J. D. Watson, *Bull. Soc. Chem. Biol.* **46,** 1399 (1964).

189. S. V. Kirillov, *Itogi Nauki Tekh.: Mol. Biol.* **18,** 5 (1980) (in Russian).
190. R. A. Grajevskaja, Y. P. Ivanov and E. M. Saminsky, *EJB* **128,** 47 (1982).
191. K. H. Nierhaus, H.-J. Rheinberger, U. Geigenmuller, A. Grinke, H. Saruyama, S. Schilling and P. Wurmbach, *in* "Structure, Function and Genetics of Ribosomes" (B. Hardesty and G. Kramer, eds.), p. 444. Springer-Verlag, New York, 1986.
192. H. Weissbach, *in* "Ribosomes: Structure, Function and Genetics" (G. Chamblis, G. R. Craven, J. Davies, K. Davis, L. Kahan and M. Nomura, eds.), p. 377, University Park Press, Baltimore, Maryland, 1980.
193. J. A. Lake, *PNAS* **74,** 1903 (1977).
194. B. Hardesty, W. Culp, and W. McKeehan, *CSHSQB* **34,** 331 (1969).
195. A. Kaji, K. Igarashi and H. Ishitsuka, *CSHSQB* **34,** 167 (1969).
196. S. Pestka, *CSHSQB* **34,** 395 (1969).
197. R. J. Harris and S. Pestka, *in* "Molecular Mechanisms of Protein Biosynthesis" (H. Weissbach and S. Pestka, eds.), p. 413. Academic Press, New York, 1977.
198. F. S. Pedersen, E. Lund and N. O. Kjeldgaard, *Nature NB* **343,** 13 (1973).
199. W. A. Haseltine and R. Block, *PNAS* **70,** 1564 (1973).
200. J. A. Lake, *in* "Ribosomes: Structure, Function and Genetics" (G. Chamblis, G. R. Craven, J. Davies, K. Davis, K. Kahan and M. Nomura, eds.), p. 207. University Park Press, Baltimore, Maryland, 1980.
201. M. Cashel, *Annu. Rev. Microbiol.* **29,** 301 (1975).
202. G. G. Abdurashidova, M. F. Turchinsky and E. I. Budowsky, *FEBS Lett.* **129,** 59 (1981).
203. G. Koch and D. Richter (eds.), "Regulation of Macromolecular Synthesis by Low Molecular Weight Mediators." Academic Press, London, 1979.
204. P. B. Moore, M. Capel, M. Kjeldgaard and D. M. Engelman, *in* "Structure, Function and Genetics of Ribosomes" (B. Hardesty and G. Kramer, eds.), p. 87. Springer-Verlag, New York, 1986.
205. J. Rychlik, O. W. Odon and B. Hardesty, *Bchem* **22,** 85 (1983).
205a. A. S. Girshovich, E. S. Bochkareva and V. D. Vasiliev, *FEBS Lett.* **197,** 192 (1986).
206. A. A. Krayevsky, M. K. Kukhanova and B. P. Gottikh, *Itogi Nauki Tekh.: Mol. Biol.* **9,** 6 (1977) (in Russian).
207. B. S. Cooperman, *in* "Ribosomes: Structure, Function and Genetics" (G. Chamblies, G. R. Craven, J. Davies, K. Davis, L. Kahan and M. Nomura, eds.), p. 331. University Park Press, Baltimore, Maryland, 1980.
209. L. Hillel and W. Wu, *Bchem* **17,** 2954 (1978).
210. B. Setyono and J. R. Greenberg, *Cell* **24,** 775 (1981).
211. J. R. Greenberg and E. Carroll, *MC Biol* **5,** 342 (1985).
212. V. Gerke and J. A. Steitz, *UCLA Symp. Mol. Cell. Biol., New Ser.* **67** (1987).
213. I. P. Schouten, *JBC* **260,** 9916 (1985).
214. G. R. Banks and S. S. Sedgwick, *Bchem* **25,** 5882 (1986).
215. F. Maley and G. Maley, *JBC* **257,** 11876 (1982).
216. J. Sperling and A. Havron, *Bchem* **15,** 1489 (1976).

Regulation of Collagen Gene Expression

Paul Bornstein* and
Helene Sage†

*Departments of *Biochemistry and
†Biological Structure
University of Washington
Seattle, Washington 98195*

The collagens are a family of related proteins characterized by the repeating tripeptide sequence Gly-X-Y, in which X is often proline and Y is often hydroxyproline, and a triple-helical structure. This minimal definition of a collagen encompasses the twelve commonly recognized collagen types,[1] a number of different collagens not yet

[1] Collagen types are structurally and genetically distinct proteins present in the same organism. Types I–XII are numbered sequentially in the order of their identifica-

accorded a type number, and several proteins whose structures and functions are sufficiently different from the collagens to warrant their exclusion from the collagen gene family. The latter include the C1q component of complement (*1*), acetylcholinesterase (*2*), several pulmonary surfactant proteins (*3, 4*), conglutinin (*5*), and at least two related mannose-binding liver proteins (*6*).

The twelve currently recognized collagen types (Table I) have characteristic tissue distributions and play different functional roles in embryogenesis, in development, and in the adult animal (*9*). A useful distinction is to assign collagens to one of two categories: fibrillar collagens, characterized by continuous or minimally interrupted triple helices, and nonfibrillar collagens, characterized by relatively frequent interruptions in the repeating triplet sequence (*10*). The former category, which includes types I, II, III, V, VI, VII, X, and XI, might be expected to function largely as fiber-forming or fiber-associated structural proteins, whereas the latter (types IV, VIII, IX, and XII) may have more diverse structural roles. The collagens have also been classified in three groups: Group 1 (types I–III, V, and XI) with chains of $M_r \sim 100{,}000$; Group 2 (types IV and VI–VIII) with chains of $M_r > 100{,}000$; Group 3 (types IX and X) with chains of $M_r < 100{,}000$ (*11*). Type XII, which was originally classified in Group 3, is now known to have an amino-terminal non-triple-helical domain of $M_r > 200{,}000$ (*12*).

It is likely that many mechanisms are employed in the transcriptional and posttranscriptional control of expression of the genes coding for these proteins, a variety that reflects their diverse biological roles and their requirement both for normal morphogenesis and for tissue injury and repair. Although the cDNA and genomic structures for the 22 or more genes coding for the 12 well-established collagen types are being elucidated, the collagens for which we have the most information are types I–IV. Studies at the molecular level of the control of collagen gene expression are therefore largely limited to these proteins. We cannot cover in this review the very extensive literature that addresses the regulation of collagen protein synthesis at the cell, tissue, or whole animal level, although in many cases these earlier studies pointed the way to the cellular and molecular biological experiments currently in progress. We also do not consider, per se,

---

tion. Each type consists of one, two, or three different chains, termed α chains and identified as $\alpha 1$, $\alpha 2$, and $\alpha 3$. The relation of a given chain to a type is indicated by placing the type number in parentheses; thus, the structure of type-I collagen is $[\alpha 1(\mathrm{I})]_2\, \alpha 2(\mathrm{I})$.

TABLE I
THE COLLAGENS

Type	Chains	Genes[a]	Chromosomal location[b]
I	α1	COL1A1	17q21–22
	α2	COL1A2	7q21–22
II	α1	COL2A1	12q13.1–13.2
III	α1	COL3A1	2q24.3–31
IV	α1	COL4A1	13q33–34
	α2	COL4A2	13q33–34
V	α1	COL5A1	
	α2	COL5A2	2q14–32
	α3	COL5A3	
VI	α1	COL6A1	21q-223[c]
	α2	COL6A2	21q-223[c]
	α3	COL6A3	2q-37[c]
VII	α1	COL7A1	
VIII	α1	COL8A1	
IX	α1	COL9A1	
	α2	COL9A2	
	α3	COL9A3	
X	α1	COL10A1	
XI	α1	COL11A1	1p21
	α2	COL11A2	
	α3	COL11A3	
XII	α1	COL12A1	

[a] A number of workers have adopted the human gene mapping workshop nomenclature for designating collagen gene loci. Thus, the α1 type-I gene is referred to as COL1A1. However, as in the hemoglobin field, the protein chain designation is also used for the gene. We adhere to the latter practice in this review.

[b] See Myers and Emanuel (7) for a recent review.

[c] From Weil *et al.* (8).

studies of collagen gene structure. Useful reviews of this topic (*13–17*) and of the structural abnormalities that occur in the heritable disorders of types-I and -III collagen (*18–21*) have appeared. We hope, at this relatively early juncture, to formulate certain general principles that underlie the regulation of collagen genes and to determine to what extent, if any, the unusual structure and function of this protein family have led to the evolution of unique genetic regulatory mechanisms.

I. Transcriptional Control

Early studies of chick embryo development showed quite clearly that increases in procollagen synthesis are directly proportional to increases in procollagen mRNA content per cell (*22, 23*). Subsequently, parallel increases in collagen content and in type-I mRNA levels were noted in CCl_4-induced liver fibrosis in rats (*24*), and several cell-culture studies showed a good correlation between mRNA levels and changes in collagen synthesis resulting from modulation by interleukin-1 (*25*), γ-interferon (*26*), a hepatic fibrogenic factor (*27*), and phorbol esters (*28*) (see Section IV). However, all of these studies were based on analyses of mRNA levels and none distinguished between changes in transcription rate and changes in mRNA processing, transport, or stability. A more detailed understanding of transcriptional control has come from the examination of a number of other experimental systems (see Sections I,A–E).

In normal fibroblasts, the ratio of $\alpha1$(I) to $\alpha2$(I) mRNA is 2 : 1 (*29–31*). Thus one need not postulate translational control to achieve the 2 : 1 ratio in $\alpha1$(I) to $\alpha2$(I) collagen-chain synthesis that has been observed by many investigators. The basis for this coordinate control of $\alpha1$(I) and $\alpha2$(I) mRNA levels, whether it be at the level of transcription rate or RNA processing or stability, is not known.

A. Viral Transformation

Initial studies of chick embryo fibroblasts (CEF) transformed with Rous sarcoma virus (RSV) demonstrated a correlation between reduced levels of translatable type-I collagen mRNA and procollagen synthesis (*32–35*). This effect is clearly at the level of transcription, as both the levels of nuclear RNA precursors of $\alpha2$(I) mRNA (*36*) and transcription rates of the $\alpha1$(I) and $\alpha2$(I) genes (*37*) were reduced in RSV-transformed CEF. However, other than an association between the activity of the protein kinase, $pp60^{src}$, and changes in collagen synthesis as revealed by the use of temperature-sensitive mutants of RSV (*37*), very little is known of the mechanisms that influence transcriptional activity of collagen genes in retroviral transformation. The process is further complicated by the finding that in some cases, as in transformation of BALB/c 3T3 cells by Kirsten murine sarcoma virus, changes in the relative proportions of different collagen types also occur (*38*); in RSV-transformed BALB/c 3T3 cells, a 4-fold increase in type-III collagen mRNA was observed in conjunction with a decrease to one-seventh in α2(I) mRNA (*39*). The possibility must also be considered that changes in collagen synthesis are secondary to other effects of retroviral transformation, such as cytoskeletal changes, rather than to the transformation process itself (*40*).

The transcription of plasmids carrying the mouse α2(I) promoter, stably introduced into NIH/3T3 cells, is reduced by transformation of cells with the mouse Moloney sarcoma and leukemia virus complex (MMSV), or by introduction of a plasmid containing the v-*mos* oncogene linked to the MMSV long-terminal-repeat sequence (*41, 42*). Thus, a common mechanism may inhibit both the endogenous and transfected collagen promoters. The same workers developed a selection system in which cells transfected with the neomycin-resistance gene fused to the α2(I) collagen promoter were transformed with MMSV or by a plasmid carrying the v-*mos* oncogene and exposed to a chemical mutagen. Since transformation caused a marked reduction in mRNA levels for the product of the neomycin-resistance gene, cells that showed resistance to Antibiotic G418[2] (an analogue of neomycin) could be selected. When mutants that failed to express the v-*mos* gene were excluded, two mutants were selected that also showed an increase in collagen mRNA. These cells may carry mutations in cellular *trans*-acting factors that regulate expression of type-I collagen genes and could be useful in the study of such factors.

Even less is known of the basis for changes in collagen synthesis and processing in transformation with DNA viruses (see *43* for a discussion). mRNA levels for GP 140, a subunit of type-VI collagen, are reduced in simian virus 40 (SV40)-transformed WI-38 human lung fibroblasts (*44*).

B. Chemical Transformation

A number of studies have documented changes in collagen synthesis in chemically transformed epithelial cells and fibroblasts (*45–50*). An underlying theme appears to be transcriptional inhibition of one or another of the two genes coding for type-I collagen, but the basis for these effects is not understood. In a rat epithelial cell line transformed with a fluorene derivative, α2(I) mRNA is absent (*46, 48*), whereas hamster embryo fibroblasts transformed by 4-nitroquinoline-1-oxide

[2] Antibiotic G418 (proprietary name, Geneticin), is a diglycoside derivative of streptamine (Chemical Abstracts Number 49863-47-0) [Eds.].

lack α1(I) mRNA (*49, 50*). Transformed hamster fibroblasts that lack hybridizable or *in vitro* translatable α1(I) mRNA do not appear to have changes in genomic DNA detected by Southern blot analysis (*50*). The heterogeneity observed in the α2(I) chain in these transformed cells (*49*) is likely to reflect posttranslational modifications or changes in processing (*50*).

C. Inactivation by Insertional Mutagenesis

Some unexpected insights into transcriptional control of collagen gene expression have come from studies of the germ-line integration of Moloney murine leukemia virus (M-MuLV) in mice. Jaenisch *et al.* (*51*) obtained 13 substrains of mice, each carrying one copy of the M-MuLV integrated at different sites in its genome. One of these substrains, *Mov*-13, was lethal at an early embryonic age in the homozygous state (*52*). Homozygous (*Mov*-13/*Mov*-13) mice were arrested in development at days 11 or 12 and died at days 12 to 14 of gestation. The locus of integration of *Mov*-13 was shown to be a 14-kb *Eco*RI fragment (*52*). When this fragment was used in Northern blots of RNA from embryos of different ages, it was observed that the locus became transcriptionally active during the second half of embryogenesis (*53*). Hybridization with a panel of cDNA clones identified the locus as the collagen α1(I) gene (*53*). Specifically, the M-MuLV was found to be integrated 19 bp downstream from the junction separating the first exon and first intron of the mouse α1(I) gene (*54*). This site of integration can account for the histological findings in homozygous *Mov*-13 mice, which include necrosis of mesenchymal cells and rupture of major blood vessels (*55*).

The basis for the failure of homozygous *Mov*-13 mice to synthesize the α1(I) chain of type-I collagen has been pursued by Breindl *et al.* (*56*), who identified two sites hypersensitive to DNase-I in the α1(I) gene present in all tissues regardless of the transcriptional activity of the gene, and a third that correlated with active transcription. The latter site, located about 200 bp upstream from the start of transcription, was absent in mutant alleles containing the viral insertion. Since sites hypersensitive to DNase-I may play important roles in developmental gene activation (*57*), the altered chromatin structure resulting from viral insertion could affect activation of the α1(I) gene during embryogenesis. It is of interest that one of the two sites hypersensitive to DNase-I that does not correlate with the transcriptional activity identified (*56*) is located in the first intron and is also found in human chromatin (*58*). This site corresponds closely to a region of substantial similarity in DNA sequence between the rat and human introns (see Fig. 2).

The failure of *Mov*-13 mice to synthesize type-I collagen results from a block in transcription of the $\alpha 1$(I) gene (*59*). This was documented by both nuclear "run-on" transcription experiments and by an S1 mapping analysis of nuclear RNA. A possible basis for altered transcriptional activity of the $\alpha 1$(I) gene was the finding (*60*) that viral insertion is associated with *de novo* methylation within 1 kb 5′ from the insertion site [encompassing the $\alpha 1$(I) promoter] in embryonic tissues. There is now substantial evidence that DNA methylation inhibits gene expression in some cases and that a possible basis for this inhibition is interference with the binding of proteins to specific *cis*-acting elements (*61*).

The abnormal collagen phenotype in *Mov*-13 fibroblast cell lines transformed by SV40 has been rescued by retroviral-mediated introduction of a full-length human $\alpha 1$(I) cDNA clone (*62*). Infected, transduced *Mov*-13 cells secrete a stable heterotrimeric type-I collagen that contains mouse $\alpha 2$(I) and human $\alpha 1$(I) chains. Both human and mouse $\alpha 1$(I) collagen genes have also been introduced into SV40-transformed *Mov*-13 cells by stable transfection (*63*). Again, the products of the introduced genes associated with the mouse $\alpha 2$(I) chain to produce stable type-I collagen. The human gene, under the control of its promoter and 1.5 kb of 5′ flanking sequence, was poorly expressed, but a high level of expression, comparable to the endogenous gene, was noted when the SV40 promoter–enhancer was used (*63*). Chains derived from hybrid mouse–human collagen molecules migrated more slowly on acrylamide gel electrophoresis than did normal chains. This behavior generally indicates an increased degree of posttranslational modification and results from a slower rate of triple-helix formation.

D. Transcriptional Control Elements

Promoter sequences have been reported for four collagen genes: the $\alpha 1$(I) (*54, 64–66*); $\alpha 2$(I) (*67–72*); $\alpha 1$(II) (*73–74*); and $\alpha 1$(III) (*39*). There is substantial conservation of sequence between mammalian species for a given collagen gene, e.g., for $\alpha 1$(I) in human and mouse (*54, 65*), for $\alpha 1$(II) in human and rat (*74*), and for $\alpha 2$(I) in human and mouse (*72*). However, the degree of conservation is far less when mammals and birds are compared (*71*), and there are few similarities between the promoters of different collagen types (*39*). A comparison of the human and mouse sequences extending 348 bases upstream from the translation initiation codon is shown in Fig. 1. All four genes contain a TATA-like sequence located 20 to 30 bases 5′ from the start of transcription. A CCAAT motif is present on the bottom strand of the $\alpha 1$(I) gene at approximately −120 (Fig. 1) and in the $\alpha 2$(I) gene at

```
     CCAGGCTGGGCTGGGGGCTGGGGAGGCAGAGCTGCGAAGAGGGGAGATGTGGGG    -173
     ** **  *  *********  *** ******* ****  ****** *  *  ***
GAATTCCCGG--GATCTGGGGGGCAAGGGCGGCAGAGTTGCG-GGAGGGGGGCGCTGGG

TGGACTCCCTTCCCTCCTCCTCCC-CCTCTCCATTCCAACTCCCAAATTGGGGCCGGGC    -114
******** ****** *   **** **** **   **      * *******************
TGGACTCCTTTCCCTTCCTTTCCCTCCTCCCCCCTCTTCGTTCCAAATTGGGGCCGGGC

CAGGCAGCTCTGATTGGCTGGGGCACGGGCGGCCGGCTCCCCCTCTCCGAGGGGCAGGGT    -54
******* ***************   ***** ** ***************** ********
CAGGCAGTTCTGATTGGCTGGGGGCCGGGCTGCTGGCTCCCCCTCTCCGAGAGGCAGGGT

TCCTCCCTGCTCTCCATCAGGACAGTATAAAAGGGGCCCGGGCCAGTCGTCGGAGCAGAC    7
******* *********** **  *************** * ******************
TCCTCCCAGCTCTCCATCAAGATGGTATAAAAGGGGCCCAG-CCAGTCGTCGGAGCAGAC

GGGAGTTTCTCCTCGGGGTCGGAGCAGGAGGCACGCGGAGTGTGAGGCCACGCATGAGCG    67
*****************  **********************   ***************
GGGAGTTTCTCCTCGGGA-CGGAGCAGGACGCCACCCCGCACTG--AGCCCACGCATGAGCC

GACGCTAA-CCCCCTCCCCAGCCACAAAGAGTCTACATGTCTAGGGTCTAGACATG    123
** ***** ***** ******** *******************************
GAAGCTAACCCCCCACCCCAGCCGCAAAGAGTCTACATGTCTAGGGTCTAGACATG
```

FIG. 1. Alignment of promoter sequences from the human (top line) and mouse (bottom line) α1(I) collagen genes. Base numbers are for the human sequence (*66*) and refer to base 1 (arrow) as the start of transcription. The mouse sequence is from Harbers *et al.* (*54*). Translation initiation codons, TATA, and inverted CCAAT motifs are boxed. Upstream ATG and in-frame stop codons are overlined. The degree of sequence identity (indicated by asterisks) is 85%.

about −80. The α1(II) and α1(III) genes do not contain typical CCAAT motifs.

Apart from the TATA and CCAAT motifs, there are few, if any, recognizable sequences in the 5′ flanking region clearly implicated in control of collagen gene transcription. The 5′ flanking sequence of the chicken α2(I) gene has the potential for forming several hairpin loops (*70*), but the significance of this observation is unclear. In the mouse and chicken α2(I) genes, a pyrimidine-rich sequence exists between −100 and −200 that is sensitive to S1 nuclease in supercoiled plasmids (*68*, *75*). In the chicken gene, this region corresponds to a previously observed S1- and DNAse-I-sensitive site in chromatin (*23*). The possibility exists that these pyrimidine-rich regions can form staggered loop structures (*75*) that could be stabilized by DNA-binding proteins or could influence nucleosome formation. However, the importance of this region is uncertain since deletions in the region do not affect the transcription of collagen promoter-chloramphenicol acetyltransferase (CAT) fusion plasmids in a major way (*67*). A similar pyrimidine-rich stretch is present in the mammalian α1(I) gene just upstream from the inverted CCAAT box (Fig. 1), but its function in transcription, if any, is also not understood. In contrast, pyrimidine-rich stretches do not occur in analogous positions in the α1(II) or α1(III) genes, or in the chicken α1(I) gene (*64*).

The chicken α1(I) gene contains unusual tandem repeats not present in mammalian homologues. At least 25 copies of a 23-bp element exist in the 5′ flanking sequence upstream of −181 (*64*). As suggested by Southern blot hybridization, this sequence is present at approximately 150 copies per haploid genome. Fifteen tandem repeats of the sequence GGGGAGA are present in the first intron, 300 bp 3′ from the first exon. This region contains two sites hypersensitive to S1 nuclease as determined *in vitro* in supercoiled plasmids (*64*). The function of these tandem repeats is not known.

A small number of deletion studies of collagen promoters have been reported. The transcription of chimeric collagen α2(I) promoter-CAT gene constructs, in which the start of translation was provided by the collagen gene, is inhibited when sequences between −979 and −502 or between −346 and −185 are deleted (*67*). However, the interpretation of these experiments is complicated by the presence of an SV40 enhancer sequence in most of the constructions examined (*67*) and by the absence of the first intron, now known to affect transcription of both the α1(I) and α2(I) genes. More recently, some of the same deletions in α2(I) that were inhibitory (*67*) were found to stimulate transcription when a 1-kb intronic sequence was added 3′ from the poly(A) addition site in collagen promoter-CAT plasmids (*76*). The transcription of a human α1(I) collagen promoter-growth hormone fusion gene was not significantly reduced by deletions extending from −625 to −161, provided the first intron was present in the constructions (*77*). Extension of the human α1(I) promoter from −253 to −950 resulted in only a doubling of transcription of collagen-α globin plasmids in *Xenopus* oocytes; extension of the promoter to −2400 increased transcription another 6-fold; the presence of far-upstream elements is therefore suggested (*78*).

An unusual feature of the regulation of collagen gene expression is the presence of regulatory elements in the first intron of several collagen genes (*66, 78–81*). There is a transcriptional enhancer in the first intron of the mouse α2(I) gene (*79*). The sequence, which has not been narrowly defined, functions in conjunction with either a collagen or SV40 early promoter sequence and in both orientations upstream or downstream from the promoter. The intronic enhancer has not, however, been placed in its natural position immediately following the start of transcription and first exon of the gene, a point that assumes importance in studies of regulatory elements in the α1(I) intron (see below). An enhancer has also been identified in the first intron of the rat α1(II) gene (*80*). The sequence was active in chondrocytes and cells synthesizing type-II collagen, but not in fibroblasts or myoblasts, or in cells down-regulated for type-II synthesis with retinoate (*80*; see Section V,C).

A transcriptional enhancer within a 782-bp intronic fragment of the human α1(I) gene has been identified (*78*). The enhancer, placed either upstream from a collagen promoter or in the first intron of a collagen α-globin fusion gene, functioned in both *Xenopus* oocytes and mouse 3T3 cells. In contrast, experiments from our laboratory (*66, 81*) suggest a more complex role for the first intron in transcriptional regulation. Deletion of much of the first intron from a collagen human growth hormone (hGH) fusion gene resulted in a reduction in transcriptional activity to 1/3 or 1/6 (*81*) but not to 1/50 or 1/100 as suggested earlier (*78*). An explanation of the differences in these results may include the presence of both positively and negatively acting elements in the α1(I) collagen gene intron (*66, 81*) and the likelihood that transcriptional activation of the α1(I) collagen gene is regulated by a complex set of interactions, mediated by DNA-binding proteins, between intronic and 5′ flanking–promoter sequences (*77*).

An indication that functional elements exist in the first intron of the α1(I) collagen gene comes from a comparison of the human and rat sequences. A matrix plot (Fig. 2) reveals a substantial degree of

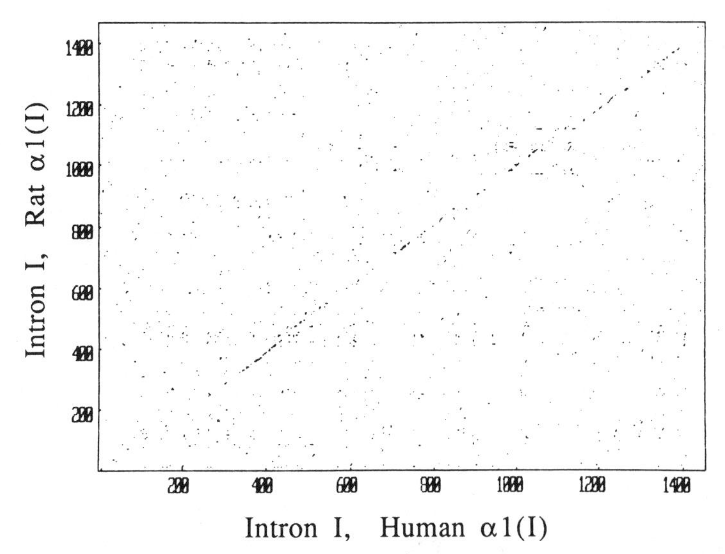

FIG. 2. Dot-plot matrix analysis of the human and rat first introns. The human sequence (bases 223–1674) is from ref. *66* and unpublished data. The rat intronic sequence was provided by D. Rowe. Parameters: Window, 10; matches, 7; ktup, 5.

identity between the introns in the two genes. A comparison of an approximately equal length of sequence from the fifth intron of the rat and human genes shows no such conservation. A more detailed analysis of a segment of the first intron (Fig. 3) shows that sequences in the human intron that are the sites of protein binding, as revealed by DNase-I protection (*66*), are also highly conserved in the rat sequence. Such homology increases the probability that these sequences harbor regulatory elements.

E. *trans*-Acting Factors

Studies of the transcription factors that regulate collagen gene expression are just beginning. Since the finite size of the genome precludes the existence of an expanding set of proteins dedicated to the transcriptional regulation of each gene, it is likely that a limited number of *cis*-acting elements, and their corresponding proteins, confer specificity as a result of a unique combinatorial arrangement of these elements. A TATA-box-binding factor (*82*) is likely to be involved, as the collagen genes that have been examined contain this motif. A CCAAT-binding factor has been identified in nuclear extracts of 3T3 cells on the basis of its ability to protect the appropriate segment of DNA in the mouse α2(I) promoter in an exonuclease-III "footprinting" assay (*83*). At least three different CCAAT-binding proteins have been identified, and each consists of heterologous

```
 AACCTGGGCTATTGC-GGGCCTGGACCCTCATCTCCAGACTAGGAAAATAAAAGCACAG
  *** *** ****** ***   ****** ** ** *    *  * ***  ***** ***
AGACCCGGGTTATTGCTGGGTGCGGACCCCCACCTGCTAGATCTGGAAAGTAAAGC-CAG   875

ATGGATGTGATAGCTTTGG--TCTTTGAAAGTGACCCAGGCCAGTAAAATTGGGTCCGCC
  ***** *  ***    *  ****  * **  *  * **** ** *  * **  *****
--GGATGGGGCAGCCCAAGCCTCTTAAAGAGGTAGTCGGGCCGGTGAGGTCGGCCCCGCC   933

CCGTG--CATTGCTTAGCGTTGCCCGCCACCTAGTGGTTG-CTTTGATCCGGTGCTAGAG
***    ******************* **********  * **  *   *   ***** *
CCGGCCCCATTGCTTAGCGTTGCCCGACACCTAGTGGCCGTCTGGGGAGC--CGCTAGCG   991

AGGTGGGCGTGGTTAACTAA--TCATTGCTATTTGTAGACTTGAGTTCTTTCTTTGGCTA
 ****** ******* ****  **    *******  *****   **  ***********
CGGTGGGAGTGGTTAGCTAACTTCTGGACTATTTGCGGACTTT--TTGGTTCTTTGGCTA  1049

AGGAGTGACCCCAAAGCTCT-ACTGGCTTTGAAGGAAGGACCGGGCAAGC
*  *******   * **  *  *********      **** *** * *
AA-AGTGACCTGGAGGCATTGGCTGGCTTTGG-----GGACTGGGGATG             1092
```

FIG. 3. Alignment of sequences from the rat (top line) and human (bottom line) α1(I) collagen first introns. Base numbers are for the human sequence and refer to base 1 as the start of transcription. Boxed sequences represent DNase-I protection footprints in the human sequence (*66*). The rat sequence was provided by D. Rowe. The degree of sequence identity (indicated by asterisks) is 66%.

subunits (*84, 85*). One of these, nuclear factor 1 (NF1), also binds to the origin of replication in adenovirus and is required for initiation of adenovirus replication. NF1, or a closely related protein, binds to the sequence TCGN_5GCCAA, located at −310 to −295 in the mouse α2(I) promoter (*86*), and this binding site mediates the transcriptional activation of the gene by transforming growth factor β (TGF-β) (*76*; see Section IV,A). The collagen promoter fragment conferred TGF-β inducibility on the SV40 early promoter, and a 3-bp mutation in the NF1 binding site abolished TGF-β inducibility as well as NF1 binding (*76*).

II. Translational Control

Evidence for translational control of type-I collagen gene expression came initially from discrepancies between collagen mRNA levels and collagen synthetic rates (*87–91*). Additional support for the existence of a translational control mechanism was derived from several observations (92–94) that fragments from the amino-terminal non-triple-helical domain of the pro-α1 chain of procollagen selectively inhibited the translation of collagen mRNA *in vitro*, as well as that of type-I collagen by fibroblasts in culture (*95*). The amino-terminal fragment did not inhibit type-II collagen synthesis by chondrocytes; thus, chondrocytes may not possess a receptor or may lack other mechanisms for recognition and transduction of the extracellular signal (*93*).

The possibility that direct translational control represents the physiological basis for the feedback regulation observed with amino-terminal propeptides raises the difficult question of how such relatively large (100–150 amino-acid) protein fragments traverse a lipid bilayer membrane to function in the cytoplasm of a cell. Additional recent findings further complicate the issue of feedback regulation. A synthetic 22-residue peptide in the carboxy-terminal propeptide of the human α2(I) procollagen chain inhibits both collagen and fibronectin synthesis by human fibroblasts in the absence of detectable changes in mRNA levels (*96*). In addition, there is evidence (*97*) that the translational inhibition observed with chick α1(I) amino-terminal propeptides is nonspecific, whereas it has been proposed (*98*) that amino- and carboxy-terminal chick propeptides function specifically at a pretranslational level.

Several investigators have raised the possibility that a highly conserved nucleotide sequence spanning the untranslated and translated regions of the first exon in several collagen genes may play a role in translational control (*39, 65, 71, 99*). A constant feature in the α1(I), α2(I), and α1(III) genes is the presence of a conserved inverted repeat sequence that contains the open reading frame ATG in its 3′ dyad of symmetry (Fig. 4). A second ATG is consistently found 14 nucleotides upstream in the 5′ axis of the inverted repeat, and a third ATG is placed 40–60 bases upstream from the second. Both upstream ATG codons are followed by in-frame "stop" codons. On the other hand, the cartilage collagen α1(II) gene, which is subject to different tissue-specific control than the genes coding for types-I and -III collagens, lacks this inverted repeat sequence (*73, 74*).

Several features of the sequence shown in Fig. 4 provide a potential basis for translational control. Highly stable stem–loop structures in mRNA can inhibit translation of preproinsulin in COS

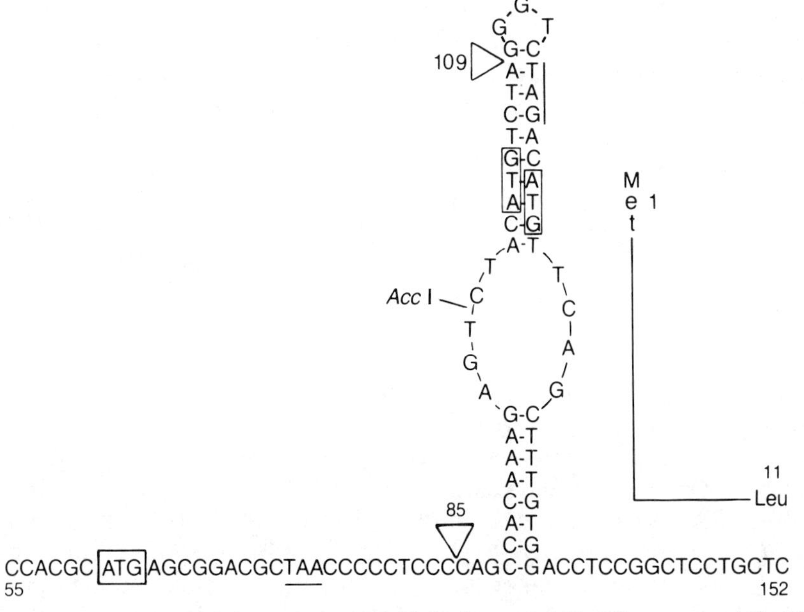

FIG. 4. Sequence of the sense strand of the human α1(I) collagen gene (*65, 66*). The sequence is written as it might exist in mRNA if the discontinuous inverted repeat were to form a stem-loop structure. The two upstream ATGs (boxed) are followed by in-frame stop codons (underlined). The third, open reading frame ATG codes for the amino-terminal residue, Met, in procollagen, and the signal sequence is shown schematically through amino acid 11, Leu. Numbers refer to the start of transcription as 1. A deletion of the upstream axis of the inverted repeat was achieved by cleavage at the unique *Acc*I site and limited *Bal*-31 digestion to remove bases 85–109 (arrowheads) (see *101* and Section II).

cells transfected by several vectors (*100*). The stability of the stem-loop structures capable of being formed in the collagen genes is less than that required to exert translational control (*100, 101*). However, it has been suggested (*102*) that an RNA dimer of the inverted repeat sequence in the mouse α2(I) gene may form, by anti-sense, intermolecular self-hybridization. The stability of such a structure would be considerably greater than that of an intramolecular stem-loop.

In a recent study, Bornstein *et al.* (*101*) constructed a collagen promoter bovine growth hormone (bGH) fusion gene and introduced a deletion in the upstream axis of the inverted repeat (Fig. 4). When the intact and deleted fusion genes were introduced into a variety of cells, by either transient or stable transfection, the efficiency of translation of bGH transcripts was unchanged (*101*). Furthermore, there was no evidence that transcriptional activity of the fusion gene was affected. The functional role of the highly conserved inverted repeat sequence therefore remains unclear.

A second feature of the 5′ untranslated region of types-I and -III collagen genes that could mediate translational control is the invariant presence of two translation initiation codons followed by in-frame stop codons upstream from the initiator ATG. Short upstream reading frames have been found in the mRNA of the yeast GCN4 gene, the transcriptional activator of amino-acid biosynthetic genes (*103*), and in the cytomegalovirus β transcript (*104*), and are thought to modulate gene expression. In accord with the scanning model for eukaryotic translational initiation, upstream translation–initiation codons can severely depress initiation of translation from the authentic start of translation (*105*). In some studies, the presence of in-frame stop codons following these upstream initiation sites modified this inhibitory effect (*105, 106*), except in the case of GCN4 (*103*).

The extent to which the several initiation codons conform to the most favorable sequence context, CCRCCATGG, will also modify the effects of multiple initiation codons. Purines at -3 and $+4$ (relative to ATG in positions $+1$ to $+3$) appear to be most important in this regard (*107, 108*). The presence of the authentic start codon in a strong sequence context and upstream start codons in weaker contexts would tend to reduce the negative influence of the latter codons. Examination of the sequence of the human α1(I) gene (Fig. 4) reveals that none of the three start sites exists in a strong sequence context. As a consequence, the potential for translational control by this region remains uncertain. In the experiments described by Bornstein *et al.* (*101*), the deletion of the 5′ dyad of symmetry of the inverted repeat, which also removes the middle ATG (Fig. 4), did not affect the translatability of the resulting transcript.

Chondrogenesis is associated with a change in synthesis from type-I to type-II collagen (*109;* Section V,A,2). Chondrocytes grown in culture can be induced to "dedifferentiate" by prolonged growth in monolayer culture, by exposure to bromodeoxyuridine or phorbol esters, or by transformation with RSV (see Section V,A,2). There is evidence both for translational control of type-II collagen in dedifferentiated cells (*110*) and for translational control of type-I collagen in differentiated chondrocytes (*89, 90, 110–112*). The basis for the poor translatability of type-I mRNAs in chondrocytes is not well understood. Translatability of RNAs *in vitro* reflects cellular levels of proteins and suggests (*90*) that intrinsic differences in mRNA modulate differences in translatability. Differences in size distribution of types-I and -III mRNA from skin, tendon, and chondrocytes have been found on Northern blot analysis (*90*). Recently, Bennett and Adams (*111*) excluded sequestration of $\alpha2$(I) mRNA into ribonucleoprotein particles as the basis for translational discrimination but suggested that there might be a block in translational elongation of this mRNA. The data could not distinguish between an intrinsic difference in mRNA structure or the presence of RNA-binding proteins that could account for a block in translational elongation.

III. mRNA Processing and Stability

Polymorphic RNA transcripts have been identified for the $\alpha1$(I) (*65*), $\alpha2$(I) (*23, 113*), $\alpha1$(II) (*114*), $\alpha1$(III) (*75, 115–117*), and $\alpha1$(IV) genes (*118, 119*). In the $\alpha1$(I) and $\alpha2$(I) genes, for which DNA sequences at the 3′ end of the gene are known (*65, 113, 120*), the utilization of different polyadenylation sites has been implicated in the generation of RNA transcripts of different size. There is currently no clear-cut evidence that different mRNA transcripts are involved in tissue-specific expression of collagen genes. It is certainly possible, however, that the 3′ untranslated segment of a collagen mRNA may influence certain properties of the transcript, including stability. The marked increase in $\alpha1$(I) collagen mRNA seen in response to treatment of 3T3 cells with TGF-β is associated with a significant increase in the ratio of the 5.7/4.7-kb forms of the mRNA (*121*). Under some conditions of cell culture, the stability of the mRNA, which normally has a half-life of about 9–10 hours (*30, 121*), is increased by TGF-β, but there is no evidence that the two different sizes of the mRNA differ in stability.

Although it would not be surprising if the complex collagen genes, consisting in some cases of 50 or more exons and introns, were subject to alternative splicing as a means of introducing structural variability, thus far no convincing evidence has been presented for alternative splicing in the processing of the genes coding for the twelve characterized collagen types. A GT→GC transition in the $\alpha2$(I) gene, resulting in the skipping of exon 6, has been described in a form of Ehlers–Danlos syndrome, type VII (*121a*). In addition, a low-molecular-weight collagen, which probably represents a new collagen type, has been shown to exist in two forms, one of which lacks a 36-bp exon (*122*).

IV. Growth Factors, Cytokines, and Other Modulators of Gene Expression

Since the synthesis and deposition of collagen (primarily types I and III) is an important element in the fibrotic response, there is considerable interest in how this process might be controlled. The effects of growth and differentiation factors on collagen expression during embryogenesis, tissue remodeling, wound repair, and normal morphogenesis are currently subjects of active study. Control of collagen synthesis in cells is complex and reflects both translational and pretranslational mechanisms that are dependent on the cellular proliferative state. For most of the mediators discussed in this section, the net effect on collagen mRNA abundance has been shown, but the specific mechanisms remain to be established.

A. Transforming Growth Factor β (TGF-β)

TGF-β is a multifunctional peptide growth factor that was first identified by its ability to transform fibroblasts *in vitro;* further studies have confirmed its presence in a variety of neoplastic and nonneoplastic cells and tissues (reviewed in *123* and *124*). There is a general consensus that TGF-β causes an increase in connective tissue synthesis that resembles fibrotic responses to injury in several respects (*125*). This idea has been strengthened by recent studies on the distribution of TGF-β protein (*126*) and mRNA (*127*) in embryonic tissues. The protein was clearly associated with mesenchymally derived tissues (connective tissue, bone, cartilage, neural crest mesenchyme), especially in areas undergoing remodeling, angiogenesis, and morphogenesis. In fetal bone, TGF-β mRNA is prominent in cells associated with type-I collagen and osteogenesis, but is considerably diminished in cells undergoing chondrogenesis and type-II collagen synthesis.

TGF-β stimulates the synthesis of extracellular matrix proteins, including types-I and -III collagen, fibronectin, and proteoglycan (*125, 128–131*). mRNA levels for type-I collagen and fibronectin are increased coordinately by TGF-β in fibroblasts and myoblasts within 2 hours after treatment (*132*). This effect is inhibited by actinomycin D but appears to be independent of protein synthesis. Similar findings are the 2- to 3-fold increase in both types-I and -III mRNAs, and a 5- to 8-fold increase in fibronectin mRNA, which persists a minimum of 72 hours after removal of TGF-β from fibroblast cultures (*133*). The mechanism(s) by which these increased mRNA levels occur are not understood.

Raghow *et al.* (*134*) were unable to demonstrate increases in nuclear transcription rates for fibronectin or α1(I) collagen genes in human dermal fibroblasts, and suggested that TGF-β functions posttranscriptionally to stabilize collagen mRNAs. In contrast, our own work shows that increases in α1(I) collagen and fibronectin mRNA levels in 3T3 cells occur in subconfluent cells in the absence of changes in mRNA stability (as measured by half-life), whereas in confluent cells a clear increase in collagen mRNA stability is observed (*121;* see also Section III). Explanations for the elevated steady-state levels of α1(I) mRNA in TGF-β-treated subconfluent cells include an increase in the rate of transcription of the gene, or an alteration in processing of the initial transcript, whereas posttranscriptional effects of this growth factor may predominate in confluent cells (*121*). At the time of this writing, only one study has addressed the former possibility. An NF1 binding site in the mouse α2(I) promoter has been shown, by deletion analysis in conjunction with transient transfection assays, to be sensitive to stimulation by TGF-β (*76;* see also Section I,E). It is likely that transcriptional activation, as well as several posttranscriptional mechanisms, can be mediated directly or indirectly by TGF-β to increase steady-state levels of types-I and -III collagen mRNAs. The complexity of this induction is evidenced by the disparate effects of TGF-β on different cell types. Within a homogeneous cellular population, these effects are also dependent on the state of differentiation, the position in the cell cycle, and the schedule of subcultivation (*124*).

B. Platelet-Derived Growth Factor (PDGF) and Epidermal Growth Factor (EGF)

Studies of mRNA induction by TGF-β *in vitro* are complicated by the secondary stimulation of other growth factors and/or their receptors by this "panregulin" (*125*). For example, TGF-β induces PDGF in AKR-2B cells (*135*) and the EGF receptor in NRK fibroblasts (*136*). Receptor transmodulation (i.e., a decrease in affinity of a receptor for its ligand by a competing heterologous ligand) also contributes to difficulties in interpreting the effects of growth factors and cytokines on gene regulation. It is perhaps for this reason that relatively few studies have directly addressed the effects of PDGF and EGF on collagen transcriptional rates. Similarly, the role of the angiogenic and broad-spectrum mitogenic FGFs in collagen gene expression is virtually unknown (*137*).

Although originally isolated from platelets, PDGF is now known to be produced by a variety of resting and "activated" normal cells, as well as by transformed cells (*138*). Its association with wound repair and several pathological conditions characterized by accumulation of connective tissue matrix is suggestive of a specific stimulatory effect on collagen transcription. However, our own studies indicate virtually constant levels of type-I collagen mRNA in smooth muscle cells exposed to PDGF (*139*). On the other hand, Narayanan and Page (*140*) observed that PDGF selectively stimulated the synthesis of type-V collagen protein by gingival fibroblasts, with a reciprocal decrease in type-III collagen. Since the primary effect of PDGF is presumably mitogenic, it is reasonable to expect that an initial diminution in extracellular matrix, with the induction of cell surface or nonfibrillar collagens, might facilitate cellular proliferation.

A similar shift in collagen type was observed in murine palatal shelves cultured in the presence of EGF (*141, 142*). Although total protein synthesis was increased 18% over controls, with the proportion of collagen remaining at 10% of total protein, a significant increase in $\alpha1(V)$ and $\alpha2(V)$ occurred under the influence of EGF. In osteoblastic cell cultures, there is a selective inhibition of type-I collagen protein synthesis in the presence of EGF, which results solely from a decrease in the synthetic rate (*143*). One interpretation of these studies is that specific induction of fibrillar collagens inhibits cellular proliferation, while other collagen types (such as V) might function to disengage cell-surface contacts from the extracellular matrix and facilitate cell movement and division.

C. Cytokines (Interleukin-1 and γ-Interferon)

Interleukin-1 (IL-1) is a product of monocyte/macrophages that, in association with the inflammatory process, acts on connective tissue cells to effect both degradation [by causing an increase in prostaglandin E_2 (PGE_2) and proteases] and repair (by fibroblast proliferation) of the extracellular matrix (*144, 145*). The availability of recombinant IL-1, and the use of such compounds as indomethacin that block production of PGE_2 (which in turn inhibits collagen synthesis), have clearly shown that this cytokine stimulates types-I and -III collagen mRNAs in several mesenchymal cells (*25*). Type-III collagen mRNA is induced to a greater extent than type I, and the levels of α1(I) and α2(I) mRNAs are not coordinately regulated by IL-1 in chondrocytes as they are in fibroblasts (*25*). Similarly, there is a 2.5-fold increase in types-I and -III collagen mRNAs in dermal fibroblasts (*146*). Both IL-1 α and β increase steady-state mRNA levels for type-I collagen mRNAs (*25, 147*). Although these experiments point to pretranslational regulation, they do not distinguish between transcription rate and mRNA stability.

The T-lymphocyte product γ-interferon (γ-IFN), in addition to its proliferative and antiviral functions, suppresses collagen synthesis in fibroblasts and chondrocytes (*148–150*). Rosenbloom and co-workers, using recombinant γ-IFN and normal fibroblasts, demonstrated a coordinate decrease in the levels of type-I collagen mRNA and protein (*149*). Moreover, scleroderma fibroblasts, which produce excessive amounts of collagen and other connective tissue proteins, exhibit a coordinate reduction in α1(I) and α2(I) mRNAs of >60% after treatment with γ-IFN (*150*). Czaja *et al.* (*151*) have recently extended these observations with interesting results. In dermal fibroblasts exposed to γ-IFN, mRNA levels for α2 type-I and type-III collagen were 23% and 7%, respectively, of control, while steady-state levels of fibronectin were 560% of control. Furthermore, transcription rates of the collagen mRNAs were unaffected by γ-IFN, whereas that of fibronectin mRNA was increased. Exposure of cells to this cytokine therefore did not result in coordinate regulation of collagen and fibronectin mRNAs, the former being regulated postranscriptionally, while the latter appeared to accumulate by a transcriptional mechanism.

D. Glucocorticoids

Glucocorticoids constitute a class of steroid hormones that are synthesized in the adrenal cortex and that have significant effects on metabolism by virtue of their interaction with cellular receptors and specific target genes. In fibroblasts, hepatocytes, and several other cell types, glucocorticoids generally inhibit DNA synthesis, cellular proliferation, and protein synthesis, which, in some instances, is specific to collagen (extensively reviewed in *152*).

Earlier studies with fibroblasts had shown a selective decrease in type-I collagen mRNAs in the presence of cortisol (*153*), leading to the postulate that this decrease was a secondary response preceded by a steroid-activated cellular regulatory pathway and was dependent on culture conditions. Subsequently, it was suggested (*154*) that the inhibition of type-I procollagen mRNA by corticosteroids is due to transcriptional regulation. Alternative explanations involving mRNA stability have also been proposed (*30, 155*). Administration of the synthetic glucocorticoid dexamethasone (DEX) to rat fibroblasts did not appreciably affect the rates of transcription of either type-I collagen or fibronectin genes; however, kinetic studies of radio-labeled collagen mRNAs showed an increased rate of mRNA turnover in the treated cells (*155*).

Hepatocytes cultured in the presence of DEX retain liver-specific functions and exhibit reduced steady-state levels of types-I and -IV collagen mRNAs (*156, 157*). This reduction was shown by nuclear "run-on" assay to result, at least in part, from lower rates of transcription of these genes (*157*). These authors also demonstrated, by transient transfection assay, a diminished utilization of the α2(I) collagen promoter in hepatocytes treated with DEX compared to controls, and therefore concluded that this glucocorticoid exerts its effect directly on the collagen promoter as well as posttranscriptionally.

DEX has also been used to study the effects of glucocorticoids on the development of the mammalian intestine. There is a differential regulation of several extracellular matrix protein mRNAs in DEX-treated rat small intestine *in vivo* (*158*). Induction of adult-specific disaccharidase activity by DEX is accompanied by decreased steady-state levels of types-I and -III collagen mRNAs, whereas mRNAs for fibronectin and the basement membrane components laminin and type-IV collagen increase. This selective regulation of interstitial versus basement membrane protein mRNAs is achieved, at least in part, by altered rates of mRNA transcription and may depend on the state of differentiation of the cells, as has recently been shown for smooth muscle cells (*159*). The effects of glucocorticoids on collagen expression are clearly complex and multifactorial, and have been complicated by the antiinflammatory and antifibrogenic properties of these compounds. Multiple cellular regulatory pathways are also likely to be involved.

E. Prostaglandins and Phorbol Esters

Prostaglandins are critical mediators of inflammation that have been associated with decreased collagen production (*144*). The effect of PGE_2 is thought to be mediated by cAMP as a second messenger; adenylate cyclase levels reflect in part the number of active cell-surface receptors for this mediator. The suppression of collagen (and fibronectin) synthesis by PGE_2 in fibroblasts was thought to be due to increased intracellular degradation of these proteins. However, the decreased protein production is correlated with a coordinate reduction in the steady-state levels of types-I and -III collagen and fibronectin mRNAs in PGE_2-treated cells (*160*). Moreover, when transcription is blocked with actinomycin D, the stabilities of these mRNAs are not affected (*160*).

Another group of compounds that could modulate the accumulation and degradation of extracellular matrix is the phorbol esters. Both phorbol myristate acetate (PMA) and 12-*O*-tetradecanoyl phorbol-13-acetate (TPA) are potent tumor promoters that, as analogues of diacylglycerol, bind to protein kinase C, although phorbol esters do not appear to initiate cellular transformation by themselves. The effects they induce on cell morphology and differentiated phenotype are thought to occur via the phosphoinositol lipid pathway.

Studies on collagen gene regulation by PMA have exploited the embryonic chick chondrocyte system, which exhibits reversible shape and biosynthetic modulation *in vitro* (*161*) (see also Section V,A,2). Differentiated chondrocytes *in vivo* and *in vitro* are spherical cells that synthesize type-II collagen and other cartilage-specific extracellular matrix proteins and proteoglycans. "Dedifferentiation" *in vitro*, which occurs as a result of subculture or in the presence of several compounds including TPA, refers to a series of changes in these homogeneous populations that include acquisition of a fibroblastic morphology, cessation of production of cartilage-specific macromolecules, and initiation of synthesis of fibronectin and type-I collagen. In an extensive study, a PMA-mediated switch in chondrocyte phenotype was accompanied by a 10-fold decrease in type-II collagen mRNA (*110*). The cells could be redifferentiated after removal of PMA up to the point at which type-I collagen mRNA were translated; mRNA levels for type-II collagen were 24 times those for type-I collagen in the redifferentiated cells, although the rates of collagen protein synthesis were similar. Further studies (*162*) showed that chondrocytes treated with PMA exhibit the shape change concomitantly with elevated levels of actin and fibronectin mRNA, prior to initiation of type-I collagen gene transcription. Both transcriptional and posttranscriptional controls are therefore operative in the processes of dedifferentiation and redifferentiation, processes that are associated with interactions between the chondrocyte cytoskeleton and matrix proteins.

The human collagenase gene has an enhancer element inducible by TPA (*163*). Phorbol esters, as well as both positive (IL-1) and negative (γ-IFN and PGE_2) regulators of collagen transcription, are therefore likely to be critical modulators in the turnover of extracellular matrix in the interstitium.

F. Other Mediators

Several recent studies have investigated the mechanisms by which parathyroid hormone (PTH) and 1,25-dihydroxyvitamin D_3 (Vit D) alter collagen synthesis in bone. In fetal rat calvaria, PTH selectively decreases the steady-state levels of type-I collagen mRNAs and has relatively little effect on collagen protein degradation (*164*). Similarly, Vit D treatment of this tissue is associated with concomitant and nearly equal decreases in type-I collagen mRNA and protein levels and is specific to differentiated osteoblasts (*165, 166*).

The marked changes in collagen types and levels that occur during embryonic development have been ascribed, in some cases, to transcriptional induction by specific factors (see also Section V). For example, a hepatic fibrogenic factor increases selectively the steady-state mRNA levels for types-I, -III, and -V collagen *in vitro* (*27*). The addition of retinoids to tissue culture fibroblasts, in contrast, reduces the levels of $\alpha2$(I) mRNA but does not affect actin or fibronectin gene expression (*167*). Further discussion related to retinoate as a morphogen and inducer of differentiation-specific collagens is presented in Section V,A.

There is a selective effect of the monokine tumor necrosis factor α (TNF-α) on collagen gene transcription (*168*). At levels of TNF-α that are not cytotoxic, the rate of type-I collagen protein production is decreased by 50% in cultured fibroblasts, with no measurable effect on noncollagenous proteins. Similarly, $\alpha1$(I) collagen mRNA levels are decreased by 50% in the presence of TNF-α. This diminution is dependent on protein synthesis, and nuclear "run-on" assays show that it occurs at the level of gene transcription. Since this monokine is associated with cachexia and impaired wound healing, its ability to modulate collagen gene transcriptional activity may be relevant to these conditions.

Collagen synthesis is remarkably sensitive to ascorbate (reviewed in *169*). In addition to the requirement of this cofactor for the posttranslational modifications contributed by prolyl and lysyl hydroxylase, ascorbate affects collagen gene transcription. In one study, the 6-fold induction of $\alpha2$(I) mRNA in ascorbate-treated primary avian tendon cells could be accounted for by a 3-fold increase in the rate of collagen gene transcription and a doubling of the mRNA half-life (*170*). Since ascorbate appears to affect several different subcellular pathways, its activity in the nucleus and the cytoplasm may be indirect, and other control mechanisms (such as feedback regulation or induction of a second messenger) may regulate collagen gene expression subsequent to ascorbate administration.

The biosynthesis of collagens is clearly regulated at several transcriptional and posttranscriptional steps. As we have attempted to summarize in this section, either positive or negative control can occur, depending on the inducer, by mechanisms which for the most part are poorly understood. The induction of collagen and matrix protein mRNAs by TGF-β and IL-1, and the generalized diminution of type-I collagen mRNA by γ-IFN, PTH, Vit D, antiinflammatory steroids, and prostaglandins, should provide researchers with many opportunities to investigate these regulatory pathways.

V. Developmental Regulation

In the previous section, we describe some effects of growth factors on collagen gene regulation. Analysis of early development in both invertebrates and vertebrates, coupled with recombinant DNA methodology, has enabled researchers to formulate growth-factor superfamilies whose members (e.g., EGF, FGF, TGF-β) function additionally as morphogenetic and differentiation molecules (*171*). A major challenge in understanding eukaryotic development concerns the temporal and tissue-specific emergence of structures and molecular pathways in embryogenesis. *cis*-Acting DNA sequences, as well as general and restricted *trans*-acting protein factors that interact with promoter and enhancer elements, have been identified (reviewed in *172*). Of the several mechanisms proposed to explain the interaction of DNA and regulatory proteins, experimental data favor the formation of transcription complexes among enhancer and promoter (and/or intronic) sequences accompanied by a looping out of intervening DNA (*173*). Although very little is known about transcriptional control of collagen genes in development, numerous studies showing differential appearance of collagen types reveal both qualitative and quantitative requirements for the expression of these molecules and identify areas for further analysis at the molecular level.

A. Models of Differentiation *in Vitro*

1. Teratocarcinoma Cells and the Early Embryo

Adamson and Ayers (*174*) were among the first to identify differential expression of the basement membrane collagen, type IV, in postimplantation mouse embryos. Early-stage embryos (day 5) show an immunopositive reaction with anti-type-IV collagen antibodies only in the primitive endoderm, while subsequent stages reveal staining in the parietal endoderm (PE), Reichart's membrane, and the mesodermal portion of the visceral yolk sac. Visceral endoderm (VE), however, produced type-I collagen. These authors also showed differential expression of interstitial versus basement membrane collagen in teratocarcinoma cells following induction of differentiated phenotypes *in vitro* (*175*). Teratocarcinoma cells have been used as model systems for the differentiation of PE and VE from the primitive endoderm of the mouse embryo (reviewed in *176*). F9 mouse teratocarcinoma cells (stem cells representative of normal inner cell mass cells), when cultured as floating aggregates in the presence of retinoate, form VE, which *in vivo* is a flat absorptive epithelium that produces α-fetoprotein. In contrast, treatment of F9 cells with retinoate and cAMP results in the formation of PE-like populations that are invasive and secrete plasminogen activator and basement membrane components *in vivo* (*176*).

Although arguments exist concerning the limits to which teratocarcinoma systems can be taken as truly representative of normal mouse embryogenesis, there is a general consensus that differential screening of VE versus PE mRNAs can yield useful probes to elucidate developmental processes. cDNA clones specific for type-IV collagen and laminin, a component of basement membranes that promotes attachment and migration of some cells, have been isolated by differential colony hybridization with uninduced and induced F9 cell RNA (*177*). A similar strategy enabled the isolation of type-IV collagen cDNAS and showed an increase in type-IV collagen mRNA, within 24–48 hours after addition of retinoate + Bt_2 cAMP, that paralleled an approximately 20-fold increase in type-IV collagen protein previously observed in these cells (*178*). Further studies have demonstrated a slow and dose-dependent induction of α1(IV) mRNA by retinoate, which was sensitive to the inhibitor of protein synthesis, cycloheximide (*179*). Although type-IV collagen mRNA is not induced in mutant cells lacking a functional retinoate-binding protein, the continued presence of retinoate is not necessary to maintain type-IV collagen mRNA expression in normal cells (*179*).

Coordinate expression among the mRNAs for laminin chains and α1(IV) collagen has recently been observed in F9 cells treated with retinoate + Bt_2 cAMP (*180*). In contrast, levels of mRNA for these components varied considerably in normal mouse tissues. The transcriptional regulation of α1(IV) mRNA is also similar to that of tissue plasminogen activator in differentiated F9 cells (*181*).

2. Chondrogenesis

The role of translational control in inhibiting type-I collagen synthesis in prechondrogenic mesenchymal cells and in generation of the chondrocyte-specific phenotype is discussed in Section II. Castagnola *et al.* (*182*) have used a dedifferentiation system to analyze the transcription of the type-I collagen gene, as well as that of three cartilage-specific collagen genes (types II, IX, and X). Dedifferentiated chondrocytes in culture exhibited an adherent, fibroblastic morphology and synthesized type-I collagen. When transferred to suspension culture, these cells reverted to the chondrocytic phenotype and synthesized types-IX and -II collagen. Redifferentiated chondrocytes also formed aggregates and matured to hypertrophic cells that produced a collagen type specific to this zone *in vivo,* type X. These changes in collagen mRNA levels were correlated with parallel changes in the relative rates of transcription of the corresponding genes.

Several agents alter the phenotypic expression of chondrocytes *in vitro*. In the presence of 5-bromo-2′-deoxyuridine, chick chondrocytes assume a polygonal shape and shift synthesis of collagen from types II, IX, X, and XI to types I and V (*112, 183*). Although both $\alpha2(I)$ and $\alpha1(II)$ genes are transcribed in chondrocyte nuclei, type-II levels are under transcriptional control; the $\alpha2(I)$ transcripts are polyadenylated but untranslated (*112*). Similar changes in collagen phenotype were observed after treatment of chondrocytes with retinoate, although induction of the interstitial types was more rapid and included type-III collagen (*183*). Yasui *et al.* (*183*) also concluded that types-II and -IX collagen are coordinately regulated, although mRNA levels were not investigated.

Studies of collagen gene regulation in other systems exhibiting differentiation *in vitro* are few. Gerstenfeld *et al.* (*184*) noted a 15-fold increase in the steady-state levels of $\alpha1(I)$ and $\alpha2(I)$ mRNAs and a 10-fold increase in $\alpha1(III)$ mRNA in cultures of differentiating chick myoblasts. These increases were maximal between days 3 and 9, when myoblast fusion and myotube formation were essentially complete, and were in agreement with a previous report of the temporal appearance and accumulation of muscle-specific collagen proteins (*185*).

B. Embryogenesis

1. Insertional Mutagenesis

Developmental mutations have been induced in mice by insertion of both recombinant DNA and retroviruses. Studies by Jaenisch and co-workers have led to isolation of a strain of mice, *Mov 13*, containing an embryonic homozygous lethal mutation in which the M-MuLV was integrated into the first intron of the α1(I) collagen gene (*54*). (Transcriptional control mechanisms elucidated by these experiments are discussed in Section I,C.) Since *Mov 13* mice failed to synthesize collagen and died 12 to 14 days postconception, an opportunity was presented to analyze midgestational development in the absence of type-I collagen. There were apparently no serious functional consequences resulting from the lack of type-I collagen gene expression in 0- to 12-day-old mutant embryos, as these animals developed normally with respect to morphogenesis and expression of other collagens (II, III, IV), fibronectin, and laminin (*55*). Normal epithelial branching was observed in the absence of type-I collagen in a variety of organs including lung, kidney, salivary glands, pancreas, and skin (*186*). Death occurred by rupture of large blood vessels, which require type-I collagen for structural integrity (*55*). Perhaps the most important finding in this study was the apparent requirement for type-I collagen in promoting and/or stabilizing cellular interactions, as *Mov 13* mice exhibited severe necrosis of hematopoietic and other mesenchymal cells.

2. Transgenic and Chimeric Mice

Jaenisch and co-workers also applied the transgenic technology successfully to create a mutant phenotype in mice that mimics a collagen-related disease (*187*). Substitutions in codons for glycine residues were engineered in the α1(I) collagen gene, and the mutant chains, which were impaired in their ability to form stable triple helices, were subsequently expressed in transgenic mice. These animals displayed a dominant lethal phenotype and characteristics of the brittle bone disease, osteogenesis imperfecta type II. A somewhat surprising result was the low level of the mutated gene product that could disrupt the synthesis and/or assembly of functionally normal collagen in the presence of normal mouse alleles for the two collagen genes.

A related study in transgenic mice was recently carried out with the α2(I) collagen promoter (*188*). Integration of this promoter (nucleotides −2000 to +54), linked to the CAT gene, into the germ lines of eight mice produced 1 to 20 copies of the gene construct per haploid genome. Six of the mice expressed CAT in a tissue-specific manner, with the highest amounts in tail tendon, a tissue particularly enriched in type-I collagen, although these levels were lower than endogenous levels. Moreover, expression of CAT was first observed in 8.5-day-old embryos, which led these investigators to conclude that the $\alpha2(I)$ collagen promoter contains the necessary information for both stage- and tissue-specific expression of type-I collagen. The disparity in onset of expression noted between these two studies may reflect differences between the mouse $\alpha1(I)$ and $\alpha2(I)$ collagen promoter sequences and their associated transcription factors.

An alternate approach was taken to analyze the expression of the human type-II collagen gene (*189, 190*). A cosmid clone containing the entire gene, including ~7 kb of flanking sequence, was transfected into mouse embryonic stem cells. Stable transformants expressing type-II collagen were subsequently injected into blastocysts to obtain chimeric mice. In this system, one can study expression of the gene *in vivo*, as the progeny of the transformed cells will populate a variety of tissues during embryogenesis. Expression of human type-II collagen was found in 14.5-day chondrocyte extracellular matrix in a patchy distribution, as expected for a chimeric animal. These very promising studies demonstrate that this approach can yield the correct temporal and spatial expression of a collagen gene.

C. Tissue- and Lineage-Specific Expression

1. Invertebrates

The cloning of collagen genes from *Drosophila melanogaster* (*191*) and *Caenorhabditis elegans* (*192*) has revealed potentially new structural classes of collagens, in addition to remarkable examples of developmental expression. Monson and her colleagues first isolated two clones, DCg 1 and DCg 2, which resemble vertebrate type-IV collagen (*191, 193*). These sequences were located by *in situ* hybridization in chromosomes 2L and X, respectively. Maximal accumulation of both transcripts occurred during the first and second larval instars coincident with the production of cuticle and basement membrane. DCg 1 mRNA expression was evident 12–15 hours after oviposition, and preceded that of DCg 2, which extended into the third larval instar (*193*).

These studies were extended by the isolation of additional cDNA clones corresponding to at least five unique collagen genes with distinct cytogenetic loci (*194*). These putative collagens were transcribed into distinct mRNAs that were in most cases stage-specific. Further characterization of DCg 1 showed it to be a single-copy gene, transcribed as two mRNA species (6.4 and 10 kb), differentially regulated throughout all stages of *Drosophila* development (*195*). The highest levels of DCg-1 mRNA were found in second-instar fat bodies, third-instar lymph glands, and adepithelial cells of third-instar imaginal discs. DCg 1 transcripts were specifically localized by *in situ* hybridization to mesodermal hemocyte-related organs and to large wandering cells (hemocytes) that accumulate during pupariation and invade larval tissues as part of dipteran metamorphosis (*196*). These authors suggested that hemocytes produce extracellular matrix component(s) critical for *Drosophila* development. The derangement in hematopoiesis seen in *Mov-13* mice (*55*) is strikingly reminiscent of the "fibroblastic" role proposed (*196*) for *Drosophila* hemocytes, although the DCg-1 probe corresponded to a collagen related more to type IV than to type I (*196*).

Kramer *et al.* (*192*) analyzed collagen gene expression in the developmental stages of *C. elegans*. The nematode cuticle, which contains a complex mixture of structurally distinct collagens, is an ideal system in which to study developmental regulation of this gene family. Five different cuticles are formed during the life cycle, and a number of mutants have been characterized in which the structure or the timing of cuticle formation was altered. By Northern blot analysis, the nematode collagen genes *col*-1 and *col*-2 were transcribed into mRNAs (1.1–1.4 kb) that exhibited differential expression (*197*). *col*-1 appeared at all developmental stages, with highest levels in the L1 and dauer cuticles, while *col*-2 occurred only during the synthesis of the dauer cuticle. The unusually large number (50–150) of collagen genes in the nematode is thought to provide both diversity and sufficiently high levels of expression during the life cycle (*197*).

An apparently novel collagen gene has been characterized from the sea urchin *Strongylocentrotus purpuratus* (*198*). As deuterostomes, these animals display several features of embryogenesis similar to those of gastrulation and induction of mesoderm in vertebrates. The *S. purpuratus* collagen clone hybridizes to a 9-kb mRNA that appears first in the morula, increases in abundance in the blastula, and decreases during later developmental stages (*198*). Transcripts accumulate specifically in mesenchyme, and become especially abundant in cells of the blastula primary mesenchyme lineage (*199*). Immunohistochemistry on pluteus-stage embryos showed this gene product to be located in the endoskeleton. Further investigation of this collagen, which in some respects resembles vertebrate type-I collagen, should provide valuable insight into lineage-specific regulation and expression of a product essential for skeletal integrity and survival.

2. Vertebrates

Early studies of tissue-specific expression noted both transcriptional and posttranscriptional control of type-I collagen synthesis during fetal development. The levels of type-I collagen mRNA and its translatability (as measured by cell-free translation) are colinear with the synthesis of type-I collagen protein in sheep lung and tendon (*88*). However, a disparity between mRNA level and protein synthetic rate in the later stages of fetal skin development led to the postulate of a translational control mechanism for this tissue. On the other hand, no evidence was found for translational control of type-I collagen mRNA in chick embryo calvaria (*22*). In this tissue, the marked increase in collagen synthesis between 10-day and 17-day embryos (from 12% to 65% of total protein) was due to an absolute increase in type-I collagen mRNA per cell and a preferential inactivation of mRNAs for other proteins. In a later study, there was observed an approximately 10-fold coordinate increase in steady-state levels of types-I, -II, and -III collagen mRNAs between day 5 and day 10 of chick embryogenesis (*23*). These increases relative to total mRNA, which were not seen for fibronectin or actin mRNAs, paralleled the general rise in collagen synthesis and developmental progress of tissues enriched for the respective collagen types.

In situ hybridization with α2(I) and α1(II) cDNA probes in chick embryos demonstrated that these genes can be expressed simultaneously within a tissue, by the same or distinct cellular populations (*200*). For example, scleral and limb bud cartilage stain positively with the type-I collagen probe in fibroblasts surrounding the cartilage and in a few chondrocytes, while labeling with the type-II collagen probe is restricted to most of the chondrocyte population and a limited number of fibroblasts. An extensive examination of developing human skeletal tissues showed that the highest levels of type-II collagen mRNA are concentrated in chondrocytes comprising the lower proliferative and upper hypertrophic zones of growth plate cartilage (*201*). Chondrocytes at the osteochondral junction exhibit little or no labeling with either probe, while osteoblasts in the mineralizing zone uniformly contain type-I collagen mRNA. These observations document the correspondence of collagen mRNA and protein in certain connective tissues and, while not strictly quantitative, provide a sensitive assay for transcripts of different collagen types in a morphologically intact context.

Collagen "type switching" at the transcriptional level has also been studied in limb cartilage differentiation. Type-II collagen mRNA is present in sterna and differentiated limb bud cultures, but is not evident in stage-24 limb buds that are not yet differentiated into cartilage (*114*). Subsequent quantitation of type-II collagen mRNA levels led to the observation (*202*) of a low expression of this mRNA at stages prior to phenotypic differentiation. mRNA levels increased from 20 copies per diploid genome in stage-24 limb mesenchyme *in vivo* to approximately 2000 copies in stage-24 limb mesenchyme after differentiation in culture (equivalent to stage 31 *in vivo*). It was concluded that translational or posttranslational controls are not involved in determining the levels of type-II collagen synthesis in cartilage, although differential regulation was apparent in distinct cartilagenous tissues.

In a related study, Kosher *et al.* (*203*) examined cytoplasmic levels of types-I and -II collagen mRNAs during limb chondrogenesis. Type-II collagen mRNA exhibited significant accumulation during the condensation stage, when prechondrogenic limb mesenchymal cells aggregate just prior to synthesis of a cartilage-specific matrix. Ensuing matrix deposition was paralleled by a steady increase in the level of type-II collagen mRNA. Two rather interesting findings were (1) that the type-II collagen gene is activated in prechondrogenic cells well before the initiation of cartilage differentiation, and (2) type-I collagen expression is regulated translationally, as mRNA was detected in differentiated chondrocytes that no longer synthesize this protein.

The cartilage system has also provided interesting, albeit limited, information about other collagens. Synthesis of type-IX collagen is initiated in developing chick limb mesenchyme at stage 26 and appears within the central cartilaginous region by stage 34, more or less concomitant with the expression of type-II collagen (*204*). In contrast, type-X collagen synthesis was restricted to the calcification region of embryonic chick sternal cartilage and was not seen until stage 43 (approximately day 17) of development (*205*). Since the levels of type-X collagen mRNA were nearly equal in several stages of chondrocytes from the calcification region, it was concluded (*206*) that expression of this collagen is translationally controlled. mRNA levels in the latter study (*206*) were estimated by measurement of cell-free translation products. The regulatory mechanisms involved in the control of this very interesting collagen gene will therefore have to be assessed as molecular probes become available.

The recently discovered type-XII collagen, which is structurally similar to type-IX collagen, has further substantiated the existence of a separate, nonfibrillar class of collagen genes that exhibit tissue-specific expression (*207*). As a member of this structurally distinct class, type-XII collagen mRNA was found in 17-day embryonic chick sternal cartilage, calvarium, tendon, and 6-day cornea (*207*). The restriction of type-IX collagen to chondrocytes and type-X collagen to hypertrophic/calcified cartilage predicts that novel promoter elements conferring temporal and histological specificity may be found in this subfamily of genes.

D. Chromatin Structure and DNA Methylation

The mechanisms for inactivating type-I collagen genes in, for example, lymphoid, hematopoietic, and neural cell lineages, are not understood. Transcriptionally active chromatin has been characterized by an increased susceptibility to digestion by nucleases as well as, in many cases, reduced levels of DNA methylation (reviewed in *61*). A DNase-I-hypersensitive site in the 5′ flanking region has been found in tissues actively transcribing the α2(I) collagen gene (*69*). Extension of these studies in developing chick embryos led to the identification of DNase-I-hypersensitive sites in the 5′ region of the α2(I) collagen gene in chromatin from cells prior to initiation of type-I collagen synthesis (*23*).

Recent studies on the chicken α1(I) collagen promoter, however, have shown that the 40-nucleotide pyrimidine tract, corresponding to an *in vivo* DNase-hypersensitive site in the chicken α2(I) and mouse α1(I) genes, is absent in this gene (*64*). There are also nuclease-sensitive sites in fibroblast chromatin that are correlated with tissue-specific usage of different polyadenylation sites of the human α1(I) collagen gene (*58*). Additional insights into possible relationships between hypersensitive sites and gene activity have come from *Mov-13* mice and are discussed in Section I,C.

Early studies indicate that transcription of the α2(I) collagen gene is independent of methylation (*208*). Furthermore, the DNAse-I sensitivity of this gene is not correlated with its methylation pattern (*75*). Type-I collagen genes are hypermethylated in fibroblasts transformed with SV40 (*209*). Demethylation of the collagen genes by treatment of the transformed cells with 5-azacytidine, a cytidine analogue that is not a substrate for DNA methyltransferase, does not reactivate type-I collagen genes. These results support earlier studies (*75*) and indicate that mechanisms other than, or in addition to, cytosine modification by methylation are important in collagen gene activation.

Similar results were obtained in a carefully executed study with chick chondrocytes (*210*). The α1(II) collagen gene was hypomethylated in embryonic chondrocytes that express high levels of type-II collagen, as compared to fibroblasts and erythrocytes in which this gene is normally inactive. However, hypomethylation of the α1(II) collagen gene was also apparent in dedifferentiated chondrocytes treated with BrdU and retinoate that produced no type-II collagen.

Although the methylation patterns, as assessed by methylation-sensitive restriction endonucleases, of the α1(II) and α2(I) collagen genes were different, no relation of the pattern to gene activation, deactivation, and tissue-specific expression was noted.

E. Possible Functions of Collagens in Development and Differentiation

1. Cellular Growth

The relationship of collagen production to cellular growth is an important question that has been addressed experimentally in a number of different systems. Solution hybridization methods showed that type-I collagen mRNAs are approximately 2-fold higher in confluent fibroblasts compared to log-phase cells *in vitro*, although overall type-I collagen production was the same for both growth states (*87*). In a later study, mRNA levels for α1(I), α2(I), and α1(III) collagens, as determined by Northern blotting, were also found to be twice as high in nonproliferating cells, and appeared to be coordinately regulated regardless of growth status (*115*). Similar results were obtained for smooth muscle cells *in vitro* (*91*), in which maximal levels of types-I and -III collagen mRNA were observed after confluence was attained; however, production of collagen protein was highest several days after the mRNA peak value was seen and was ascribed to an increase in translational activity of collagen mRNAs.

Data have been presented (*211*) supporting the contention that the chick α2(I) collagen promoter, when transfected as part of an expression vector into growing or growth-retarded cells, is growth-regulated and displays reduced activity in slowly proliferating 3T6 and CV1 cells. Since most cells in adult animals are not undergoing continuous replication, the question was raised as to how collagen genes are activated in quiescent or slowly dividing cellular populations. When a polyoma enhancer sequence was inserted 5′ or 3′ to the collagen promoter, the activity of the promoter in slowly growing, serum-starved cells was increased significantly.

On the other hand, no differences were detected in mRNA levels for α1(I) collagen in human fetal lung fibroblasts as a function of density in culture (*212*). Sucrose gradient profiles of sparse, rapidly growing, and confluent, quiescent cells showed that the distribution of mRNA in ribonucleoprotein particles (mRNP) and polysomes remained constant, even though RNA and poly(A)-rich mRNA were shifted from polysomes to the mRNP pool in quiescent cells.

Regulation of collagen expression *in vitro* is further complicated by the presence of serum at some stages of cellular subculture.

Although serum stimulation of collagen production occurred by a mechanism of transcriptional control, similar levels of $\alpha 1$(I) collagen mRNA were found in both S and G_1 phases of the cell cycle in fibroblasts (*213*). It was concluded that stimulation of collagen gene transcription is not contingent upon completion of the cell cycle and mitogenesis.

In summary, all studies to date indicate that regulation of collagen production in cells is a complex process that does not depend on the cell cycle, is stimulated by exogenous factors, and varies according to confluence and rate of cellular growth.

2. Collagen as an Adhesive Macromolecule in Directing Cellular Migration and Condensation

As a secreted protein that forms a significant portion of the extracellular matrix, the role of collagens in directing cellular migration, attachment, proliferation, differentiation, and specific gene expression has been demonstrated in a variety of systems (reviewed in *214*). It is reasonable to assume that most of these interactions occur or are initiated at the cell surface. The tripeptide sequence, Arg-Gly-Asp—which has been shown to determine the adhesion of cells to some extracellular proteins—occurs several times in most types of collagen chains. Several cell-surface receptors for collagen have recently been isolated (*215–218*). Collagens also interact with at least two major attachment proteins, fibronectin and vitronectin (*219*).

There is increasing evidence that expression of several adhesive macromolecules determines in part, or is coincident with, the status of cellular growth and phenotypic differentiation. The various types of collagen in the extracellular space or in apposition to the cell surface direct several aspects of cellular behavior. It is likely that these secreted products are regulated in a coordinate manner and that their expression at specific stages of development and cellular differentiation results from utilization of common transcription factors or structural elements within their genes.

Acknowledgments

We thank numerous colleagues for providing us with manuscripts prior to publication, and Kathleen Doehring for preparation of the manuscript. We are especially indebted to David Rowe (University of Connecticut) for sharing unpublished sequence data for the rat $\alpha 1$(I) gene. Original work from our laboratories was supported by National Institutes of Health Grants HL18645, AM11248, and DE08229.

References

1. K. B. M. Reid and R. R. Porter, *ARB* **50**, 433 (1981).
2. C. Mays and T. L. Rosenberry, *Bchem* **20**, 2810 (1981).
3. J. Floros, R. Steinbrink, K. Jacobs, D. Phelps, D. Driz, M. Reany, L. Sultzman, J. Jones, H.-W. Taeusch, H. A. Frank and E. F. Fritsch, *JBC* **261**, 9029 (1986).
4. P. L. Ballard, S. Hagwood, H. Liley, G. Wellenstein, L. W. Gonzales, B. Benson, B. Cordell and R. T. White, *PNAS* **83**, 9527 (1986).
5. A. E. Davis and P. J. Lachmann, *Bchem* **23**, 2139 (1984).
6. K. Drickamer and V. McCreary, *JBC* **262**, 2582 (1987).
7. J. C. Myers and B. S. Emanuel, *Collagen Relat. Res.* **7**, 149 (1987).
8. D. Weil, M.-G. Mattei, E. Passage, N. Van Cong, D. Pribula-Conway, K. Mann, R. Deutzmann, R. Timpl and M. L. Chu, *Am. J. Hum. Genet.* **42**, 435 (1988).
9. R. Mayne and R. E. Burgeson (eds.), "Structure and Function of Collagen Types." Academic Press, New York, 1987.
10. E. J. Miller, *Ann. N.Y. Acad. Sci.* **460**, 1 (1985).
11. E. J. Miller and S. Gay, *Methods Enzymol.* **144**, 3 (1987).
12. B. R. Olsen, personal communication.
13. H. Boedtker, M. Finer and S. Aho, *Ann. N.Y. Acad. Sci.* **460**, 85 (1985).
14. F. Ramirez, M. Bernard, M.-L., Chu, L. Dickson, F. Sangiorgi, D. Weil, W. de Wet, C. Junien and M. Sobel, *Ann. N.Y. Acad. Sci.* **460**, 117 (1985).
15. W. B. Upholt, C. M. Strom and L. J. Sandell, *Ann. N.Y. Acad. Sci* **460**, 130 (1985).
16. B. R. Olsen, Y. Ninomiya, G. Lozano, H. Konomi, M. Gordon, G. Green, J. Parsons, J. Seyer, H. Thompson and G. Vasios, *Ann. N.Y. Acad. Sci.* **460**, 141 (1985).
17. B. de Crombrugghe, A. Schmidt, G. Liau, C. Setoyama, M. Mudryj, Y. Yamada and C. McKeon, *Ann. N.Y. Acad. Sci.* **460**, 154 (1985).
18. D. J. Prockop and K. I. Kivirikko, *New Engl. J. Med.* **311**, 376 (1984).
19. K. S. E. Cheah, *BJ* **229**, 287 (1985).
20. P. H. Byers, *in* "The Metabolic Basis of Inherited Disease" (C. R. Scriver, A. L. Baudet, W. S. Sly and D. Valle, eds.), 6th Ed. McGraw-Hill, New York, 1988.
21. B. Sykes, *Oxford Surv. Eukaryotic Genes* **4**, 1 (1987).
22. R. C. Moen, D. W. Rowe and R. D. Palmiter, *JBC* **254**, 3526 (1979).
23. G. T. Merlino, C. McKeon, B. de Crombrugghe and I. Pastan, *JBC* **258**, 10041 (1983).
24. R. A. Pierce, M. R. Glaug, R. S. Greco, J. W. MacKenzie, C. D. Boyd and S. B. Deak, *JBC* **262**, 1652 (1987).
25. M. B. Goldring and S. M. Krane, *JBC* **262**, 16724 (1987).
26. M. B. Goldring, L. J. Sandell, M. L. Stephenson and S. M. Krane, *JBC* **261**, 9049 (1986).
27. I. Choe, R. S. Aycock, R. Raghow, J. C. Myers, J. M. Seyer and A. H. Kang, *JBC* **262**, 5408 (1987).
28. M. E. Sobel, L. D. Dion, J. Vuust and N. H. Colburn, *MCBiol* **3**, 1527 (1983).
29. W. J. de Wet, M.-L. Chu and D. J. Prockop, *JBC* **258**, 14385 (1983).
30. L. Hämäläinen, J. Oikarinen and K. I. Kivirikko, *JBC* **260**, 720 (1985).
31. A. Olsen and D. J. Prockop, *Methods Enzymol.* **144**, 74 (1987).
32. S. L. Adams, M. E. Sobel, B. H. Howard, K. Olden, K. M. Yamada, B. de Crombrugghe and I. Pastan, *PNAS* **74**, 3399 (1977).
33. M. E. Sobel, T. Yamamoto, B. de Crombrugghe and I. Pastan, *Bchem* **20**, 2678 (1981).
34. D. W. Rowe, R. C. Moen, J. M. Davidson, P. H. Byers, P. Bornstein and R. D. Palmiter, *Bchem* **17**, 1581 (1978).

35. S. Sandmeyer and P. Bornstein, *JBC* **254,** 4950 (1979).
36. E. Avvedimento, Y. Yamada, E. Lovelace, G. Vogeli, B. de Crombrugghe and I. Pastan, *NARes* **9,** 1123 (1981).
37. S. Sandmeyer, B. Gallis and P. Bornstein, *JBC* **256,** 5022 (1981).
38. J. F. Bateman and B. Peterkofsky, *PNAS* **76,** 6028 (1981).
39. G. Liau, M. Mudryj and B. de Crombrugghe, *JBC* **260,** 3773 (1985).
40. S. L. Adams, M. Pacifici, R. J. Focht, E. S. Allebech and D. Boettiger, *Ann. N.Y. Acad Sci.* **460,** 202 (1985).
41. A. Schmidt, C. Setoyama and B. de Crombrugghe, *Nature* **314,** 286 (1985).
42. C. Setoyama, G. Liau and B. de Crombrugghe, *Cell* **41,** 201 (1985).
43. W. G. Carter, *JBC* **257,** 13805 (1982).
44. B. Trüeb, J. B. Lewis and W. G Carter, *J. Cell Biol.* **100,** 638 (1985).
45. R.-I. Hata and B. Peterkofsky, *PNAS* **74,** 2933 (1977).
46. B. D. Smith and R. Niles, *Bchem* **19,** 1820 (1980).
47. S. Sandmeyer, R. Smith, D. Kiehn and P. Bornstein, *Cancer Res.* **41,** 830 (1981).
48. E. Marsilio, M. E. Sobel and B. D. Smith, *JBC* **259,** 1401 (1984).
49. B. Peterkofsky and W. Prather, *JBC* **261,** 16818 (1986).
50. G. Majmudar, E. Schalk, J. Bateman and B. Peterkofsky, *JBC* **263,** 5555 (1988).
51. R. Jaenisch, D. Jähner, P. Nobis, I. Simon, J. Löhler, K. Harbers and D. Grotkopp, *Cell* **24,** 519 (1981).
52. R. Jaenisch, K. Harbers, A. Schnieke, J. Löhler, I. Chumakov, D. Jähner, D. Grotkopp and E. Hoffman, *Cell* **32,** 209 (1983).
53. A. Schnieke, K. Harbers and R. Jaenisch, *Nature* **304,** 315 (1983).
54. K. Harbers, M. Kuehn, H. Delius and R. Jaenisch, *PNAS* **81,** 1504 (1984).
55. J. Löhler, R. Timpl and R. Jaenisch, *Cell* **38,** 597 (1984).
56. M. Breindl, K. Harbers and R. Jaenisch, *Cell* **38,** 9 (1984).
57. H. Weintraub, *Cell* **42,** 705 (1985).
58. G. S. Barsh, C. L. Roush and R. E. Gelinas, *JBC* **259,** 14906 (1984).
59. S. Hartung, R. Jaenisch and M. Breindl, *Nature* **320,** 365 (1986).
60. D. Jähner and R. Jaenisch, *Nature* **315,** 594 (1986).
61. H. Cedar, *Cell* **53,** 3 (1988).
62. A. Stacey, R. Mulligan and R. Jaenisch, *J. Virol.* **61,** 2549 (1987).
63. A. Schnieke, M. Dziadek, J. Bateman, T. Tascara, K. Harbers, R. Gelinas and R. Jaenisch, *PNAS* **84,** 764 (1987).
64. M. H. Finer, S. Aho, L. C. Gerstenfeld, H. Boedtker and P. Doty, *JBC* **262,** 13323 (1987).
65. M.-L. Chu, W. de Wet, M. Bernard and F. Ramirez, *JBC* **260,** 2315 (1985).
66. P. Bornstein, J. McKay, J. K. Morishima, S. Devarayalu and R. E. Gelinas, *PNAS* **84,** 8869 (1987).
67. A. Schmidt, P. Rossi and B. de Crombrugghe, *MCBiol* **6,** 347 (1986).
68. M. H. Finer, E. J. B. Fodor, H. Boedtker and P. Doty, *PNAS* **81,** 1659 (1984).
69. C. McKeon, A. Schmidt and B. de Crombrugghe, *JBC* **259,** 6636 (1984).
70. G. Vogeli, H. Ohkubo, M. E. Sobel, Y. Yamada, I. Pastan and B. de Crombrugghe, *PNAS* **78,** 5334 (1981).
71. A. Schmidt, Y. Yamada and B. de Crombrugghe, *JBC* **259,** 7411 (1984).
72. L. A. Dickson, W. de Wet, M. DiLiberto, D. Weil and F. Ramirez, *NARes* **13,** 3427 (1985).
73. K. Kohno, M. Sullivan and Y. Yamada, *JBC* **260,** 4441 (1985).
74. A. M. Nunez, K. Kohno, G. R. Martin and Y. Yamada, *Gene* **44,** 11 (1986).
75. C. McKeon, I. Pastan and B. de Crombrugghe, *NARes* **12,** 3491 (1984).

76. P. Rossi, G. Karseny, A. B. Roberts, N. S. Roche, M. B. Sporn and B. de Crombrugghe, *Cell* **52,** 405 (1988).
77. P. Bornstein, J. McKay, D. J. Liska, S. Apone and S. Devarayalu, *MCBiol* **8,** 4851 (1988).
78. C. M. S. Rossouw, W. P. Vergar, S. J. du Plooz, M. P. Bernard, F. Ramirez and W. J. de Wet, *JBC* **262,** 15151 (1987).
79. P. Rossi and B. de Crombrugghe, *PNAS* **84,** 5590 (1987).
80. W. Horton, T. Miyashita, K. Kohno, J. R. Hassell and Y. Yamada, *PNAS* **84,** 8864 (1987).
81. P. Bornstein and J. McKay, *JBC* **263,** 1603 (1988).
82. J. L. Workman and R. G. Roeder, *Cell* **51,** 613 (1987).
83. A. Hatamochi, B. Paterson and B. de Crombrugghe, *JBC* **261,** 11310 (1986).
84. A. Hatamochi, P. T. Golumbek, E. Van Schflingen and B. de Crombrugghe, *JBC* **263,** 5940 (1988).
85. L. A. Chodash, A. S. Balwin, R. W. Carthew and P. A. Sharp, *Cell* **53,** 11 (1988).
86. J. Oikarinen, A. Hatamochi and B. de Crombrugghe, *JBC* **262,** 11064 (1987).
87. P. Tolstoshev, R. A. Berg, S. I. Rennard, K. H. Bradley, B. C. Trapnell and R. G. Crystal, *JBC* **256,** 3135 (1981).
88. P. Tolstoshev, R. Haber, B. C. Trapnell and R. G. Crystal, *JBC* **256,** 9672 (1981).
89. E. S. Allebach, D. Boettiger, M. Pacifici and S. L. Adams, *MCBiol* **5,** 1002 (1985).
90. R. J. Focht and S. L. Adams, *MCBiol* **4,** 1843 (1984).
91. M. A. Stepp, M. S. Kindy, C. Franzblau and G. E. Sonenshein, *JBC* **261,** 6542 (1986).
92. L. M. Paglia, *Bchem* **18,** 5030 (1979).
93. L. M. Paglia, M. Wiestner, M. Duclene, L. A. Ouellete, D. Hörlein, G. R. Martin and P. K. Müller, *Bchem* **20,** 3523 (1981).
94. D. Hörlein, J. McPherson, S. H. Goh and P. Bornstein, *PNAS* **78,** 6163 (1981).
95. M. Wiestner, T. Krieg, D. Hörlein, R. W. Glanville, P. Fietzek and P. Müller, *JBC* **254,** 7016 (1979).
96. R. S. Aycock, R. Raghow, G. P. Stricklin, J. M. Seyer and A. H. Kang, *JBC* **261,** 14355 (1986).
97. R. Goldenberg and R. E. Fine, *BBA* **826,** 101 (1986).
98. C. H. Wu, C. B. Donovan and G. Y. Wu, *JBC* **261,** 10462 (1986).
99. Y. Yamada, M. Mudryj and B. de Crombrugghe, *JBC* **258,** 14914 (1983).
100. M. Kozak, *PNAS* **83,** 2850 (1986).
101. P. Bornstein, J. McKay, S. Devarayalu and S. C. Cook, *NARes* **16,** 9721 (1988).
102. P. Rossi and B. de Crombrugghe, *NARes* **15,** 8935 (1987).
103. P. P. Mueller and A. G. Hinnebusch, *Cell* **45,** 201 (1986).
104. A. P. Geballe, R. R. Spaete and E. S. Mocarksi, *Cell* **46,** 865 (1986).
105. M. Kozak, *NARes* **12,** 3873 (1984).
106. C.-C. Liu, C. S. Simonsen and A. D. Levinson, *Nature* **309,** 82 (1984).
107. M. Kozak, *Cell* **44,** 283 (1986).
108. D. S. Peabody, *JBC* **262,** 11847 (1987).
109. K. von der Mark, *Curr. Top. Dev. Biol.* **14,** 199 (1980).
110. M. H. Finer, L. C. Gerstenfeld, D. Young, P. Doty and H. Boedtker, *MCBiol* **5,** 1415 (1985).
111. V. D. Bennet and S. L. Adams, *JBC* **262,** 14806 (1987).
112. S. A. Saxe, L. N. Lukens and P. J. Pawlowski, *JBC* **260,** 3812 (1985).
113. J. C. Myers, L. A. Dickson, W. J. de Wet, M. P. Bernard, M.-L. Chu, M. DiLiberto, G. Pepe, F. O. Sangiorgi and F. Ramirez, *JBC* **258,** 10128 (1985).

114. E. Vuorio, L. Sandell, D. Kravis, V. C. Scheffield, T. Vuorio, A. Dorfman and W. B. Upholt, *NARes* **10**, 1175 (1982).
115. M. Miskulin, R. Dalgleish, B. Kluve-Beckerman, S. I. Rennard, P. Tolstoshev, M. Brantly and R. G. Crystal, *Bchem* **25**, 1408 (1986).
116. H. R. Loidl, J. M. Brinker, M. May. T. Pihlajaniemi, S. Morrow, J. Rosenbloom and J. C. Myers, *NARes* **24**, 9383 (1984).
117. M.-L. Chu, D. Weil. W. de Wet, M. Bernard, M. Sippola and F. Ramirez, *JBC* **260**, 4357 (1985).
118. T. Pihlajaniemi, K. Tryggvason, J. C. Myers, M. Kurkinen, R. Lebo, M.-C. Cheung, D. J. Prockop and C. D. Boyd, *JBC* **260**, 7681 (1985).
119. J. M. Brinker, L. J. Gudas, H. R. Loidl, S.-Y. Wang, J. Rosenbloom, N. A. Kefalides and J. C. Myers, *PNAS* **82**, 3649 (1985).
120. S. Aho, V. Tate and H. Boedtker, *NARes* **11**, 5443 (1983).
121. R. P. Penttinen, S. Kobayashi and P. Bornstein, *PNAS* **85**, 1105 (1988).
121a. D. Weil, M. Bernard, N. Combates, M. K. Wietz, D. W. Hollister, B. Steinmann and F. Ramirez, *JBC* **263**, 8561 (1988).
122. T. Pihlajaniemi, R. Myllylä, J. Seyer, M. Kurkinen and D. J. Prockop, *PNAS* **84**, 940 (1987).
123. M. B. Sporn, A. B. Roberts, L. M. Wakefield and B. de Crombrugghe, *J. Cell Biol.* **105**, 1039 (1987).
124. M. B. Sporn and A. B. Roberts, *Nature* **332**, 217 (1988).
125. A. B. Roberts, M. B. Sporn, R. K. Assoian, J. M. Smith, N. S. Roche, L. M. Wakefield, V. I. Heine, L. A. Liotta, V. Falanga, J. H. Kehrl and A. S. Fauci, *PNAS* **83**, 4167 (1986).
126. U. I. Heine, E. F. Munoz, K. C. Flanders, L. R. Ellingsworth, H.-Y. P. Lam, N. L. Thompson, A. B. Roberts and M. B. Sporn, *J. Cell Biol.* **105**, 2861 (1987).
127. M. Sandberg, T. Vuorio, H. Hirvonen, K. Alitalo and E. Vuorio, *Development* **102**, 461 (1988).
128. R. A. Ignotz and J. Massagué, *JBC* **261**, 4337 (1986).
129. A. Fine and R. H. Goldstein, *JBC* **262**, 3897 (1987).
130. S. M. Seyedin, A. Y. Thompson, H. Bentz, D. M. Rosen, J. M. McPherson, A. Conti, N. R. Siegel, G. R. Galluppi and K. A. Piez, *JBC* **261**, 5693 (1986).
131. M. Centrella, T. L. McCarthy and E. Canalis, *JBC* **262**, 2869 (1987).
132. R. A. Ignotz, T. Endo and J. Massagué, *JBC* **262**, 6443 (1987).
133. J. Varga, J. Rosenbloom and S. A. Jimenez, *BJ* **247**, 597 (1987).
134. R. Raghow, A. E. Postlethwaite, J. Keski-Oja, H. L. Moses and A. H. Kang, *J. Clin. Invest.* **79**, 1285 (1987).
135. E. B. Leof, J. A. Proper, A. S. Goustin, G. D. Shipley, P. E. DiCorleto and H. L. Moses, *PNAS* **83**, 2453 (1986).
136. R. K. Assoian, C. A. Frolik, A. B. Roberts, D. M. Miller and M. B. Sporn, *Cell* **36**, 35 (1984).
137. K. A. Thomas and G. Gimenez-Gallego, *TIBS* **11**, 81 (1986).
138. R. Ross, E. W. Raines and D. F. Bowen-Pope, *Cell* **46**, 155 (1986).
139. S. Kobayashi, R. Penttinen and P. Bornstein, unpublished.
140. A. S. Narayanan and R. C. Page, *JBC* **258**, 11694 (1983).
141. G. Carpenter and S. Cohen, *ARB* **48**, 193 (1979).
142. M. H. Silver, J. C. Murray and R. M. Pratt, *Differentiation* **27**, 205 (1984).
143. R.-I. Hata, H. Hori, Y. Nagai, S. Tanaka, M. Kondo, M. Hiramatsu, N. Utsumi and M. Kumegawa, *Endocrinology* **115**, 867 (1984).
144. S. M. Krane, J.-M. Dayer, L. S. Simon and M. S. Byrne, *Collagen Relat. Res.* **5**, 99 (1985).

145. S. M. Krane, E. P. Amento, M. B. Goldring, S. R. Goldring, M. L. Stephenson, B. Polla, S. Arai, A. K. Bahn and J. T. Kurnick, *in* "Articular Cartilage Biochemistry" (K. E. Kuettner, R. Schleyerbach and V. C. Hascall, eds.), p. 413. Raven, New York, 1986.
146. V.-M. Kähäri, J. Heino and E. Vuorio, *BBA* **929**, 142 (1987).
147. A. E. Postlethwaite, R. Raghow, G. P. Stricklin, H. Poppleton, J. Seyer and A. H. Kang, *J. Cell Biol.* **106**, 311 (1988).
148. M. L. Stephenson, S. M. Krane, E. P. Amento, P. A. McCroskery and M. Byrne, *FEBS Lett.* **180**, 43 (1985).
149. J. Rosenbloom, G. Feldman, B. Freundlich and S. A. Jimenez, *BBRC* **123**, 365 (1984).
150. J. Rosenbloom, G. Feldman, B. Freundlich and S. A. Jimenez, *Arthritis Rheum.* **29**, 851 (1986).
151. M. J. Czaja, F. R. Weiner, M. Eghbali, M.-A. Giambrone, M. Eghbali and M. A. Zern, *JBC* **262**, 13348 (1987).
152. K. R. Cutroneo, K. M. Sterling and S. Shull, *in* "Regulation of Matrix Accumulation" (R. P. Mecham, ed.), p. 119. Academic Press, Orlando, Florida, 1986.
153. J. Oikarinen, T. Pihlajaniemi, L. Hämäläinen and K. Kivirikko, *BBA* **741**, 297 (1983).
154. D. Cockayne, K. M. Sterling, S. Shull, K. P. Mintz, S. Illeyne and K. R. Cutroneo, *Bchem* **25**, 3202 (1986).
155. R. Raghow, D. Gossage and A. H. Kang, *JBC* **261**, 4677 (1986).
156. D. M. Jefferson, L. M. Reid, M.-A. Giambrone, D. A. Shafritz and M. A. Zern, *Hepatology* **5**, 14 (1985).
157. F. R. Weiner, M. J. Czaja, D. M. Jefferson, M.-A. Giambrone, R. Tur-Kaspa, L. M. Reid and M. A. Zern, *JBC* **262**, 6955 (1987).
158. M. J. Walsh, N. S. LeLeiko and K. M. Sterling, *JBC* **262**, 10814 (1987).
159. D. C. Leitman, S. C. Benson and L. K. Johnson, *J. Cell Biol.* **98**, 541 (1984).
160. J. Varga, A. Diaz-Perez, J. Rosenbloom and S. Jimenez, *BBRC* **147**, 1282 (1987).
161. M. E. Lowe, M. Pacifici and H. Holtzer, *Cancer Res.* **38**, 2350 (1978).
162. L. C. Gerstenfeld, M. H. Finer and H. Boedtker *MCBiol* **5**, 1425 (1985).
163. P. Angel, I. Baumann, B. Stein, H. Delius, H. J. Rahmsdorf and P. Herrlich, *MCBiol* **7**, 2256 (1987).
164. B. E. Kream, D. W. Rowe, S. C. Gworek and L. G. Raisz, *PNAS* **77**, 5654 (1980).
165. D. W. Rowe and B. E. Kream, *JBC* **257**, 8009 (1982).
166. C. Genovese, D. Rowe and B. Kream, *Bchem* **23**, 6210 (1984).
167. H. Oikarinen, A. Oikarinen, E. M. L. Tan, R. P. Abergel, C. A. Meeker, M.-L. Chu, D. J. Prockop and J. Uitto, *J. Clin. Invest.* **75**, 1545 (1985).
168. J. A. Solis-Herruzo, D. A. Brenner and M. Chojkier, *JBC* **263**, 5841 (1988).
169. S. R. Pinnell, *Yale J. Biol. Med.* **58**, 553 (1985).
170. B. L. Lyons and R. I. Schwarz, *NARes* **12**, 2569 (1984).
171. M. Mercola and C. D. Stiles, *Development* **102**, 451 (1988).
172. T. Maniatis, S. Goodbourn and J. A. Fischer, *Science* **236**, 1237 (1987).
173. M. Ptashne, *Nature* **322**, 697 (1986).
174. E. D. Adamson and S. E. Ayers, *Cell* **16**, 953 (1979).
175. E. D. Adamson, S. J. Gaunt and C. F. Graham, *Cell* **17**, 469 (1979).
176. B. L. M. Hogan, D. P. Barlow and R. Tilly, *Cancer Surv.* **2**, 115 (1983).
177. S.-Y. Wang and L. J. Gudas, *PNAS* **80**, 5880 (1983).
178. M. Kurkinen, D. P. Barlow, D. M. Helfman, J. G. Williams and B. L. M. Hogan, *NARes* **11**, 6199 (1983).

179. L. Gudas and S.-Y. Wang, *Prog. Clin. Biol. Res.* **226,** 181 (1986) .
180. H. K. Kleinman, I. Ebihara, P. D. Killen, M. Sasaki, F. B. Cannon, Y. Yamada and G. R. Martin, *Dev. Biol.* **122,** 373 (1987).
181. R. J. Rickles, A. L. Darrow and S. Strickland, *JBC* **263,** 1563 (1988).
182. P. Castagnola, B. Dozin, G. Moro and R. Cancedda, *J. Cell Biol.* **106,** 461 (1988).
183. N. Yasui, P. D. Benya and M. E. Nimni, *JBC* **261,** 7997 (1986).
184. L. C. Gerstenfeld, D. R. Crawford, H. Boedtker and P. Doty, *MCBiol* **4,** 1483 (1984).
185. J. H. Sasse, H. von der Mark, U. Kühl, W. Desson and K. von der Mark, *Dev. Biol.* **83,** 79 (1981).
186. K. Kratochwil, M. Dziadek, J. Löhler, K. Harbers and R. Jaenisch, *Dev. Biol.* **117,** 596 (1986).
187. A. Stacey, J. Bateman, T. Choi, T. Mascara, W. Cole and R. Jaenisch, *Nature* **332,** 131 (1988).
188. J. S. Khillan, A. Schmidt, P. A. Overbeek, B. de Crombrugghe and H. Westphal, *PNAS* **83,** 725 (1986).
189. R. H. Lovell-Badge, A. E. Bygrave, A. Bradley, E. Robertson, M. J. Evans and K. S. E. Cheah, *CSHSQB* **50,** 707 (1985).
190. R. H. Lovell-Badge, A. Bygrave, A. Bradley, E. Robertson, R. Tilly and K. S. E. Cheah, *PNAS* **84,** 2803 (1987).
191. J. M. Monson, J. Natzle, J. Friedman and B. J. McCarthy, *PNAS* **79,** 1761 (1982).
192. J. M. Kramer, G. N. Cox and D. Hirsh, *Cell* **30,** 599 (1982).
193. J. E. Natzle, J. M. Monson, and B. J. McCarthy, *Nature* **296,** 368 (1982).
194. Y. LeParco, J.-P. Cecchini, B. Knibiehler and C. Mirre, *Biol. Cell* **56,** 217 (1986).
195. Y. LeParco, B. Knibiehler, J.-P. Cecchini and C. Mirre, *Exp. Cell Res.* **163,** 405 (1986).
196. B. Knibiehler, B. C. Mirre, J.-P. Cecchini and Y. LeParco, *Wilhelm Roux's Arch. Dev. Biol.* **196,** 243 (1987).
197. J. M. Kramer, G. N. Cox and D. Hirsh, *JBC* **260,** 1945 (1985).
198. M. Venkatesan, F. de Pablo, G. Vogeli and R. T. Simpson, *PNAS* **83,** 3351 (1986).
199. L. M. Angerer, S. A. Chambers, Q. Yang, M. Venkatesan, R. C. Angerer and R. T. Simpson, *Genes Dev.* **2,** 239 (1988).
200. M. Hayashi, Y. Ninomiya, J. Parsons, K. Hayashi, B. R. Olsen and R. L. Trelstad, *J. Cell Biol.* **102,** 2302 (1986).
201. M. Sandberg and E. Vuorio, *J. Cell Biol.* **104,** 1077 (1987).
202. D. Kravis and W. B. Upholt, *Dev. Biol.* **108,** 164 (1985).
203. R. A. Kosher, W. M. Kulyk and S. W. Gay, *J. Cell Biol.* **102,** 1151 (1986).
204. T. Kimura, N. Yasui, S. Ohsawa and K. Ono, *BBRC* **130,** 746 (1985).
205. G. J. Gibson and M. H. Flint, *J. Cell Biol.* **101,** 277 (1985).
206. A. M. Reginato, J. W. Lash and S. A. Jimenez, *BBRC* **137,** 1125 (1986).
207. M. K. Gordon, D. R. Gerecke and B. R. Olsen, *PNAS* **84,** 6040 (1987).
208. C. McKeon, H. Ohkubo, I. Pastan and B. de Crombrugghe, *Cell* **29,** 203 (1982).
209. M. I. Parker and W. Gevers, *BBRC* **124,** 236 (1984).
210. M. P. Fernandez, M. F. Young and M. E. Sobel, *JBC* **260,** 2374 (1985).
211. G. Moore and M. Yaniv, *EJB* **162,** 333 (1987).
212. T. Voss and P. Bornstein, *EJB* **157,** 433 (1986).
213. A. S. Narayanan and R. C. Page, *BBRC* **145,** 639 (1987).
214. E. D. Hay, *in* "Cell Biology of the Extracellular Matrix" (E. Hay, ed.), p. 379. Plenum, New York, 1981.
215. S. Dedhar, E. Ruoslahti and M. D. Pierschbacher, *J. Cell Biol.* **104,** 585 (1987).

216. T. M. Chiang and A. H. Kang, *JBC* **257,** 7581 (1982).
217. J. Mollenhauer and K. von der Mark, *EMBO J.* **2,** 45 (1983).
218. E. A. Wayner and W. G. Carter, *J. Cell Biol.* **105,** 1873 (1987).
219. C. Gebb, E. G. Hayman, E. Engvall and E. Ruoslahti, *JBC* **261,** 16698 (1986).

Left-Handed Z-DNA and Genetic Recombination

John A. Blaho[1] and
Robert D. Wells

Department of Biochemistry
Schools of Medicine and Dentistry
The University of Alabama at Birmingham
Birmingham, Alabama 35294

I. Historical Perspective

The processes through which living organisms rearrange their genetic information have been extensively studied biochemically and genetically. However, surprisingly little is known on the molecular level about the structures of the DNA intermediates involved in these mechanisms. It is generally assumed that during genetic rearrangements, DNA adopts non-B conformations. Recent theoretical and experimental studies indicate that left-handed DNA may be involved.

[1] Present address: Marjorie B. Kovler Viral Oncology Laboratories, University of Chicago, Chicago, Illinois 60637.

Progress in Nucleic Acid Research
and Molecular Biology, Vol. 37

A. Genetic Recombination

Genetic recombination is the process by which DNA is translocated from one region of the chromosome to another, as shown in Fig. 1. The first model proposed to describe this phenomenon was by Holliday (1). Recombinational events are now classified into two basic groups: homologous (general) recombination and site-specific recombination. The integration of bacteriophage lambda into the *Escherichia coli* chromosome best illustrates site-specific recombination (reviewed in *2–4*). Several other site-specific recombination systems have been described (*5–9*). Homologous recombination (reviewed in *10–13*) occurs between regions of extensive homology essentially anywhere on the chromosome. Recombination events involving neither specific sequences nor extensive homology are termed illegitimate recombination. This review focuses on mechanisms of homologous recombination.

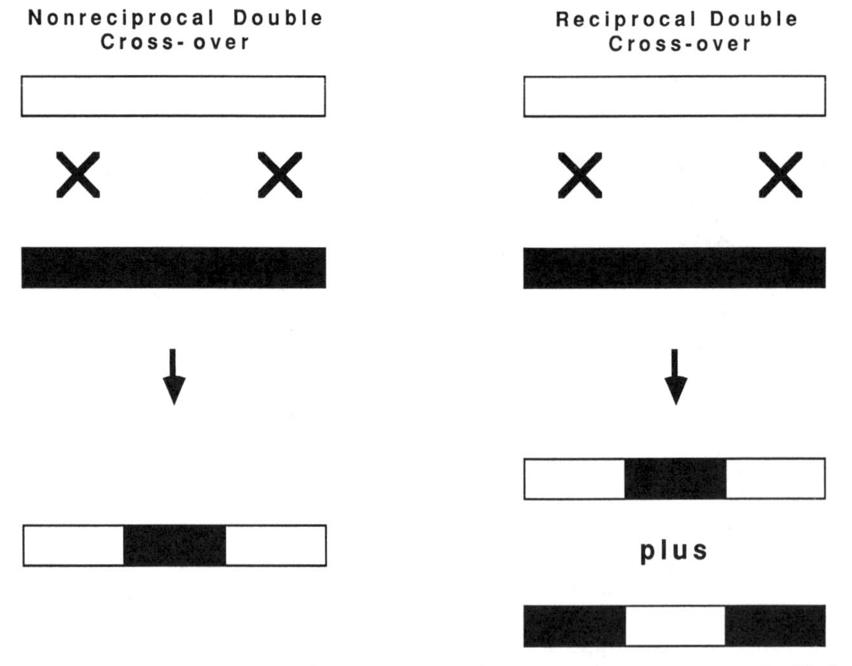

FIG. 1. Representation of genetic recombination events. Open and filled rectangles (upper portions) are genomic segments. The lower portions show the products of double cross-over events that result in nonreciprocal and reciprocal exchanges. The plasmid multimerizations and sequence-specific deletions discussed in the text could proceed through similar mechanisms.

B. Left-Handed Z-DNA

DNA microheterogeneity—neighboring sequences with different secondary structures—is the focus of much attention (reviewed in *14–16*). The most dramatic deviation in secondary structure from the orthodox right-handed B-DNA is that of left-handed Z-DNA (reviewed in *17–20*). Figure 2 shows van der Waals models of B- and Z-DNA. Both of these right-handed and left-handed conformations are double-helices containing antiparallel sugar–phosphate backbones. In Z-DNA, the phosphate groups (shown as a heavy line) are positioned along the helix in a zig-zag fashion; it is this irregularity in the backbone for which the conformation is named. Z-DNA is generally defined as a left-handed structure in which there are 12 bp per helical turn and the successive nucleotide residues alternate between the *syn* and *anti* conformations.

An equilibrium exists between B- and Z-DNA. Since Z-DNA is a higher energy form than B-DNA, it requires the presence of a stabilizing agent such as high salt, divalent metal ions, base modifica-

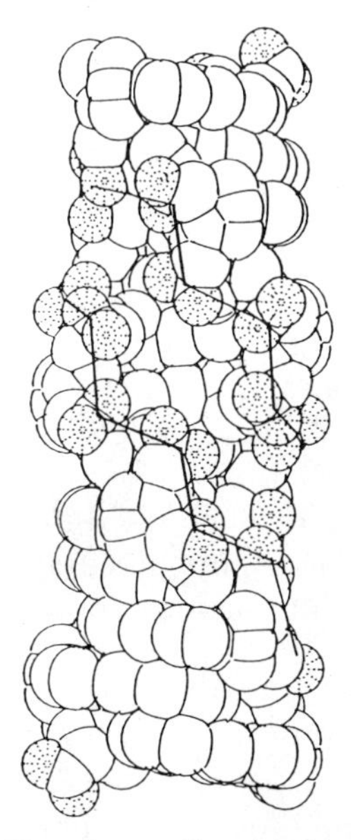

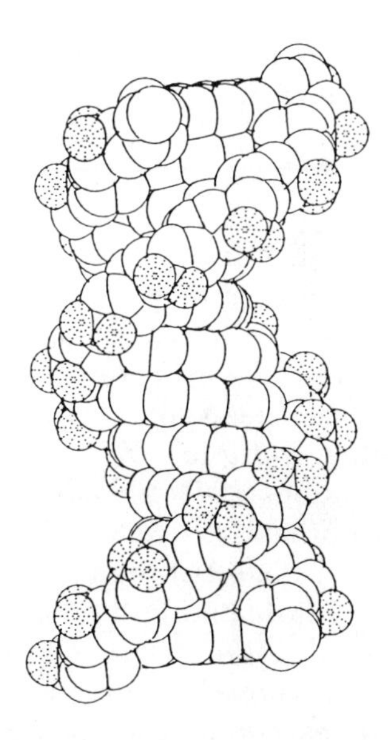

FIG. 2. van der Waals representations of left-handed Z-DNA (left side) and right-handed B-DNA (right side). The dark lines on the Z-DNA model emphasize the zigzag nature of its phosphodiester backbone. These computer diagrams were generated by Dr. Stephen C. Harvey, The University of Alabama at Birmingham, and are reproduced with permission from *103*.

tions (for example, halogenation and methylation), or negative supercoiling. Z-DNA exists in oligonucleotides, DNA polymers, restriction fragments, and recombinant plasmids. Tracts of left-handed Z-DNA can be contiguous with regions of right-handed B-DNA, thus necessitating the presence of B–Z junctions. Numerous physical, chemical, and enzymatic analyses have been used to characterize this unique DNA conformation; the interested reader is directed to reviews of the topic (*14–20*). Recently, left-handed DNA was shown to exist and to elicit biological responses *in vivo (21, 22)*.

II. Models for the Role of Left-Handed Z-DNA in Genetic Recombination

In 1967, based on optical rotary dispersion data with T4 and calf thymus DNA, Pohl (*23*) proposed a model for DNA structure in which the topological constraints resulting from replication, transcription, and recombination could be alleviated by the presence of both right-handed and left-handed double-helices. This was, perhaps, the first indication that left-handed DNA could be involved in genetic recombination. Since that time, our understanding of left-handed DNA has become much more sophisticated, and several models for the role of Z-DNA as a recombination participant have been brought forward.

A. Initiation of Recombination at Regions of Z-DNA

Left-handed Z-DNA duplexes possess base residues that are in a *syn* conformation and are therefore exposed to the environment. For example, in the case of alternating G-C sequences, the guanosines are *syn* while the cytosines remain *anti*. As a result, the exposed N7 and C8 of the guanosines of one Z-duplex are available to interact with another Z-duplex. Z-DNA aggregates in the presence of Mg^{2+} and polyamines (*24*) and sodium acetate (*25*). Thus any process, like recombination, that involves duplex–duplex interactions should be easily initiated at regions of left-handed Z-DNA.

Haniford and Pulleyblank (*26*) expanded this theory to explain the physical requirements for correct chromosomal pairing during meiosis. The presence of numerous repeated sequences in eukaryotic genomes could lead to nonhomologous recombinations if chromosomal pairing is simply based on DNA sequence homology. Since Z-DNA is "inherently sticky" (*26*) and potential Z-DNA-forming sequences are dispersed in eukaryotic genomes (see Section V), these segments provide a simpler mode for homologous chromosomal binding than those involving other intermediates or proteins. Also, Z-helix formation requires a change in the helical winding of DNA. Since recombination requires an unwinding of participating DNA molecules, left-handed Z-DNA formation could be a precursor. Figure 3 shows a model of how the initiation of recombination might begin at regions of Z-DNA.

Recently, Pugh and Cox (27) proposed a general mechanism for the binding of duplex DNA by the *recA* protein (discussed in detail below). *recA*, which is the major recombinase of *E. coli*, rapidly associates with highly negatively supercoiled DNA, suggesting that it initiates binding at locally underwound regions. Local structural variations in DNA, such as Z-helices or B–Z junctions, are likely to cause faster nucleation of binding by *recA* than most other naturally occurring DNAs.

B. Synaptic Paranemic Joints and Left-Handed Z-DNA

Synapsis is the step in homologous recombination by which DNA molecules are brought together and complementary sequences are aligned. "Nicks" or breaks are not introduced in the participating

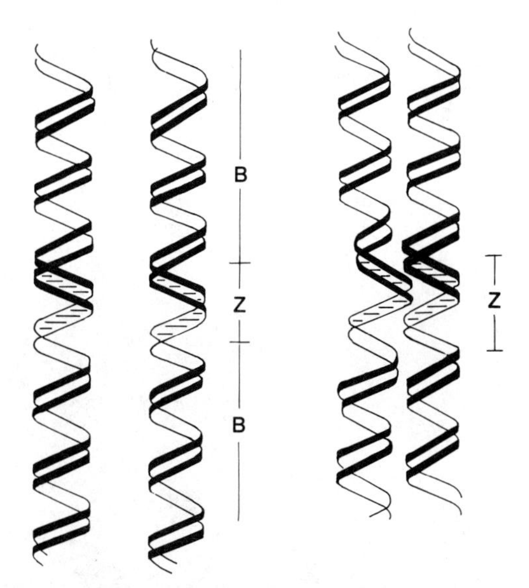

FIG. 3. Model for initiation of recombination at Z-DNA regions. Due the nature of left-handed Z-DNA helices (see text), parallel tracts might interact and precipitate the initial pairing associated with genetic exchanges. The diagram was redrawn similar to a figure previously described (*26*).

strands. Reversal of the process therefore results in separation of the paired strands without breaking them. The nascent heteroduplex formed during synapsis has been termed a paranemic joint (*28*) to emphasize its non-intertwined character. Paranemic joints form without creating new topological constraints in the DNA (i.e., the linking number between the two associated strands is zero). This results by either aligning parallel strands or alternating right- and left-handed helical turns. Recent electron-microscopic work (*29*) argues that the duplex segment of *recA*-mediated paranemic joints exist in alternating left-handed and right-handed turns. A model for a paired but nonintertwined paranemic joint is shown in Fig. 4. Recently, a similar four-stranded paranemic joint model was proposed for the general role of Z-DNA in general genetic recombination (*30*).

Kmiec and Holloman (*31*) found that the *rec1* protein of *Ustilago maydis* (discussed below) promotes synaptic paranemic joint formation. Envisioning that the joints contain alternating right- and left-

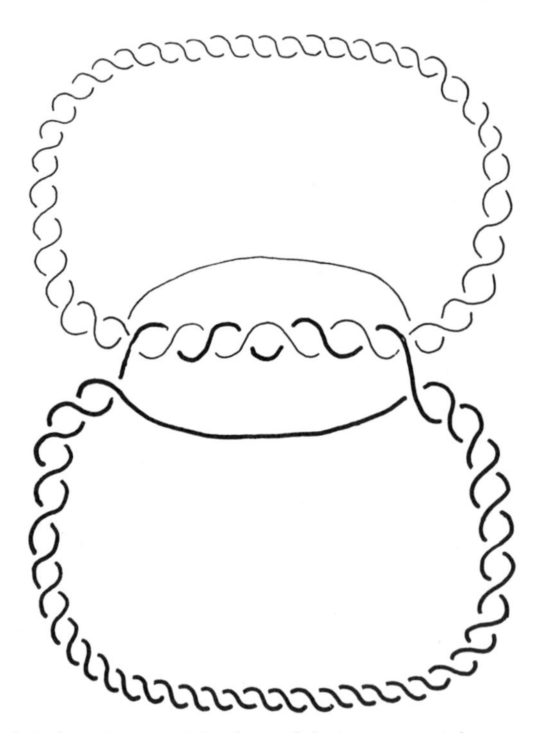

FIG. 4. Model of a paranemic joint formed during synapsis between two covalently closed circular DNAs. The heteroduplex at the joint is a non-intertwined paired structure containing equal right-handed and left-handed turns of DNA. For simplicity, only one of the possible heteroduplexes is shown; the other is drawn as a single-strand.

handed DNA, antibodies to left-handed Z-DNA were used to probe for such structures in this system. Reaction of the Z-DNA antibody with the nascent heteroduplex indicated that left-handed DNA was forming in the paranemic joints.

Thaler, Stahl, and co-workers (*32, 33*) have suggested that the formation of analogous paranemic joints might be involved in Chi-enhanced recombination. It was observed (D. Thaler and F. W. Stahl, personal communication) for certain Chi, that stimulated exchange in the *recBCD* pathway was not maximal at Chi but, rather, to the left at a sequence theoretically capable of forming Z-DNA. Recent exchange studies (C. Shurvinton and F. W. Stahl, personal communication) using recombinant lambda phage containing potential Z-DNA inserts are compatible with these ideas. Thus, it seems likely that left-handed Z-DNA may function in *recBCD*-mediated homologous recombination.

III. Left-Handed Z-DNA and the *recA* Protein of *Escherichia coli*

The *recA* protein of *E. coli* is a relatively small protein (monomer of 37,800 Da) that is vital to the cell for recombination, postreplication repair and the "SOS" response (reviewed in *10–13, 34–38*). *recA* has been cloned (*39, 40*), thus facilitating production of large amounts of pure protein and enabling extensive characterization of its *in vitro* properties.

recA is capable of renaturing complementary single strands of DNA to form double-stranded DNA. It can hybridize single-stranded DNA to double-stranded DNA, thus forming a three-stranded D-loop. The requirements for the formation of both single-stranded and double-stranded DNA complexes with *recA* are well established, and the binding and hydrolysis of ATP significantly influences these interactions (*41–44*). ATP hydrolysis by *recA* is generally assumed to coincide with dissociation of the nucleoprotein complex. *recA*-mediated homologous pairing of double-stranded DNA with single-stranded DNA proceeds in three distinct sequential phases: presynaptic polymerization, synapsis, and postsynaptic strand exchange. Left-handed DNA may occur in the heteroduplex of a synaptic paranemic joint (discussed above).

A. *recA* Binding to Left-Handed Z-DNA

If left-handed DNA occurs as a synaptic intermediate in *recA*-mediated homologous recombination, *recA* should interact with Z-DNA. Specific binding to left-handed Z-DNA by the *recA* protein was first measured using nitrocellulose filter binding assays (*45, 46;* C. Lagravere and J. Laval, personal communication). Two- to 7-fold preferential binding to Z-DNA polymers over B-DNA was observed, as in Fig. 5. Left-handed Z-DNA polymer binding by *recA* required ATP or its nonhydrolyzable analogue, ATP-[γ-S], while ADP inhibited binding. The extent of binding to Z-DNA was found to be dependent on the length of the polymer, since *recA* bound longer polymers better (*46*). The kinetics of *recA* protein binding to Z-DNA is relatively slower than the binding by anti-Z-specific immunoglobulin-G antibodies (*45*).

Nitrocellulose filter binding assays were also used in studies with recombinant plasmids (*46*). *recA* binding to plasmids containing supercoil-stabilized left-handed Z-DNA was essentially similar to that found for control vectors; thus, no preferential binding of recA to the Z-form was observed. Nevertheless, these results did not eliminate

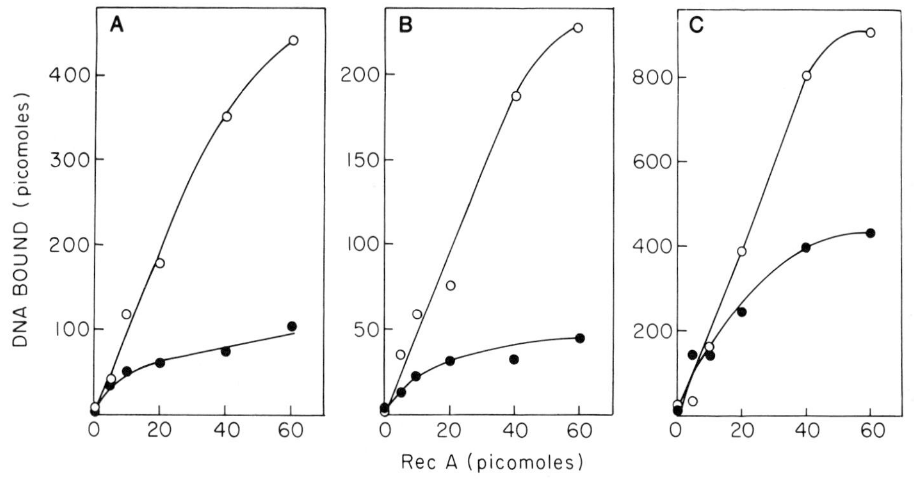

FIG. 5. Left-handed Z-DNA polymer binding by the *recA* protein of *Escherichia coli.* The extent of *recA* binding was measured using nitrocellulose filter binding assays under optimal conditions for *recA* binding to double-stranded DNA. In all cases, *recA* preferentially bound the left-handed Z-DNA polymer. In panel A, poly(dG-m^{5}dC) · poly(dG-m^{5}dC) was the Z-DNA polymer (open circles) and poly(dG-dC) · poly(dG-dC) was the B-DNA polymer (filled circles). In panel B, poly(dG-dC) · poly(dG-dC) was the B-polymer (filled circles), but Br-poly (dG-dC) · poly(dG-dC) was the Z-polymer (open circles). In panel C, left-handed poly(dG-dC) · poly(dG-dC) (open circles) was induced by heat (55°C, 10 minutes) in 4-mM $MnCl_2$. Right-handed poly(dG-dC) · poly(dG-dC) (filled circles) was unheated in 4-m*M* $MnCl_2$. The polymer lengths were 3000 base-pairs in panels A and C and 750 in panel B. This figure is reproduced with permission from *46*.

the possibility that *recA* binds Z-DNA in plasmids because *recA* preferentially binds supercoiled plasmids two to three times better than the relaxed form. Since the amounts of Z-DNA in the plasmids were only ~1% of the total DNA, the filter binding assay may not have been sensitive enough to detect preferential Z-DNA binding (as described with DNA polymers) from binding to the vector.

Direct visualization of *recA* binding to left-handed Z-DNA in supercoiled recombinant plasmids and DNA polymers has been obtained using electron microscopy. Nucleation of plasmid binding by *recA* seems to be enhanced two-fold at regions of Z-DNA compared to binding elsewhere in the vector (J. Pinsence, J. A. Blaho, R. D. Wells, and J. D. Griffith, unpublished observations). *recA* formed right-handed filaments on left-handed (G-C) polymers stabilized in the Z-form by either methylation (J. Heuser and M. M. Cox, personal communication) or bromination (B. Muller, J. A. Blaho, R. D. Wells, and A. Stasiak, unpublished observations) at the 5-position of cytosine. Left-handed (G-C), stabilized as Z-DNA by interstrand diepoxybutane cross-linking, formed *recA* nucleoprotein complexes with regions without visible helical structure (B. Muller, J. A. Blaho, R. D. Wells, and A. Stasiak, unpublished observations). These results may be interpreted in at least two ways. First, the *recA* may bind to the DNA, causing a DNA conformational change. Second, the protein may use the DNA only as a scaffold, with the helical sense of the nucleoprotein complex determined solely by protein–protein associations. Nevertheless, these studies corroborate the filter-binding data that indicate left-handed Z-DNA binding by *recA*.

B. Stimulation of ATPase of *recA* by Left-Handed Z-DNA

Since Z-DNA polymer binding by *recA* requires ATP, it seemed probable that *recA* contains a Z-DNA-dependent ATPase activity. *recA* utilizes left-handed Z-DNA polymers in the hydrolysis of ATP (*46*). This Z-DNA-stimulated activity differs from the B-DNA-dependent activity since it is less sensitive to increased pH. The kinetics of ATP hydrolysis in mixtures of B-DNA and Z-DNA showed that the turnover of the Z-DNA–*recA* complex is slower than for B-DNA. The data, shown in Table I, suggest that left-handed Z-DNA is more stably bound by *recA* in the active complex, supporting the binding data above.

Under optimal conditions, a lag of about 5 minutes is observed for the Z-DNA-enhanced ATPase activity compared to a 2 to 3 minute lag with B-DNA. As the percentage of Z-DNA in the mixing experiments

TABLE I
ATP HYDROLYSIS BY *recA* IN MIXTURES OF RIGHT-HANDED B-DNA AND LEFT-HANDED Z-DNA[a]

Z-DNA (pmol)	B-DNA (pmol)	V (pmol ADP/min)	V/E (ADP/*recA* monomer-min)
—	466	257	7.8
116	350	244	7.4
233	233	220	6.7
350	116	171	5.2
466	—	58	1.8

[a] The left-handed Z-DNA polymer was poly(dG-br^{5}dC) · poly(dG-br^{5}dC) and the right-handed B-DNA polymer was poly(dG-dC) · poly(dG-dC). Velocities of ATP hydrolysis by *recA* were determined by measuring the time course of the reaction. These data are reproduced with permission from *46*.

increased, the lag increased accordingly, suggesting that Z-DNA can compete with B-DNA for the pool of available *recA* monomers. The binding to and unwinding of DNA by *recA* prior to stable complex formation seem responsible for the lags (*12*). These data indicate that stable Z-DNA–*recA* complexes require more time for formation than B-DNA–*recA* complexes.

Taken together, the filter binding, electron microscope, and ATPase data reveal that *recA* rapidly binds (nucleates) Z-DNA and then slowly forms a stable complex that does not dissociate as fast as that with B-DNA. The significance of these observations remains to be elucidated.

C. Potential Z-DNA Regions and Anomalous Recombinations *In Vivo*

1. Z-DNA-SPECIFIC DELETION EVENTS

The first indication that left-handed Z-DNA may be involved in *recA*-mediated recombination occurred during the initial cloning attempts of potential left-handed DNA sequences (*47*). Segments of (C-G), longer than about 50 bp, were not stable and suffered deletions when cloned into pBR322. The (C-G) tracts seemed to enhance *recA*-mediated recombination when they were of suitable length and cloned into certain sites in the recombinants. Similar anomalous deletions mapping directly to potential left-handed Z-DNA sequences were observed (*48–50*).

It was recently demonstrated (*21*) that insertions of a 56-bp insert, capable of left-handed helix formation *in vivo*, into the *Bam*HI site of pBR322 are readily deleted and are generally unstable. However, when the same insert is cloned into the *Eco*RI site, it is quite stable.

This behavior is presumably due to the existence *in vivo* of left-handed DNA and may be a consequence of transcription (A. Jaworski, J. A. Blaho, and R. D. Wells, unpublished observations) since the *Bam*HI site is in the tetracycline-resistance gene.

Similar results were obtained (*51*) when a series of plasmids containing defined numbers of (C-G) units were cloned within the *E. coli lacZ* gene. Sequences capable of adopting a Z-conformation were subject to high rates of spontaneous deletions. These deletions reduced the lengths of potential Z-DNA stretches such that they could no longer form *in vivo*. Additionally, it was observed (*51*) that frameshift mutations induced by the carcinogen 2-acetamidofluorene (AAF) are localized at Z-like conformations that had been stabilized by the AAF. Interestingly, this latter phenomenon does not require active *recA* protein but, rather, depends on an unidentified SOS function(s). These results lead one to question whether such DNA structures are mutagenic.

2. Z-DNA-Enhanced Plasmid Multimerizations

During the course of numerous isolations of the original (C-G)-containing plasmids, the formation of plasmid oligomers was observed (*47*). Certain plasmids, transformed as monomers and then isolated from *E. coli* strains MO *recA*$^+$ or C600 SF8, were 20–50% monomer, 50–75% dimer and higher multimers. The distributions of these forms were stable over several generations. Transformation of the monomer forms into *E. coli* MO *recA*$^-$ cells were maintained indefinitely, indicating that the recombinations were *recA*-mediated.

These experiments were recently repeated and quantitated using a series of recombinant plasmids capable of *in vivo* left-handed helix formation (J. A. Blaho, A. Jaworski, and R. D. Wells, unpublished observations). Overall concatemer production increased linearly with increasing lengths of (C-G). Multimer copy numbers (i.e., dimers, trimers, tetramers, etc.) also increased with increasing (C-G) content. The multimerizations were invariant throughout the cell cycle and required the presence of *recA*.

Such plasmid–plasmid concatemerizations result from homologous recombination events occurring throughout the entire molecule. If left-handed helix formation is intermediate in the process (as described above), one might assume that this step is energetically unfavorable and therefore rate-limiting. Accordingly, if sequences are present that require less energy for left-handed helix formation, by virtue of their base composition, recombination might proceed through these regions preferentially. Since longer Z-DNA tracts require less energy for formation (*14–20*), this model could explain these *in vivo* observations.

This model assumes that all DNA sequences have the capacity to adopt left-handed conformations, albeit the transitions may be protein-induced with the required energy derived from ATP hydrolysis. Supporting this requirement are the demonstrations that (i) any natural sequence can form a left-handed helix during the generation of form-V DNA (*52–55*) and (ii) consecutive A · T pairs, generally thought to form cruciform structures (*14–17*), as well as interruptions in alternating purine-pyrimidine sequences, can form left-handed Z-DNA (*49, 50, 56,* reviewed in *57*).

IV. Eukaryotic Recombinase Interactions with Left-Handed Z-DNA

A. Left-Handed Z-DNA and the *rec1* Protein of *Ustilago maydis*

The *rec1* protein (*58*) was isolated from mitotic *Ustilago maydis* cells based on its ability to reanneal complementary single strands of DNA. *rec1* can catalyze the uptake of linear single-stranded DNA by homologous supercoiled DNA, as well as efficiently pairing circular single-strands with linear double-stranded DNA. It possesses an ATPase activity, which is required for efficient pairing. Complexes formed between *rec1* and homologous single- and double-stranded DNA are synaptic structures in which the nascent heteroduplex formed is a paranemic joint. As described above, *rec1*-mediated paranemic joint formation is accompanied by the genesis of left-handed Z-DNA (*31*). *rec1* catalyzes several reactions similar to *recA*, its more widely studied prokaryotic counterpart (*31, 58–60,* reviewed in *10–13*).

Since a considerable amount of unwinding accompanies the formation of a paranemic joint, Kmiec *et al.* (*59*) presumed that Z-DNA or a left-handed Z-DNA conformer is generated to compensate for the right-handed turns produced as pairing proceeds (Fig. 3). If left-handed Z-DNA were involved in or could mimic the productive intermediate of the pairing reaction, it was thought that *rec1* might bind Z-DNA. Nitrocellulose filter binding experiments, as in Fig. 6, demonstrated that *rec1* binds Z-DNA tightly in DNA polymers and recombinant plasmids and this binding parallels the pairing reaction in several respects. Binding and pairing occur at similar rates and are similarly influenced by ATP and ADP. *rec1* binds Z-DNA about 75

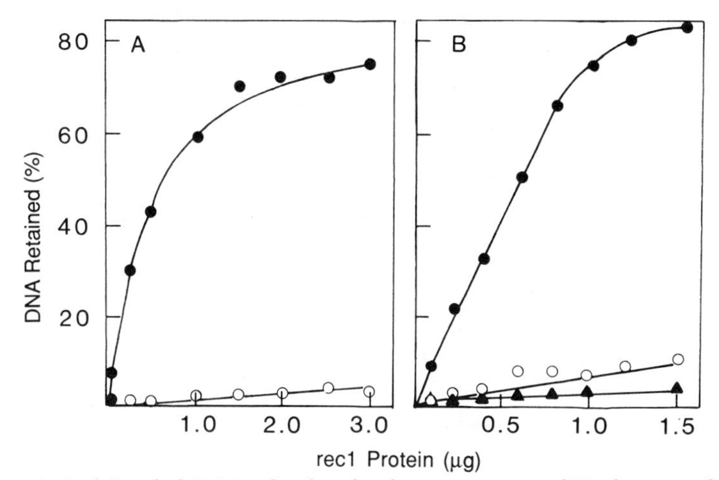

FIG. 6. Left-handed Z-DNA binding by the *rec1* protein of *Ustilago maydis*. *rec1* binding to DNA polymers and recombinant plasmids was measured using a nitrocellulose filter binding assay. Panel A shows that *rec1* preferentially binds left-handed Br-poly(dG-dC) · poly(dG-dC) (filled circles) better than right-handed poly(dG-dC) · poly(dG-dC) (open circles). Panel B shows that *rec1* binds a plasmid containing a $(C\text{-}G)_{13}$ tract, stablized as Z-DNA by negative supercoiling (filled circles), better than the relaxed form of the plasmid (filled triangle) or the supercoiled control vector (open circle). These data are redrawn with permission from 59.

times better than B-DNA and is able to compete with anti-Z-DNA antibodies for binding. Since Z-DNA is a potent inhibitor of the natural pairing reaction, which precedes paranemic joint formation, it was proposed that left-handed Z-DNA closely resembles the DNA structure at the transition state in *rec1*-promoted synapsis.

To test this theory, Kmiec and Holloman (*60*) performed experiments that showed that plasmids containing left-handed Z-DNA sequences could be paired and linked by the combined action of *rec1* and topoisomerase. The pairing began within the Z-DNA regions, and anti-Z-DNA antibodies inhibited, the reaction. These results suggest that *rec1* can initiate homologous pairing at left-handed Z-DNA helices.

Due to the provocative nature of this work, comparative analyses were performed. However, attempts to compare results directly with the *recA* and *rec1* proteins (*46;* J. A. Blaho, R. D. Wells, and W. K. Holloman, unpublished observations) were thwarted by difficulties in obtaining highly purified and stable *rec1*. These problems should be alleviated now that a working transformation system has been set up in *U. maydis* (W. K. Holloman, personal communication).

B. Left-Handed Z-DNA and Human Strand Transfer Recombinase

Complementing the observations reported for *rec1* and *recA*, Fishel *et al.* (*61*) developed a Z-DNA affinity chromatography system, based on an avidin–biotin linkage, to identify proteins associated with recombination. A homologous pairing and strand-exchange activity was purified from a human tumor cell line based on this technique (*62*). This activity was purified to near homogeneity (S. P. Moore, C. J. Harris, and R. Fishel, personal communication); it can transfer a complementary strand from a linear duplex to a single-stranded circular DNA and strongly binds Z-DNA better than B-DNA. The human protein appears capable of linking duplex circular molecules through the combined action of a second Z-DNA chromotography-enriched topoisomerase I activity. In addition to the strand-transfer and topoisomerase activities, two DNA-dependent ATPases and a DNA ligase activity were also eluted from the Z-DNA affinity column. These observations suggest that recombination proteins may be recognized based on their affinity for left-handed Z-DNA.

V. Left-Handed Z-DNA at Recombinational "Hotspots"

Potential left-handed Z-DNA sequences are found in a variety of organisms including bacteria (*63–66*), yeast (*67, 68*), salmon (*69*), trout (*70*), chicken (*71*), mouse (*72, 73*), rat (*74*), and human (*75*). Extensive computer analyses of DNA-sequence data-bases indicate that unusually high frequencies of certain alternating purine–pyrimidine runs occur in natural genomes (*76, 77*).

Numerous recombinational hotspots contain sequences that either adopt or have the potential to form left-handed Z-DNA helices. The proposed roles of certain of these sites are presented in Table II, which is representative rather than exhaustive and emphasizes that most naturally occurring potential Z-DNA tracts are located in regions associated with recombinational events.

Recent analyses of human recombination hotspots (W. P. Wahls and P. D. Moore, personal communication) indicate that potential left-handed Z-DNA tracts enhance homologous chromosomal rearrangements in EJ cells. These results further corroborate the notion that left-handed Z-DNA participates in recombinational processes.

TABLE II
LOCATION OF LEFT-HANDED Z-DNA AT RECOMBINATIONAL HOTSPOTS

DNA	Comments	Reference
Mouse immunoglobulin V_H genes	Z-DNA-forming sequences border regions that appear to be recombined	*78*
Large intervening sequences of human fetal globin genes	Sequences involved in genetic recombination and gene conversion and adopt Z-helices *in vitro*	*79–81*
Human α2 immunoglobulin heavy-chain constant region	Z-DNA-forming sequences border regions that appear to be recombined	*82*
Simian virus 40 (SV40) constructs	Recombination occurs between potential left-handed Z-DNA sequences in SV40-infected cultured cells	*83, 84*
Metaphase chromosomes of *Cebus*	Sequence that adopts Z-DNA *in vitro* seems to be an active recombination region	*85*
poly(G-T) sequences in pBR322	Potential Z-DNA sequences enhance *recA*-independent recombination in *Escherichia coli*	*86*
SV40 integration site into rat genome	Recombination site is potential Z-DNA-forming sequence	*87*
Bovine satellite DNA	Recombination hotspots are associated with potential Z-DNA sequences	*88*
poly(T-G) inserted into yeast chromosome	Potential Z-DNA inserts promote reciprocal exchange and generate unusual tetrads during meiosis	*89*
poly(T-G) and poly(C-G) inserts in SV40	Potential left-handed Z-DNA sequences enhance homologous recombination in somatic cells 5- to 15-fold	*90*
Murine major histocompatibilty complex	Recombination site is a potential Z-DNA-forming sequence	*91*
Murine IgG2b and IgG2a heavy-chain constant region gene	Site of unequal sister chromatid exchange contains a potential Z-DNA-forming tract	*92*
Rous sarcoma virus	Recombination site of cellular oncogene transduction contains potential Z-DNA sequences	*93*

VI. Other Non-B-DNA Structures and Genetic Recombination

The involvement of left-handed Z-DNA in recombination remains circumstantial and further work is clearly necessary. Several recent studies indicate that other types of non-B-DNA structures may be involved in genetic recombination. Greaves and Patient (*94*) observed that potential cruciform formation, at regions of inverted repetition, stimulates *recA*-independent recombination in *E. coli*. Hyrien *et al.* (*95, 96*) found that potential cruciform-forming sequences are located at sites of amplification-associated rearrangements in Chinese hamster fibroblasts. Cruciforms were implicated in other recombinational processes (*97–99*).

Non-B-DNA structures adopted by sequences possessing purines on one strand and pyrimidines on the other also exist at recombination sites. The site of segment inversion in herpes simplex virus type 1 forms a novel conformation, termed anisomorphic DNA (*100*). Homopurine · homopyrimidine sequences located at the site of switching in murine immunoglobulin genes (*101;* A. Weinreb, D. A. Collier, R. D. Wells, and B. K. Birshtein, unpublished observations) may adopt an intramolecular triple-stranded structure (reviewed in *102*).

Also, DNA bending has been implicated in certain site-specific recombination mechanisms (reviewed in *6*). These findings clearly support the notion that the non-B-DNA conformations may be intermediates during genetic recombination events.

VII. Prospects for the Future

The prospects for further elucidating the role of non-B-DNA in genetic recombination in the near future are excellent. The reasons for this are two-fold. First, substantial advancements were made in the past 10 years in our comprehension of the genetics and biochemistry of recombination. Improved methodologies for the isolation, characterization, and determination of the biological roles of recombination proteins have been important. The capacity to overproduce these proteins will enable a sharper understanding of their functions.

Second, enormous progress was made in this time period in determining the properties of unusual DNA structures (such as left-handed Z-DNA, cruciforms, anisomorphic DNA, DNA triplexes, and bent DNA). A number of chemical, enzymatic, and physical probes (reviewed in *14–20, 102–105*) have been developed to study these unorthodox conformations. These probes will be useful for analyzing the role of non-canonical structures in recombination. Also, the demonstration of the existence and behaviors of left-handed DNA and cruciforms *in vivo* (*21, 22, 26, 106–108*) will enhance future investigations.

Appropriate *in vivo* systems must be developed in the future for elucidating the role of non-B-DNA structures in genetic recombination. An understanding at the molecular level of the types of DNA conformations adopted during genetic recombination is necessary for our complete comprehension of this key biological process.

Acknowledgments

We express our gratitude to B. K. Birshstein, S. K. Brahmachari, M. M. Cox, R. A. Fishel, R. P. P. Fuchs, J. D. Griffith, W. K. Holloman, S. C. Kowalczykowski, J. Laval, M. Leng, M. J. McLean, P. D. Moore, A. Nordheim, G. R. Smith, F. W. Stahl, A. Stasiak, B. D. Stollar, J. R. Stringer, and D. S. Thaler for providing helpful information and for graciously communicating data and manuscripts prior to publication. This work was supported by National Institutes of Health Grants GM 30822 and CA 13148 and by National Science Foundation Grant 08644. J.A.B. was supported in part by the Bertram Marx Award for Research in Cancer-Related Areas, Comprehensive Cancer Center, The University of Alabama at Birmingham.

References

1. R. Holliday, *Genet. Res.* **5,** 282 (1964).
2. H. Nash, *ARGen* **15,** 143 (1981).
3. R. A. Weisberg and A. Landy, *in* "Lambda II" (R. W. Hendrix, J. W. Roberts, F. W. Stahl and R. A. Weisberg, eds.), p. 211. Cold Spring Harbor Lab., Cold Spring Harbor, New York, 1983.
4. P. Sadowski, *JBact* **165,** 341 (1986).
5. R. H. A. Plasterk and P. van de Putte, *CSHSQB* **49,** 295 (1984).
6. N. D. F. Grindley and R. R. Reed, *ARB* **54,** 863 (1985).
7. N. L. Craig, *Cell* **41,** 649 (1985).
8. S. A. Wasserman and N. R. Cozzarelli, *Science* **232,** 951 (1986).
9. M. Gellert and H. Nash, *Nature* **325,** 401 (1987).
10. C. M. Radding, *ARB* **47,** 847 (1978).
11. D. Dressler and H. Potter, *ARB* **51,** 727 (1982).
12. M. M. Cox and I. R. Lehman, *ARB* **56,** 229 (1987).
13. G. Smith, *Microbiol. Rev.* **52,** 1 (1988).
14. "Structures of DNA," *CSHSQB* **47** (1983).
15. R. D. Wells and S. C. Harvey, "Unusual DNA Structures." Springer-Verlag, New York, 1988.
16. R. D. Wells, *JBC* **263,** 1095 (1988).
17. A. Rich, A. Nordheim and A. H.-J. Wang, *ARB* **53,** 791 (1984).
18. R. D. Wells, B. F. Erlanger, H. B. Gray, L. H. Hanau, T. M. Jovin, J. E. Larson, J. C. Martin, J. J. Migletta, C. K. Singleton, S. M. Stirdivant, C. M. Veneziale, R. M. Wartell, C. F. Wei, W. Zacharias and D. Zarling, *UCLA Symp. Mol. Cell. Biol. New Ser.* **8,** 3 (1983).

19. M. Leng, *BBA* **825,** 339 (1985).
20. P. K. Latha and S. K. Brahmachari, *J. Sci. Ind. Res.* **45,** 521 (1986).
21. A. Jaworski, W.-T. Hsieh, J. A. Blaho, J. E. Larson and R. D. Wells, *Science* **238,** 773 (1987).
22. W. Zacharias, A. Jaworski, J. E. Larson and R. D. Wells, *PNAS* **85,** 7069 (1988).
23. F. Pohl, *Naturwissenschaften* **23,** 616 (1967).
24. J. H. van de Sande and T. M. Jovin, *EMBO J.* **1,** 115 (1982).
25. W. Zacharias, J. C. Martin and R. D. Wells, *Bchem* **22,** 2398 (1983).
26. D. B. Haniford and D. E. Pulleyblank, *J. Mol. Struct. Dyn.* **1,** 593 (1983).
27. B. F. Pugh and M. M. Cox *JMB* **203,** 479 (1988).
28. M. Bianchi, C. DasGupta, and C. M. Radding, *Cell* **34,** 931 (1983).
29. G. Christiansen and J. Griffith, *PNAS* **83,** 2066 (1986).
30. R. Fishel and A. Rich, *ICN/UCLA Symp.* in press (1989).
31. E. B. Kmiec and W. K. Holloman, *Cell* **36,** 593 (1984).
32. D. S. Thaler and F. W. Stahl, *ARGen* **22,** 169 (1988).
33. D. S. Thaler, E. Sampson, I. Siddiqi, S. M. Rosenberg, F. W. Stahl and M. Stahl, *UCLA Symp. Mol. Cell. Biol. New Ser.* **83** (1988).
34. A. J. Clark, *ARGen* **7,** 67 (1973).
35. E. M. Wilkin, *Bacteriol. Rev.* **40,** 869 (1976).
36. K. McEntee and G. M. Weinstock, *Enzymes* **14,** 445 (1981).
37. J. W. Little and D. W. Mount, *Cell* **29,** 11 (1981).
38. J. D. Griffith, *CRC Crit. Rev. Biochem.* **23** (Suppl. 1), S43 (1988).
39. K. McEntee and W. Epstein, *Virology* **77,** 306 (1977).
40. A. Sancar, C. Stachelek, W. Konigsberg and W. D. Rupp, *PNAS* **77,** 2611 (1980).
41. G. M. Weinstock, K. McEntee, and I. R. Lehman, *JBC* **256,** 8829 (1981).
42. K. McEntee, G. M. Weinstock and I. R. Lehman, *JBC* **256,** 8835 (1981).
43. G. M. Weinstock, K. McEntee and I. R. Lehman, *JBC* **256,** 8845 (1981).
44. S. C. Kowalczykowski, J. Chow and R. A. Krupp, *PNAS* **84,** 3127 (1987).
45. D. Zarling and R. Carlisle, *Fed. Proc.* **45,** 1778 (1986).
46. J. A. Blaho and R. D. Wells, *JBC* **262,** 6082 (1987).
47. J. Klysik, S. M. Stirdivant and R. D. Wells, *JBC* **257,** 10152 (1982).
48. L. J. Peck, A. Nordheim, A. Rich and J. C. Wang, *PNAS* **79,** 4560 (1982).
49. M. J. McLean, J. A. Blaho, M. W. Kilpatrick and R. D. Wells, *PNAS* **83,** 5884 (1986).
50. M. J. McLean and R. D. Wells, *JBC* **263,** 7370 (1988).
51. R. P. P. Fuchs, A.-M. Freund, M. Bichara and N. Koffel-Schwartz, *in* "DNA Replication and Mutagenesis" (W. Summers and R. Moses, eds.), Chap. 27, p. 263. American Society for Microbiology, Washington, D.C., 1988.
52. U. H. Settler, H. Weber, T. Koller and C. Weissman, *JMB* **131,** 21 (1979).
53. F. M. Pohl, R. Thomae and E. DiCapua, *Nature* **300,** 545 (1982).
54. S. Brahms, J. Vergne, J. G. Brahms, E. DiCapua and P. Bucher, *JMB* **162,** 473 (1982).
55. S. K. Brahmachari, Y. S. Shouche, C. R. Cantor and M. McClelland, *JMB* **193,** 201 (1987).
56. M. J. Ellison, R. J. Kelleher III, A. H.-J. Wang, J. F. Habener and A. Rich, *PNAS* **82,** 8320 (1985).
57. M. J. McLean and R. D. Wells, *BBA* in press (1989).
58. E. Kmiec and W. K. Holloman, *Cell* **29,** 367 (1982).
59. E. B. Kmiec, K. J. Angelides and W. K. Holloman, *Cell* **40,** 139 (1985).
60. E. B. Kmiec and W. K. Holloman, *Cell* **44,** 545 (1986).
61. R. Fishel, P. Anziano and A. Rich, *Methods Enzymol.* in press (1989).

62. R. A. Fishel, K. Detmer and A. Rich, *PNAS* **85,** 36 (1988).
63. E. DiCapua, A. Stasiak, T. Koller, S. Brahms, R. Thomae and F. M. Pohl, *EMBO J.* **2,** 1531 (1983).
64. A. Nordheim, E. M. Lafer, L. J. Peck, J. C. Wang, B. D. Stollar and A. Rich, *Cell* **31,** 309 (1982).
65. B. Revet, D. A. Zarling, T. M. Jovin and E. Delain, *EMBO J.* **3,** 3353 (1984).
66. J. D. Hoheisel and F. M. Pohl, *JMB* **193,** 447 (1987).
67. R. M. Walmsley, J. W. Szostak and T. D. Petes, *Nature* **302,** 84 (1983).
68. R. W. Walmsey, C. S. M. Chan, B.-K. Tye and T. D. Petes, *Nature* **310,** 157 (1984).
69. H. Hamada, M. G. Petrino and T. Kakunaga, *PNAS* **79,** 6465 (1982).
70. J. M. Aiken, F. D. Miller, F. Haegn, D. I. McKensie, S. A. Krawetz, J. H. van de Sande, J. B. Rattner and G. H. Dixon, *Bchem* **24,** 6268 (1985).
71. L. Runkel and A. Nordheim, *NARes* **14,** 7143 (1986).
72. J. R. Thomas, R. I. Bolla, J. S. Rumbyrt and D. Schlessinger, *PNAS* **82,** 7595 (1985).
73. Y. Nishioka and P. Leder, *JBC* **255,** 3691 (1980).
74. T. E. Hayes and J. E. Dixon, *JBC* **260,** 8145 (1985).
75. H. Hamada and T. Kakunaga, *Nature* **298,** *396* (1982).
76. E. N. Trifonov, A. K. Konopka and T. M. Jovin, *FEBS Lett.* **185,** 197 (1985).
77. P. S. Ho, M. J. Ellison, G. J. Quigley and A. Rich, *EMBO J.* **5,** 2737 (1986).
78. J. B. Cohen, K. Effron, G. Rechavi, Y. Ben-Neriah, R. Zakut and D. Givol, *NARes* **10,** 3353 (1982).
79. J. L. Slightom, A. E. Blechl and O. Smithies, *Cell* **21,** 627 (1980).
80. S. Shen, J. L. Slightom and O. Smithies, *Cell* **26,** 191 (1981).
81. M. W. Kilpatrick, J. Klysik, C. K. Singleton, D. A. Zarling, T. M. Jovin, L. H. Hanua, B. F. Erlanger and R. D. Wells, *JBC* **259,** 7268 (1984).
82. J. G. Flanagan, M. P. Lafranc, and T. H. Rabbits, *Cell* **36,** 681 (1984).
83. J. R. Stringer, *J. Virol.* **53,** 698 (1985).
84. J. R. Stringer, *MCBiol* **5,** 1247 (1985).
85. B. Malfoy, N. Rousseau, N. Vogt, E. Viegas-Pequignot, B. Dutrillaux and M. Leng, *NARes* **14,** 3197 (1986).
86. K. E. Murphy and J. R. Stringer, *NARes* **14,** 7325 (1986).
87. J. R. Stringer, *Nature* **296,** 363 (1982).
88. J. Skowronski, A. Plucienniczak, A. Bednarek and J. Jaworski, *JMB* **177,** 399 (1984).
89. D. Treco and N. Arnheim, *MCBiol* **6,** 3934 (1986).
90. P. Bullock, J. Miller and M. Botchan, *MCBiol* **6,** 3948 (1986).
91. M. Steinmetz, D. Stephan and K. F. Lindahl, *Cell* **44,** 895 (1986).
92. A. Weinreb, D. R. Katzenberg, G. L. Gilmore and B. K. Birshstein, *PNAS* **85,** 529 (1988).
93. R. Swanstrom, R. C. Parker, H. E. Varmus and J. M. Bishop, *PNAS* **80,** 2519 (1983).
94. D. R. Greaves and R. K. Patient, *NARes* **14,** 4147 (1986).
95. O. Hyrien, M. Debaisse, G. Buttin and B. R. de Saint Vincent, *EMBO J.* **6,** 2401 (1987).
96. O. Hyrien, M. Debatisse, G. Buttin and B. R. de Saint Vincent, *EMBO J.* **7,** 407 (1988).
97. B. W. Glickman and L. S. Ripley, *PNAS* **81,** 512 (1984).
98. U. Krawinkel, G. Zoebelein and A. L. M. Bothwell, *NARes* **14,** 3871 (1986).
99. J. Nalbantoglu and M. Meuth, *NARes* **14,** 8361 (1986).
100. F. Wohlrab, M. J. McLean and R. D. Wells, *JBC* **262,** 6407 (1987).
101. D. A. Collier, J. A. Griffith and R. D. Wells, *JBC* **263,** 7397 (1988).

102. R. D. Wells, D. A. Collier, J. C. Hanvey, M. Shimizu and F. Wohlrab, *FASEB J.* **2,** 2939 (1988).
103. F. Wohlrab and R. D. Wells, *in* "Gene Amplification and Analysis" (J. G. Chirikijian, ed.), Vol. 5, p. 247. Elsevier, Amsterdam, 1987.
104. B. D. Stollar, *CRC Crit. Rev. Biochem.* **20,** 1 (1986).
105. E. H. Egelman and A. Stasiak, *JMB* **191,** 677 (1986).
106. D. B. Haniford and D. E. Pulleyblank, *NARes* **13,** 4343 (1985).
107. N. Panayotatos and A. Fontaine, *JBC* **262,** 11364 (1987).
108. J. A. Blaho, J. E. Larson, M. J. McLean and R. D. Wells, *JBC* **263,** 1446 (1988).

Polycistronic Animal Virus mRNAs

CHARLES E. SAMUEL

Section of Biochemistry and Molecular Biology
Department of Biological Sciences
University of California
Santa Barbara, California 93106

Of central importance to the understanding of biological processes is the elucidation of the mechanisms by which the products of genes are produced in a regulated fashion. The production of a particular protein in an animal cell may be regulated, in principle, at any of the many steps of mRNA formation (*1, 2*) and translation (*3*). At the level of translation, most animal virus and animal cell mRNAs analyzed in detail appear to be functionally monocistronic; that is, a single primary translational product is produced from a single mature mRNA transcript (*4, 5*). In striking contrast, many bacteriophage and bacterial cell mRNAs are polycistronic; that is, multiple primary translational products are specified by a single mature mRNA transcript using independent initiation and/or termination codons for protein synthesis (*4*).

Although the organization of animal virus genomes and the mechanisms of viral genome replication and transcription can differ substantially among the major families of animal viruses, the translation of viral mRNA by the host's protein-synthesizing machinery is a central step common to all viral multiplication strategies (*6*). Historically, molecular genetic studies of animal viruses have yielded important new insights into our understanding of the mechanisms by which

Progress in Nucleic Acid Research
and Molecular Biology, Vol. 37

proteins are produced in animal cells (*1–3*, *7–9*). Now, detailed analyses have revealed that, in some cases, the synthesis of animal virus proteins involves the translation of functionally polycistronic mRNAs to yield multiple primary translation products. Such fundamentally important conclusions have been made possible, in part, by the availability of nucleotide and peptide sequences of a number of animal virus genomes and their encoded products, together with complete cDNA clones of animal virus genomes and mRNAs that can be manipulated by recombinant DNA techniques to generate synthetic mRNAs or mutated versions of authentic mRNAs. Careful analyses of various viral mRNAs have now made it clear that some viral mRNAs are indeed recognized and translated as functionally polycistronic mRNAs by the protein-synthesizing machinery of animal cells. Furthermore, many different biochemical mechanisms may exist in animal cells to permit the expression of functionally polycistronic viral mRNAs.

In this essay, I review some observations concerning the natural occurrence and structural organization of polycistronic animal virus mRNAs, and the mechanisms by which they may be translated to yield two or more unique polypeptide products.

I. Natural Occurrence of Polycistronic Animal Virus mRNAs

Most animal virus mRNAs characterized so far are functionally monocistronic. However, functionally polycistronic animal virus mRNAs also exist. Initiation of translation in monocistronic viral mRNAs most commonly occurs at AUG and, furthermore, at the 5′-proximal AUG (*5–7*, *10*, *11*). In most polycistronic viral mRNAs, initiation of translation of both the 5′-proximal, upstream cistron and the internal, downstream cistron(s) likewise occurs at an AUG codon (*12–79* inclusive). Animal viruses encoding polycistronic mRNAs in which translation-initiation occurs alternatively at one or more AUG initiation sites include members of several virus families that utilize a variety of different replication strategies as parts of their life cycles (Table I). They include: viruses with DNA genomes and viruses with RNA genomes; viruses with circular genomes and viruses with linear genomes; viruses whose genomes are constituted by a single piece of nucleic acid as well as viruses with segmented genomes; and viruses that utilize the cell nucleus as the site for mRNA biogenesis as well as viruses whose mRNA is synthesized in the cytoplasm.

The identification and characterization of polycistronic mRNAs have been greatly facilitated by the availability of nucleotide sequences derived from cDNA clones of viral mRNAs. From these cDNA sequences, it has been possible to identify potential initiation and termination codons and open reading frames capable of encoding polypeptide products of a predicted size. The availability of these cDNA clones has also permitted the construction of various types of vectors for translational studies, including *in vivo* expression vectors and *in vitro* transcription vectors. Eukaryotic expression vectors using various high-efficiency promoters have made possible the analysis of protein synthesis *in vivo* using constructions that encode chimeric mRNAs possessing defined mutations in regions of potential regulatory importance, and/or chimeric mRNAs specifying "reporter" genes that facilitate assays for translational expression. *In vitro* transcription vectors with SP6 and T7 promoters have made possible the efficient synthesis of wild-type and mutant mRNAs that can be examined for their translational activity with a variety of different cell-free protein-synthesizing systems. In addition, the ability to identify and quantitate the expression within infected or transfected cells of proteins predicted from cDNA sequences has been facilitated by the production of antibodies against synthetic peptides having the amino-acid sequences deduced from open reading frames revealed by cDNA sequence data, and antibodies produced against recombinant fusion proteins produced in bacteria. Primer extension analysis, S1-protection analysis, and RNA "blotting" and hybridization have been used to demonstrate the presence of mRNAs *in vivo* that possess a particular 5′ region in which potential translation initiation codons are localized. Thus, much experimental evidence exists for the occurrence of functionally polycistronic mRNAs, specific examples of which are considered in the following section, which emphasizes their possible mechanisms of expression.

II. Mechanisms of Expression of Polycistronic Viral mRNAs

A. Initiation by 5′ Ribosome-Binding and Leaky Scanning

The scanning model provides a conceptual framework for understanding the process of initiation of mRNA translation in animal cells. According to the scanning model, a 40-S ribosomal subunit binds at or near the 5′ end of an mRNA and advances linearly until an initiation codon in a favorable context is reached, at which point assembly of a complete 80-S ribosomes takes place and initiation of polypeptide synthesis occurs (*5, 10, 12, 80*). The efficiency of initiation of transla-

TABLE I
POLYCISTRONIC ANIMAL VIRUS mRNAs

Virus	mRNA	Products (5′; 3′)	Proposed expression mechanism[a]	Reference
I. DNA Viruses				
A. Adenoviridae				
Ad12, Ad5	E1b	p19 (p21); p54 (p55)	1c	*27*
Ad2	E3a	p6.7; gp19	1c	*28*
Ad2	DNA pol	p120; p62	2	*56*
B. Herpesviridae				
Herpes simplex virus	Thymidine kinase	p43; p39; p38	1a	*16, 17*
C. Papovaviridae				
Simian virus 40	19-S late	VP2; VP3	1a	*13–15*
	16-S late	LP1; VP1	1, 5?	*78, 135*
	19-S early	SELP; T, t	1, 5?	*69*
D. Parvoviridae				
AAV2	B/C	p70 (B); p60 (C)	1b	*24, 25*
B19	VP1	VP1; VP2	5?	*70*
II. RNA Viruses				
A. Bunyaviridae				
Bunyavirus	Segment S	N; NS_s	1c	*37–39*
B. Coronaviridae				
Mouse hepatitis virus	Gene 5	p13; p9.6	1c	*49*
C. Orthomyxoviridae				
Influenza B	Segment 6	NA; NB	1c?	*47, 48*

D. Hepadnaviridae				
Hepatitis B virus	Pregenomic	core; *pol*	2	*141*
E. Paramyxoviridae				
Sendai virus	P/C	P; C, C′; Y, Y′	1b, 1c	*21–23*
Parainfluenza	P/C	P; C	1c	*34–35*
Measles	P/C	P; C	1c	*36*
Newcastle disease virus	P/C	P; (p38,p29) C	1a	*20*
F. Picornaviridae				
Poliovirus	Genomic	Polyprotein	2	*101, 104*
Encephalomyocarditis virus	Genomic	Polyprotein	2	*102, 105*
G. Reoviridae				
Reovirus T1, T3	Segment 1	$\sigma1$; $\sigma1_{ns}$ (p12, p14)	1c	*40–46*
Rotavirus SA11	Segment 9	(VP7) p38; p35	1a	*19*
H. Retroviridae				
Rous sarcoma virus	Genomic	*gag; pol*	3	*61*
	Subgenomic	*src*	5?	*72*
Human T-cell leukemia virus	Genomic	*gag; pol*	3	*62*
	Subgenomic	*rex; tax*	1a	*18*
Human immunodeficiency virus	Genomic	*gag; pol*	3	*64*
Mouse mammary tumor virus	Genomic	*gag*-X; *pro*	3	*65, 66*
		x-pro; pol		
Bovine leukemia virus	Genomic	*gag; pol*	3	*63*
Murine leukemia virus	Genomic	*gag; pol*	4	*67*
Feline leukemia virus	Genomic	*gag; pol*	4	*68*
I. Rhabdoviridae				
Vesicular stomatitis virus	NS	NS (P); p7	2	*55*

[a] Candidate expression mechanisms. 1, Leaky scanning: 1a, alternative in-frame AUG initiation codon; 1b, alternative in-frame non-AUG initiation codon; 1c, alternative AUG initiation codon in a different, overlapping reading frame. 2, Initiation by internal binding of ribosomes. 3, Ribosomal frameshifting during elongation. 4, Suppression of in-frame UAG termination codon. 5, Reinitiation following termination.

tion is thought to be modulated, at least in part, by the position and context of the initiation codon (*5, 10, 12, 80–83*). If the 5′-proximal initiation codon is in an optimal context, initiation at that codon is efficient. If the context of the 5′-proximal initiation codon is suboptimal, initiation at that codon is inefficient. Although a fraction of the scanning ribosomes may indeed initiate translation at a suboptimal 5′-proximal initiation codon, a portion of the 40-S ribosomal subunits would presumably bypass the suboptimal codon and initiate downstream at an internal AUG codon positioned in a more favorable context. This "leaky scanning" model provides a mechanism for the initiation of translation at internal initiation codons that would be present in polycistronic mRNAs (*12*).

Considerable evidence consistent with the leaky scanning model for initiation of translation has accumulated. This evidence includes the unique 5′-terminal $m^7G(5')ppp(5')N$ . . . cap structure characteristic of most animal virus and cell mRNAs, a structure that often significantly stimulates translation in many types of cell-free protein-synthesizing systems (*84–86*); the inability of ribosomes to bind and translate synthetic circular mRNAs (*87, 88*); the ability of 40-S ribosomal subunits to migrate along mRNA (*89, 90*); the inhibitory effect on translation-initiation of synthetic secondary structures inserted upstream from an "authentic" initiation codon (*91–95*); and the inhibitory effect on translation of AUG initiation codons inserted upstream from an "authentic" initiation codon (*83, 96–99*).

By inspection of sequences near the 5′ ends of animal cell and viral mRNAs (*5, 11, 80, 81*) and by site-directed mutagenesis of a cDNA copy of a cellular gene (*82, 83*), two nucleotide postions relative to the initiation codon were found to have a dominant effect on the efficiency of initiation: position −3 (three nucleotides upstream from the initiation codon), and position +4 (immediately following the initiation codon). A purine in position −3 and a guanine in position +4 were optimal for efficient initiation at 5′-proximal AUG codons. Furthermore, the efficiency of initiation was, to a large extent, modulated by the context—that is, the −3/+4 nucleotide positions flanking the 5′-proximal AUG—rather than by the relative distance of the AUG from the 5′ end of the mRNA (*5, 10–12, 80–83*).

For several of the polycistronic animal virus mRNAs, the initiation codon of the 5′-proximal upstream cistron is present in a suboptimal context with regard to the −3/+4 flanking nucleotides, whereas the initiation codon of the internal downstream cistron is often, but not always, present in an optimal −3/+4 context. Thus, some of the scanning 40-S ribosomal subunits would presumably be able to "leak" past the suboptimal initiation site of the upstream cistron and continue scanning until reaching the optimal initiation site of the downstream cistron, present in either the same or a different reading frame. The relative efficiency of utilization of the upstream and downstream cistron initiation codons by ribosomes scanning the same mRNA is of possible regulatory importance, as certain polycistronic mRNAs appear to be organized so that the relative abundance of the proteins encoded by different cistrons correlates with the relative strength of the nucleotides flanking the initiation sites of the cistrons.

One mechanism by which one gene can encode two forms of a protein that differ only in the amino-terminal region is the production of one mRNA transcript from which translation initiation can occur alternatively at one of two in-frame initiation codons. A number of animal virus genes encode mRNA species in which different in-frame AUG initiation sites are utilized in a manner such as to generate related protein products that differ in their amino-terminal sequences. For example, the following viral polycistronic mRNAs each encode two or more proteins in the same reading frame: simian virus 40 (SV40) late 19-S mRNAs encoding capsid proteins VP2 and VP3 (*13–15*); the herpes simplex virus thymidine kinase gene mRNA encoding 43-kDa, 39-kDa, and 38-kDa proteins (*16, 17*); the human T-cell leukemia virus subgenomic mRNA encoding the *rex* and *tax* nonstructural regulatory proteins (*18*); the rotavirus segment-9 mRNA encoding 38-kDa and 35-kDa VP7 capsid glycoproteins (*19*); and the Newcastle disease virus mRNA encoding capsid phosphoprotein P and nonstructural 38- and 29-kDa C-like nonstructural proteins (*20*). In each of these cases (*13–20*), the upstream or 5′-proximal AUG initiation site is in a suboptimal −3/+4 context relative to the downstream AUG initiation site.

The relative translational efficiency of the upstream and downstream cistrons and the abundance of their encoded products may largely be determined by the −3/+4 sequence context of these viral mRNAs. For example, the less abundant VP2 capsid protein of SV40 is synthesized from the 5′-proximal AUG, which is in a suboptimal context, and the 3-fold more abundant VP3 capsid protein is synthesized from a downstream AUG positioned in a more favorable −3/+4 flanking nucleotide context. The results of genetic and biochemical studies of deletion and insertion mutants of SV40, and studies of SV40 transient expression vectors utilizing the bacterial chloramphenicol acetyltransferase (CAT; EC 2.3.1.28) as a reporter activity, provide strong evidence consistent with the conclusion that both VP2 and VP3 are synthesized from each of the alternatively spliced late 19-S mRNAs in a regulated manner by independent initiation of translation via a leaky scanning mechanism (*13–15*).

The Sendai P/C mRNA (*21–23*), the adeno-associated virus B/C mRNA (*24, 25*), and the human c-*myc* mRNA (*26*) also each encode two or more proteins in the same reading frame. Different in-frame initiation sites are utilized by these mRNAs to generate two protein products that differ in their N-terminal sequence. However, in these cases, the most upstream 5′-proximal initiation site is not AUG, but rather is either ACG (*21–25*) in the Sendai P/C and AAV B/C mRNAs, or CUG (*26*) in the c-*myc* mRNA. These 5′-proximal non-AUG initiation codons are in a favorable −3/+4 flanking nucleotide context. However, as discussed under Section II, C, the non-AUG codons are not as efficient as is AUG for initiation, and thus may possibly facilitate a leaky scanning of the 40-S subunit and subsequent initiation at the downstream AUG initiation codon at a frequency sufficient to permit the required levels of synthesis of the encoded proteins.

Several polycistronic viral mRNAs encode two proteins in different, overlapping reading frames. Examples include: the adenovirus (Ad12 and Ad5) 2.2-kb E1b mRNA encoding the 19/21-kDa and 54/55-kDa tumor antigens (*27*); the adenovirus E3a mRNA encoding a 6.7-kDa protein and a 19-kDa glycoprotein (*28*); the paramyxovirus P/C mRNAs of Sendai virus (*21–23, 29–33, 140*), parainfluenza virus 3 (*34, 35*) and measles virus (*36*), each of which encodes the phosphoprotein P and one or more nonstructural proteins designated C, C′, Y, and Y′; the bunyavirus-S segment mRNA encoding the nucleoprotein N and the nonstructural protein NS_s (*37–39*); and the reovirus segment-S1 mRNA encoding the minor capsid protein sigma-1 and the nonstructural protein sigma-1_{NS} (*40–46*). The upstream AUG initiation site in each of these polycistronic viral mRNAs possessing overlapping reading frames is in a suboptimal −3/+4 context, and the relative abundance of the upstream and downstream cistron products may be controlled by the −3/+4 context of their initiation codons, both *in vivo* and *in vitro*. For example, the adenovirus E3 downstream AUG codon is strong and the downstream cistron gp19K protein is the most abundantly synthesized E3 protein *in vivo* (*28*); the synthesis of the Sendai virus C-protein is about five times more efficient than that of P protein and the P/C mRNA initiation codon for C protein synthesis has an adenine at −3 whereas the P protein AUG has a cytosine at −3 (*21, 32*); and the reovirus S1 mRNA 5′-proximal AUG encoding the less abundant sigma-1 protein is in a weak −3/+4 context as compared to the internal AUG encoding sigma-1_{NS}, which is in a strong context and is more efficiently synthesized *in vitro* than is sigma-1 (*40–44, 103*).

Although the apparent efficiencies of utilization of the AUG codons for the first and second cistrons of polycistronic RNAs expressed *in vivo* or translated *in vitro* often agree with the predictions of the leaky scanning model when the context of the −3/+4 flanking nucleotides is considered, exceptions do exist. For example, the influenza virus B segment 6 NB/NA mRNA encodes two glycoproteins in different overlapping reading frames, the neuraminidase NA protein and a nonstructural glycoprotein NB (*47, 48*). The −3/+4 context of the 5′-proximal AUG initiation site used for NB synthesis is more favorable than the context of the downstream AUG initiation codon used for NA synthesis, yet NA and NB accumulate to approximately equal amounts in infected cells (*48, 138*). Furthermore, mutations in the sequence immediately around the 5′-proximal AUG codon do not make a large difference in the amounts of NB and NA that accumulate in transfected cells, but when the 5′ AUG is displaced from its normal position it becomes efficient at preventing downstream initiation events (*138*). The coronavirus gene 5 mRNA of mouse hepatitis virus (MHV) also possesses two open reading frames. Although MHV polypeptides corresponding in size to both gene-5 mRNA reading frames are synthesized *in vitro* from pGEM transcripts, so far only the second open reading frame product has been shown to be expressed within MHV-infected cells, and the −3/+4 context of the AUG initiation site for this reading frame is not optimal (*49–51*).

The analyses of certain mutant mRNAs—for example, reovirus s1 and s4 mRNAs (*52*), influenza virus NB/NA mRNA (*138*), avian retrovirus mRNAs (*53*), and hepatitis-B surface antigen mRNA (*54*)—suggest that nucleotides other than the previously identified consensus context nucleotides flanking the initiator AUG codon at the −3 (purine) and +4 (guanine) positions are also important and perhaps even play a dominant role in determining the efficiency of translation in animal cells. Conceivably, mRNA primary or higher ordered structures responsible for the differential ability of mRNAs to interact with mRNA binding initiation factors such as eIF-4A and eIF-4F may, in some cases, be a major determinant of the translational efficiency of 5′-capped mRNAs (*86, 100*). Thus, the selection of an "authentic" initiation codon with an appropriate efficiency may well be the consequence of collective contributions of many parameters, including both *cis*-acting sequences such as the −3/+4 context nucleotides, and *trans*-acting components such as protein synthesis-initiation factors that may discriminate among different mRNAs by virtue of different mRNA binding affinities.

B. Initiation of Translation by Internal Binding of Ribosomes

Binding of 40-S ribosomal subunits to internal sequences within the body of a polycistronic mRNA, rather than binding to the 5′ end of the mRNA as proposed by the modified scanning model, would provide a mechanism by which translation-initiation could occur from initiation codons positioned downstream from the 5′-proximal AUG codons. Among the best evidence for internal initiation of translation is that obtained with picornavirus mRNAs, both *in vitro* and *in vivo* (*101, 102*).

Picornaviruses possess nonsegmented, plus-strand RNA genomes of about 7500 nucleotides (*6*). Although picornavirus mRNAs appear functionally monocistronic in that a single large polypeptide precursor is synthesized that undergoes posttranslational proteolytic cleavages to generate the mature protein products (*9*), studies with poliovirus (*101, 104*) and encephalomyocarditis (EMC) virus (*102, 105*) suggest that initiation of translation occurs by a cap-independent mechanism that involves internal binding of ribosomes.

The 5′-noncoding region of poliovirus mRNA is unusually long, about 750 nucleotides, and contains several AUG codons upstream from the major initiator AUG located at nucleotide 743 (*106–109*). Unlike most animal virus mRNAs, poliovirus mRNA does not contain a 5′-terminal $m^7G(5')ppp(5')N$ . . . cap structure. Rather, poliovirus mRNA terminates in pUp . . . and is translated by a cap-independent mechanism (*110*). Furthermore, the initiation of translation of 5′-capped cellular and viral mRNAs is inhibited in poliovirus-infected cells. The inhibition mechanism appears to involve a proteolytic inactivation of the p220 subunit of eIF-4F, the cap-binding complex initiation factor required for ribosomes to bind to the 5′ terminus and subsequently to initiate the translation of capped mRNAs (*110, 111*).

Biochemical and genetic analysis of poliovirus mutants generated using a cDNA copy of the viral genome revealed regions within the 5′-noncoding region of poliovirus mRNA important for the efficient translation of the viral mRNA (*101, 112–115*). Surprisingly, mutational analysis of the seven upstream AUG codons present in the 5′-noncoding region of poliovirus mRNA has revealed that the upstream short open reading frames are not essential for virus replication and do not act as barriers to the translation of poliovirus mRNA (*136*). By contrast, analysis of a series of poliovirus deletion mutants identified a functional *cis*-acting element within the 5′-untranslated sequences of poliovirus mRNA that enables it to translate in a cap-independent manner (*112, 113, 115*). The major determinant of the polio cap-independent translational element maps between nucleotides 320 and 631 from the 5′ end of the poliovirus mRNA and is functional both *in vivo* (*113*) and *in vitro* (*112, 113*). Translation *in vitro* of SP6 transcripts containing additional deletions in the 5′-noncoding region more narrowly focused the *cis*-acting element responsible for cap-independent translation, as measured in a mixed rabbit reticulocyte–HeLa system, to a 60-nucleotide sequence located between positions 567 and 627 (*115*). The poliovirus 5′-noncoding mRNA sequences responsible for the cap-independent synthesis of poliovirus polypeptides can also confer cap-independent translation on heterologous chimeric mRNAs encoding either the bacterial CAT or the herpes virus thymidine kinase as a reporter enzymic activity (*112, 113*). Expression of the reporter enzyme from the chimeric mRNA is extensively augmented by poliovirus-mediated inhibition of cap-dependent protein synthesis.

The cap-independent translation-initiation on poliovirus (*101, 104, 115*) and EMC virus (*102, 105*) mRNAs appears to occur by a mechanism that involves binding of ribosomes to an internal sequence within the 5′-noncoding region. A bicistronic plasmid containing the herpes simplex thymidine kinase gene as the first cistron and the CAT gene as the second cistron does not express the first cistron in poliovirus-infected COS cells but does express the second cistron when the poliovirus 5′-untranslated region is inserted as the intercistronic spacer (*101*). The second cistron is not expressed when the CAT 5′-untranslated region is the intercistronic spacer (*101*). Deletion analysis of the poliovirus 5′-untranslated region suggests that the internal ribosome binding site occupies several hundred nucleotides located between nucleotides 140 and 630 of the poliovirus 5′-untranslated region (*101*). It is unclear whether internal ribosome binding is directed toward a specific AUG or whether the internal binding is followed by scanning. However, because the introduction of a hairpin secondary structure ($\Delta G° = -30$ kcal/mol) at position 631 of the polio RNA dramatically inhibits translation initiation from the downstream AUG, it has been proposed that the ribosomes, following internal binding, are translocated by scanning until they reach the initiator AUG located at position 745 of the poliovirus mRNA (*101*).

Internal ribosome binding can also occur to picornavirus mRNAs *in vitro* (*131*). In HeLa cell extracts, internal binding to the 5′-noncoding region of poliovirus mRNA in a bicistronic context is independent of an upstream open reading frame (*139*). Data obtained with EMC virus RNA are also consistent with an internal binding of ribosomes to mRNA that is followed by scanning toward the initiator AUG site (*102, 105*). Hybridization of complementary cDNA fragments to different sites within the first 338 nucleotides of the 5′-noncoding region of EMC RNA did not affect translation of the viral mRNA *in vitro*, whereas the binding of cDNA fragments to eight different sites located between nucleotides 450 and the initiator AUG codon at postion 834 caused high degrees of translation inhibition in reticulocyte lysates (*105*). These findings were extended and confirmed by the analysis of artificial bicistronic mRNAs that contained, in order from 5′ to 3′, the 5′-noncoding region of poliovirus connected to the coding region of the *sea* oncogene as the first indicator gene followed by truncated versions of the 5′-noncoding region of EMC virus connected to the poliovirus 2A coding region as the second indicator gene (*102*). The translation *in vitro* of run-off T7 polymerase transcripts of the chimeric polio(*sea*)–EMC (*2A*) constructs revealed that a specific, internal ribosome entry site probably exists within the 5′-noncoding region of EMC virus RNA. The translational efficiency of the second 2A cistron was not reduced in the presence of a poorly translated first *sea* cistron as long as the second reporter cistron remained under the control of the 5′-noncoding region of EMC virus RNA. Deletion analysis revealed that the EMC RNA 5′-noncoding sequence between nucleotides 260 and 484 plays a critical role in the efficient translation of adjacent coding sequences, in both mono- and bicistronic mRNAs, presumably because the internal ribosomal entry site resides within this region of the EMC RNA (*102*).

It appears that internal initiation sites are also utilized during the translation *in vitro* of vesicular stomatitis virus (VSV) NS mRNA encoding the phosphoprotein NS(P) (*55*), adenovirus mRNA encoding the viral DNA polymerase (*56*), and hepatitis B virus (HBV) pregenomic mRNA encoding the viral reverse transcriptase (*141*). In the case of HBV, combined genetic and biochemical studies both in cell culture and in ducks reveal that the reverse transcriptase is synthesized by a mechanism involving translation initiation at an internal *pol* AUG codon rather than by ribosomal frameshifting within the core-*pol* overlap (*141*). In these cases, the internal initiation occurs at an AUG positioned several hundred nucleotides downstream from, but in-frame with, the 5′-proximal AUG. Synthesis of the VSV 7-kDa (*55*) and the adenovirus 62-kDa (*56*) protein products initiated from the internal AUG is unaffected by hybrid-arrest translation conditions using cDNA fragments complementary to 5′ region viral sequences that inhibit the synthesis of the 5′-proximal AUG-initiated protein products (NS protein for the VSV mRNA, and the 120-kDa DNA *pol* protein for the adenovirus mRNA). The hybrid-arrest translation results suggest that leaky scanning of ribosomes from the 5′ end of these viral mRNAs to the respective internal initiation sites does not occur to an appreciable extent. However, additional studies are necessary further to support this interpretation. The formation of RNA · DNA duplex structures may inhibit translation by two mechanisms: the duplex may exert a direct steric effect that affects the binding of factors or the movement of ribosomes; alternatively, the mRNA may be cleaved at the site of the duplex by RNase H present in the reticulocyte lysate (*116*).

A highly significant sequence similarity extends through the 5′-noncoding region of the three poliovirus serotypes, PV1, PV2, and PV3 (*108, 117*). Furthermore, a comparative sequence analysis of the 5′-noncoding region of several picornaviruses, including coxsackie B3, human rhinoviruses HRV2 and HRV14, and polioviruses PV1, PV2, PV2S, PV3, and PV3S, revealed the conservation of secondary structure predicted to encompass the entire 5′-noncoding regions of the picornaviruses (*117*). The fact that divergence of picornavirus 5′-noncoding sequences occurs in a manner that permits conservation of certain overall structural features—including over 20 stem and loop structures, two pyrimidine-rich regions and long stretches of conserved sequence—suggests important functional roles for the 5′-noncoding region, possibly mediated as much by overall structure as by specific primary sequence. Undoubtedly, the conserved 5′-noncoding structure of picornavirus mRNAs plays important roles in several stages of the virus multiplication cycle, perhaps including protein synthesis. The conserved 5′-noncoding structure may define a region recognized by the protein synthesis initiation factors and ribosome subunits that results in the internal, cap-independent initiation of translation.

Biochemical evidence suggests that protein-synthesis initiation factors eIF-4A and eIF-4B may play important roles in cap-independent, internal initiation of eukaryotic mRNA translation (*118*). These two factors, together with ATP, are normally specifically required for the binding of 5′-capped mRNA to the 43-S ribosome complex (*119–121*). The eIF-4A is an ATP-dependent single-stranded RNA-binding protein that displays mRNA-dependent ATPase activity (*119*); eIF-4F is the three-subunit cap-binding protein (CBP) complex that includes eIF-4A, the CBP eIF-4E, and p220 (*120*); and, eIF-4B is a factor whose exact function remains unknown, but which appears to stimulate the activities of eIF-4A and eIF-4F and to function in the binding of mRNA to the 43-S complex. Both eIF-4A and eIF-4F can function as RNA-unwinding proteins (*121*); however, eIF-4F; is not required for the cap-independent initiation of translation observed in poliovirus-infected cells (*110*).

In the absence of eIF-4F, both eIF-4A and eIF-4B can bind to an uncapped synthetic mRNA lacking secondary structure with essentially the same degree of effectiveness and affinity observed for capped natural mRNA in the presence of all three factors (*118*). Perhaps the conserved structural elements within the 5′-noncoding sequence of picornaviruses include a feature that allows eIF-4A and eIF-4B to bind to uncapped picornavirus mRNA in the absence of eIF-4F, thereby permitting internal, cap-independent translation-initiation. It should be noted that the efficiency of internal initiation *in vitro* can vary with the nature of the cell-free protein-synthesizing system, as revealed from studies with rabbit reticulocyte and HeLa cell-free systems (*102, 104*). Possibly different kinds of cell-free extracts contain varying concentrations of a *trans*-acting protein-synthesis factor(s) which play(s) an important role in affecting the relative efficiency of ribosomal entry at or near the 5′ end of an mRNA as compared to entry at internal sites within the mRNA. Undoubtedly, the efficiency of internal initiation of translation will be modulated by multiple parameters, including both the degree of optimization of *cis*-acting mRNA sequences and/or structures and the relative concentration and form of the *trans*-acting components of the protein-synthesis machinery.

C. Non-AUG Initiation of Translation

It is apparent that codons other than AUG may initiate the synthesis of proteins in animal cells, albeit so far rarely and generally at a reduced efficiency relative to AUG. Utilization of ACG as an initiator codon has been described for polycistronic mRNAs of two animal viruses, adeno-associated virus (AAV) (*24*) and Sendai virus (*21, 22*). In addition, utilization of CUG as an initiation codon has been described for c-*myc* mRNA (*26*). The adeno-associated virus, Sendai virus, and c-*myc* mRNAs that utilize a non-AUG initiation codon display certain similarities: two or more independent translation-initiation sites are utilized on the same mRNA; the non-AUG initiation codon is located upstream from the AUG initiation codon(s); and, the non-AUG initiation codon is generally utilized less efficiently than the downstream AUG initiation codon (*22, 24–26*).

Adeno-associated virus is a defective parvovirus that replicates in the nucleus of human cells in culture coinfected with adenovirus. The genome of AAV is single-stranded DNA, either plus or minus (*6*). The AAV capsid protein synthesis was the first of non-AUG initiation described for animal cells. From the sequence of the AAV2 genome (*57*), the structure of the mRNAs that encode the capsid proteins (*58, 59*), and the amino-terminal sequence of the AAV capsid protein B (*24*), it has been concluded that the synthesis of the AAV capsid protein B is initiated at an ACG codon (*24, 25*).

The ACG codon responsible for the initiation of translation of the AAV 70-kDa capsid protein B occurs upstream from, and in the same reading frame as, the AUG codon utilized for initiation of translation of the 60-kDa capsid protein C (*24*). The capsid protein C is synthesized in amounts about 10 to 20 times greater than the capsid protein B (*24, 25, 80*). Translation *in vitro* of a synthetic SP6 AAV transcript has definitively established that the AAV capsid proteins B and C are indeed synthesized from a single mRNA species by alternative use of their respective in-frame initiation codons; the main source of the B and C proteins *in vivo* is probably the known spliced 2.3-kb RNA (*25*). The coordinated synthesis of B and C from the same mRNA may be due to leaky scanning through the non-AUG initiation codon for B. The less efficient ACG initiation of B-protein synthesis as compared to the more efficient AUG initiation of C-protein synthesis would provide a mechanism for the regulation of the amount of B- and C-protein synthesis in a fixed ratio independent of mRNA concentration (*24, 25*).

Interestingly, in the case of the cellular dihydrofolate reductase (DHFR) mRNA, substitution of ACG for the normal AUG translation codon leads to the synthesis of a normal DHFR protein both *in vivo* and *in vitro* (*122*). In addition, a truncated form of DHFR is also produced, apparently by initiation at the next in-frame AUG located downstream from the ACG. Initiation of DHFR mRNA translation at the ACG codon depends upon a favorable $-3/+4$ sequence context (*122*).

Sendai virus possesses a nonsegmented, negative-strand RNA genome and replicates in the cytoplasm of infected cells (*6*). The Sendai virus P/C mRNA is polycistronic. Deletion and site-directed point mutants of the P/C mRNA indicate that it codes for five proteins in two overlapping reading frames, utilizing both ACG and AUG as the initiation codons (*21–23, 29–33*). The C reading frame is responsible for the synthesis of four proteins, C′, C, Y1, and Y2, each of which is initiated at an independent site, but all of which appear to terminate at the same site. The P reading frame is responsible for the synthesis of a single product, the P protein (*21, 22*). The P protein is a phosphoprotein that appears to be associated with the virion-associated RNA-dependent RNA polymerase activity (*57*); the C, C′, Y1, and Y2 proteins all appear to be nonstructural proteins and their functions are not well established (*21–23, 32*).

The 5′-proximal initiation codon utilized in the Sendai virus P/C mRNA is an ACG in the C reading frame (*21–23*). The ACG codon at nucleotide position 81 is used to initiate the synthesis of the C′ protein. The succeeding initiation codons of the P/C mRNA are all AUG codons. The first AUG downstream from the ACG is in a different frame, the P reading-frame, and is used to initiate the synthesis of P protein at nucleotide position 104. The further downstream initiation AUG codons are all in the C reading-frame and are used to initiate the synthesis of C protein at nucleotide 114, Y1 protein at 181, and Y2 protein at 175 (*21, 22*).

The ACG codon responsible for the initiation of the Sendai virus C′ protein is in a context similar to that of the ACG codon responsible for the initiation of the AAV capsid protein B (*21, 22, 24*). Both ACG codons are in a favored context for efficient ribosome initiation with a purine at position −3 and a guanine at position +4; in addition, positions +5 to +10 are also identical with the exception of +7 which is a uracil in the Sendai mRNA and a cytosine in the AAV mRNA (*21, 22, 24*).

The efficiency of synthesis *in vivo* of the C′ protein from the 5′-proximal ACG differs less than 2-fold from that of the P protein expressed from the downstream AUG, whereas the amount of C protein synthesized is four to five times that of P protein (*21, 32*). These results suggest that an ACG in an otherwise favorable −3/+4 context can function almost as efficiently for initiation of translation as an AUG codon in a less favorable −3/+4 context, by only about 10 to 20% as efficiently as an AUG in a more favored context. The ACG initiator codon of the P/C mRNA appears to be even more efficiently utilized *in vitro* than it is *in vivo* (*32, 60*). Proteins Y1 and Y2 are both expressed *in vivo,* although less efficiently than P protein (N. Gupta, personal communication). Neither the Y1 nor the Y2 AUG initiation codon is in a favorable context for efficient initiation, as pyrimidines are present in the −3 and +4 positions of both AUGs (*21, 22*).

A non-AUG translational initiation also is utilized by the cellular mRNA that encodes c-myc proteins. The c-*myc* gene comprises three exons with a single large AUG-initiated open reading frame extending from exon 2 through exon 3; exon 1 lacks any AUG codons. Two major forms of C-myc proteins have been identified that, depending upon the species of cell analyzed, differ by 2 to 4 kDa in apparent mass (*123,*

124). These two proteins are derived from alternative translational initiations in the same reading frame in exons 1 and 2. Site-specific mutagenesis results show that initiation of translation of capped SP6 or T7 transcripts *in vitro* occurs at an AUG codon in exon 2 and at a CUG codon near the 3′ end of exon 1. The initiation of translation from the exon-1 CUG and from the downstream exon-2 AUG results in the production of c-myc proteins with distinct amino-termini (*26*). Analysis of c-myc proteins synthesized in Burkitt's lymphoma cell lines containing different chromosomal translocations, and in Epstein–Barr virus (EBV)-immortalized lymphoblastoid cell lines that do not have a rearranged c-*myc* locus, suggests that alternative CUG and AUG translation-initiation sites are used *in vivo* as well as *in vitro* (*26*).

D. Ribosomal Frameshifting during Elongation

Ribosomal frameshifting during the elongation stage of mRNA translation provides a mechanism for the synthesis of a single protein from two different reading frames on an RNA template. The coupling of the reading frames requires that the ribosome shifts correctly from one reading frame to the other at a discrete position on the mRNA so as to avoid termination of polypeptide synthesis (*125*). Frameshifting permits the production of two unique polypeptide products from a single mRNA by initiation of translation at a single site, but subsequent termination of translation at different sites in different reading frames. Termination of protein synthesis at different sites is the result of translational frameshifting that causes a fraction of the ribosomes to change reading frame at a discrete position on the mRNA. Several retroviruses utilize such a frameshifting strategy (*61–68*).

Most, if not all, retroviruses express the *gag*-encoded viral core proteins in the form of a *gag*–polyprotein precursor that is processed by a virus-encoded protease; the *pol*-encoded reverse transcriptase and integrase proteins are normally expressed at much lower levels by a similar proteolytic processing of a large, fused *gag–pol* precursor polypeptide (*103, 126*). However, the genetic structure of the *gag-pol* domains of the Rous sarcoma virus (RSV) genome (*61*), the human T-cell leukemia virus (HTLV-I and -II) genomes (*62*), the bovine leukemia virus genome (*63*), the human immunodeficiency virus (HIV-1) genome (*64*), and the mouse mammary tumor virus (MMTV) genome (*65*) seemingly would preclude the synthesis of a *gag–pol* fusion protein, because the *gag* and *pol* genes are in different reading frames. The *pol* gene of RSV and HIV is in the −1 reading frame with respect to the *gag* gene; however, the products of *pol* do not arise from independent translation-initiation events or from the translation of a spliced mRNA, but rather because of a ribosomal frameshifting event during *gag–pol* expression (*61, 64*). In MMTV, two −1 frameshift events are required for the synthesis of the *gag–pol* fusion protein, one at the *gag–x/pro* overlap and the other at the *x/pro–pol* overlap (*65, 66*). The frameshifting model is supported by synthesis of both the *gag* protein and the *gag–pol* fusion protein in an animal cell-free protein-synthesizing system *in vitro* using a single RSV, HIV, or MMTV mRNA template synthesized *in vitro* from cloned viral cDNA by SP6 polymerase.

Site-directed mutagenesis and amino-acid sequencing located the site of HIV *gag–pol* frameshifting at a leucine UUA codon within a U-UUA sequence near the 5′ end of the *gag–pol* overlap region (*64*). The same sequence also appears in the *gag–pol* overlap of RSV (*61*) and the *pro–pol* overlap of MMTV (*65, 66*). The exact molecular mechanism of translational frameshifting in animal cell systems is not yet known, but in RSV, HIV and MMTV, the frameshifting events may involve a slippage of the leucyl-tRNA reading the UUA codon back to the −1-frame UUU codon (*64, 137*). However, a larger sequence context, possibly including secondary structure within the region flanking the site of the frameshift event, also appears necessary for frameshifting, because synthetic oligonucleotides containing either of the MMTV overlap regions inserted into novel contexts do not induce frameshifting (*66*). Indeed, the effects of deletion and site-directed mutations best correlate with the potential to form an RNA stem–loop structure adjacent to the frame-shift site. A 147-nucleotide sequence from RSV RNA containing the frameshift site and stem–loop structure is sufficient to direct frameshifting in a novel genetic context (*137*). The efficiency of ribosomal frameshifting during translation of the mRNA encoding *gag* and the *gag–pol* fusion is about 5% within the RSV *gag–pol* overlap (*61*) and about 10% within the HIV *gag–pol* overlap (*64*). By contrast, in MMTV, about 25% of the ribosomes traversing the *gag–x/pro* overlap and 10% traversing the *x-pro/pol* overlap frameshift in the −1 direction (*66*).

E. Suppression of Termination

The genetic structure of the *gag–pol* domains of the murine leukemia virus (MuLV) (*67*) and the feline leukemia virus (FeLV) (*68*) genomes precludes the synthesis of the *gag–pol* fusion protein because *gag* and *pol* are separated by an in-frame nonsense codon. To circumvent this apparent block to synthesis of the *gag–pol* leukemia virus fusion proteins, MuLV and FeLV utilize a strategy of termination codon suppression in which ribosomes "read-through" the in-frame stop codon. The MuLV protease is located at the 5′ end of the *pol* gene and is synthesized within the precursor gag–pol fusion protein by suppression of an amber (UAG) termination codon located at the 3′ end of the *gag* gene (*67*). The first four amino-terminal amino acids of the protease are derived from the *gag* gene; the fifth residue, glutamine, is derived through suppression of the in-frame amber termination codon located at the *gag–pol* junction (*67*). Similar to MuLV (*67*), the FeLV protease is likewise synthesized through in-frame suppression of the *gag* amber termination codon by insertion of a glutamine residue at the UAG codon (*68*). Based on precedents from prokaryotic systems, natural suppression of the nonsense termination codon present in the MuLV and FeLV mRNAs presumably occurs by a mechanism involving suppressor tRNAs that possess altered anticodons (*127*). Normal mouse cells contain a natural UAG suppressor glutamine tRNA, representing about 1–2% of the total glutamine tRNA (*133, 134*). The supressor glutamine tRNA may be increased in some cells infected with MuLV (*133*) and selectively decreased in cells treated with Avarol,[1] a sesquiterpenoid hydroquinone displaying anti-MuLV activity (*134*). Indeed, nonsense suppressor tRNAs active in mammalian cells have been constructed using site-specific mutagenesis of cloned tRNA genes. The biological activity of these recombinant suppressor tRNAs was directly demonstrated by suppression of termination of nonsense codons introduced into cDNA clones encoding animal virus mRNAs (*128–130*). However, in principle, either the primary or a higher ordered structure of the MuLV or FeLV mRNA templates could also affect the efficiency of suppression of termination by affecting the competition between the protein-synthesis release factors promoting termination and the tRNA species mediating suppression by promoting read-through. Analysis of recombinant fusion genes consisting of the viral *gag–pol* junction upstream and in the same reading frame as the *E. coli lacZ* gene revealed that *gag* amber codon suppression in these constructs does not require augmented levels of suppressor tRNA species (*98*). Rather, suppression is caused by an intrinsic *cis*-acting component of the viral mRNA (*98*).

[1] Avarol is a highly reduced, highly methylated nephthalenylmethyl-1,4-benzenediol (Chem. Abstr. No. 55303-98-5) [Eds.].

F. Reinitiation following Termination

In polycistronic mRNAs with open reading frames or cistrons that do not overlap, reinitiation without prior dissociation of the 40-S subunit from the mRNA following termination of translation of the upstream cistron would provide a possible mechanism for the subsequent initiation of translation of the downstream cistron. Such reinitiation following termination by animal cell ribosomes has been shown to occur with several viral and cellular mRNAs (*69–77, 79*). Furthermore, termination-reinitiation of translation has been implicated as a possible negative regulatory mechanism in the expression of certain viral mRNAs [for example, the B19 parvovirus VP1 RNA (*70*), a cytomegalovirus (CMV) β mRNA (*79*), SV40 early mRNA (*69*), and the RSV *src* mRNA (*72*)].

The B19 parvovirus produces two capsid proteins in strikingly different quantities—VP1 (<4%) and VP2 (>96%)—from overlapping RNAs that are derived from the same transcriptional unit (*71*). Translation of the VP1 RNA is very inefficient compared to VP2 RNA in the reticulocyte cell-free system. The region about 250 nucleotides upstream from the VP1 initiation AUG codon contains seven AUG codons followed by in-frame termination codons that create multiple minicistrons; this region is removed by splicing and is not present in the efficiently translated VP2 RNA. The multiple upstream AUG codons of the VP1 RNA negatively regulate VP1 synthesis (*70*). Removal of the upstream AUG codons from VP1 RNA greatly increases the efficiency of VP1 RNA translation. Conversely, the addition of the same VP1 RNA upstream sequence containing multiple AUG codons to a position upstream from the initiation codon of VP2 markedly decreases VP2 translation to a level of about 5% of the original level (*70*). Although the upstream AUG-rich region of the VP1 RNA behaves as a negative regulatory element in translational control, no significant difference is observed when the termination site of the last minicistron precedes rather than overlaps the authentic VP1 AUG initiation codon (*70*). Because some of the upstream AUG codons in the B19 VP1 RNA are present in an appropriate −3/+4 context for translation-initiation (*71*), it was proposed that reinitiation rather than leaky scanning is mainly responsible for the reduced efficiency of VP1 translation (*70*).

SV40 (*69*) and CMV (*79*), similar to B19 papovavirus (*70*), also possess upstream "leader" minicistrons that display a negative regulatory activity for mRNA translation. The SV40 early region encodes two proteins, "large T" and "small t," from mRNAs that share common 5′-noncoding regions (*132*). During the course of SV40 infection of permissive cells, a change occurs in the position of the start sites for the synthesis of SV40 early RNA. Prior to DNA replication, the start site of early–early RNA is at map position 5235–5237, whereas following DNA replication two start sites are observed, one for late–early RNA located at m.p. 22–28 and the other for far upstream late–early RNA at m.p. 35–43. During the switch from early–early to late–early transcription, the shift in the start site of the SV40 early RNAs results in the introduction of AUG initiation codons in the 5′-noncoding region of the late–early mRNA. The introduction of the upstream AUG codons contributes to a reduced translational efficiency of the mRNAs encoding the tumor antigens. When SP6 transcripts are prepared and translated in a wheat system, the early–early RNA is a 3- to 6-fold more efficient mRNA for synthesis of t antigen, and presumably also T antigen, than is the late–early mRNA; furthermore, the early–early RNA is about a 10-fold better mRNA than the far upstream late–early mRNA. If AUG initiation codons followed by in-frame termination codons occur in the 5′-noncoding region upstream from the authentic initiation codon, minicistrons are created that, in principle, may be translated. Such a situation appears to occur in the case of the SV40 late–early RNA. A 23-residue polypeptide product of the SV40 late–early minicistron has been identified in infected cells; its function is unknown (*69*). In the case of the human CMV β gene transcript, two short minicistrons of 7 and 35 codons in length occur within the 5′-noncoding leader region of a 170-codon cistron. Transcripts carrying the 5′ leader are inefficiently translated as a consequence of the two upstream minicistrons, and the presence of either of the two minicistrons reduces expression of a downstream cistron during viral infection (*79*).

The three RSV mRNAs encoding *gag* and *gag–pol*, *env*, and *src* contain a common 5′-noncoding leather segment (*72*, *126*). With the spliced RSV *src* mRNA, the AUG codon used to initiate *src* protein synthesis is positioned 90 nucleotides downstream from the AUG codon present in the common 5′-leader region used to initiate *gag* and *gag–pol* protein synthesis. In addition, an in-frame UGA termination codon lies between the *gag* and *gag–pol* AUG and the *src* AUG in the spliced *src* mRNA. Mutation of the upstream minicistron termination UGA codon to CGA virtually eliminates initiation of translation at the downstream "authentic" initiator AUG used for synthesis of the 60-kDa *src* oncoprotein (*72*). Perhaps the absence of 60-kDa *src* synthesis is caused by eliminating the possibility of reinitiation because of the absence of minicistron termination. Conceptually similar results were obtained when the expression of the hepatitis-B surface antigen (HB_sAg) gene was studied in transfected COS cells (*76*). The insertion of AUG initiation codons upstream from the "authentic" AUG used to initiate translation of the downstream HB_sAg gene could severely depress the initiation of translation at the "authentic" AUG codon. Such inhibition could, however, be at least partially suppressed by the presence of a translational termination codon in-frame with the upstream AUG (*76*).

These results are consistent with the possibility that animal cell ribosomes can reinitiate translation at an AUG codon after previously initiating, and subsequently terminating, at an upstream site. The fact that the presence of an upstream termination codon can be shown to modulate directly the efficiency of initiation of translation at the "authentic" downstream internal AUG codons of *src* mRNA (*72*) and HB_2Ag mRNA (*76*) strongly suggests that the internal initiation is by a reinitiation mechanism and is not mediated by ribosomes that escaped initiation at the upstream site.

The translational analysis in stably transfected Chinese hamster ovary (CHO) cells and transiently transfected COS cells of polycistronic mRNAs constructed to contain two or more cellular genes in non-overlapping open reading frames likewise demonstrate that an upstream open reading frame usually (*74, 75, 77*), but not always (*75, 77*), markedly reduces the efficiency of translation of a downstream cistron. Furthermore, the relative position of the terminator codon of the upstream cistron's open reading frame appears to be a critical parameter in determining expression from the downstream, internal cistron. Some evidence suggests that efficient reinitiation of translation can occur at an internal downstream AUG if the translation that initiated from the upstream AUG codon terminates about 80 to 150 nucleotides before the downstream AUG (*77*). However, other evidence suggests that reinitiation can occur at an internal AUG if the translation that initiated from the upstream AUG codon terminates within about 20 to 50 nucleotides either before or after the downstream AUG (*74, 75*). Although the maximum intercistronic distance over which an animal cell ribosome may reinitiate has not been precisely resolved (*75, 77*), these latter results (*74, 75*) imply that ribosomes may also be able to "reach back" or scan bidirectionally.

In summary, the results of many studies designed to assess the effects of upstream initiation and termination codons on downstream translation-initiation are consistent collectively with the notion that internal initiation of translation may occur by a mechanism involving translation termination-reinitiation. Translation termination-reinitiation may occur when the termination codon of the upstream cistron either precedes or is in the close vicinity of the initiation site of the downstream second cistron. The evidence supporting reinitiation following termination is derived from a number of different translational systems, both *in vitro* and *in vivo*, and for both viral and cellular mRNAs (*69–77, 79*).

III. Conclusions

Among the best-characterized examples of naturally occurring polycistronic mRNAs are those specified by animal viruses. Careful genetic and biochemical analyses have firmly established that the protein-synthesizing machinery of animal cells can indeed recognize and translate polycistronic viral mRNAs. The translation of polycistronic mRNAs involves multiple initiation and/or termination codons in a manner that permits the synthesis of multiple primary translational products from a single, mature mRNA. Several different biochemical mechanisms appear to allow the translation of polycistronic mRNAs by the protein-synthesizing machinery of animal cells. Indeed, different mechanisms can best account for the translational expression of different polycistronic mRNAs. For example, for some polycistronic viral mRNAs, translation-initiation appears to occur by the binding of the 40-S ribosomal subunit at or near the mRNA 5′ terminus and a subsequent "leaky" scanning in which alternative initiation codons are recognized as functional initiation sites. For other mRNAs, translation-initiation appears to occur by the binding of ribosomal subunits to internal sequences, or by reinitiation following termination without prior dissociation of the ribosomal subunits from the mRNA. Certain polycistronic mRNAs utilize a non-AUG codon, for example, ACG, to initiate translation of one of the cistrons. Finally, the translation of polycistronic mRNAs, in some cases, is by mechanisms in which alternative termination codons are used to terminate translation, either as the result of ribosomal frameshifting during elongation, or by suppression of an in-frame termination signal.

Many of the molecular genetic mechanisms of eukaryotic gene expression first elucidated from studies of animal viruses have subsequently been shown to be used for the expression of cellular genes. Hopefully, future studies will establish which of the multiple mechanisms so far described for the translation of polycistronic animal virus mRNAs are used likewise for the translation of polycistronic cellular mRNAs. Future studies may also better identify and characterize the parameters important in the regulated translation of polycistronic mRNAs, parameters including *cis*-acting primary sequences or higher ordered structures of the mRNA, and *trans*-acting protein factors of the protein-synthesis machinery.

Acknowledgments

I would like to thank the many authors who generously provided preprints and reprints of their work. The work from the author's laboratory was supported in part by research grants from the National Institute of Allergy and Infectious Diseases, National Institutes of Health.

References

1. J. R. Nevins, *ARB* **52,** 441 (1983).
2. R. A. Padgett, P. J. Grabowski, M. M. Konarska, S. Seiler and P. A. Sharp, *ARB* **55,** 1119 (1986).
3. K. Moldave, *ARB* **54,** 1109 (1985).
4. M. Kozak, *Microbiol. Rev.* **47,** 1 (1983).
5. M. Kozak, *NARes* **12,** 857 (1984).
6. B. N. Fields, D. M. Knipe, R. M. Chanock, J. L. Melnick, B. Roizman and R. E. Shope (eds.), "Virology." Raven, New York, 1985.
7. M. Kozak, *Adv. Virus Res.* **31,** 229 (1986).
8. R. J. Schneider and T. Shenk, *ARB* **56,** 317 (1987).
9. H.-G. Kräusslich and E. Wimmer, *ARB* **57,** 701 (1988).
10. M. Kozak, *Curr. Top. Microbiol. Immunol.* **93,** 81 (1981).
11. H. A. Lutcke, K. C. Chow, F. S. Mickel, K. A. Moss, H. F. Kern and G. A. Scheele, *EMBO J.* **6,** 43 (1987).
12. M. Kozak, *Cell* **47,** 481 (1986).
13. S. A. Sedman and J. E. Mertz, *J. Virol.* **62,** 954 (1988).
14. C. Dabrowski and J. C. Alwine, *J. Virol.* **62,** 3182 (1988).
15. P. J. Good, R. C. Welch, A. Barkan, M. B. Somasekhar and J. E. Mertz, *J. Virol.* **62,** 944 (1988).
16. C. M. Preston and D. J. McGeoch, *J. Virol.* **38,** 593 (1981).
17. L. Haarr, H. S. Marsden, C. M. Preston, J. R. Smiley, W. C. Summers and W. P. Summers, *J. Virol.* **56,** 512 (1985).
18. J. D. Rosenblatt, A. J. Cann, D. J. Slamon, I. S. Smalberg, N. P. Shah, J. Fujii, W. Wachsman and I. S. Y. Chen, *Science* **240,** 916 (1988).
19. W.-K. Chan, M. E. Penaranda, S. E. Crawford and M. K. Estes, *Virology* **151,** 243 (1986).
20. L. McGinnes, C. McQuain and T. Morrison, *Virology* **164,** 256 (1988).
21. J. Curran and D. Kolakofsky, *EMBO J.* **7,** 245 (1988).
22. K. C. Gupta and S. Patwardhan, *JBC* **263,** 8553 (1988).
23. S. Patwardhan and K. C. Gupta, *JBC* **263,** 4907 (1988).
24. S. P. Becerra, J. A. Rose, M. Hardy, B. M. Baroudy and C. W. Anderson, *PNAS* **82,** 7919 (1985).
25. S. P. Becerra, F. Koczot, P. Fabisch and J. A. Rose, *J. Virol.* **62,** 2745 (1988).
26. S. R. Hann, M. W. King, D. L. Bentley, C. W. Anderson and R. N. Eisenman, *Cell* **52,** 185 (1988).
27. J. L. Bos, L. J. Polder, R. Bernards, P. I. Schrier, P. J. van den Elsen, A. J. van der Eb and H. van Ormondt, *Cell* **27,** 121 (1981).

28. W. S. M. Wold, S. L. Deutscher, N. Takemori, B. M. Bhat and S. C. Magie, *Virology* **148,** 168 (1986).
29. R. A. Lamb, B. W. J. Mahy and P. W. Choppin, *Virology* **69,** 116 (1976).
30. C. Giorgi, B. M. Blumberg and D. Kolakofsky, *Cell* **35,** 829 (1983).
31. L. Dethlefsen and D. Kolakofsky, *J. Virol.* **46,** 321 (1983).
32. J. A. Curran, C. Richardson and D. Kolakofsky, *J. Virol.* **57,** 684 (1986).
33. J. A. Curran and D. Kolakofsky, *J. Gen. Virol.* **68,** 2515 (1987).
34. D. Luk, A. Sańchez and A. Banerjee, *Virology* **153,** 318 (1986).
35. M. S. Galinski, M. A. Mink, D. M. Lambert, S. L. Wechsler and M. W. Pons, *Virology* **155,** 46 (1986).
36. W. J. Bellini, G. Englund, S. Rozenblatt, H. Arnheiter and C. D. Richardson, *J. Virol.* **53,** 908 (1985).
37. F. Fuller, A. S. Bhown and D. H. L. Bishop, *J. Gen. Virol.* **64,** 1705 (1983).
38. M. Bouloy, P. Vialat, M. Girard and N. Pardigon, *J. Virol.* **49,** 717 (1984).
39. S. Gerbaud, P. Vialat, N. Pardigon, C. Wychowski, M. Girard and M. Bouloy, *Virus Res.* **8,** 1 (1987).
40. L. Nagata, S. A. Masri, D. C. W. Mah and P. W. K. Lee, *NARes* **12,** 8699 (1984).
41. L. W. Cashdollar, R. A. Chmelo, J. R. Wiener and W. K. Joklik, *PNAS* **82,** 24 (1985).
42. S. M. Munemitsu, J. A. Atwater and C. E. Samuel, *BBRC* **140,** 508 (1986).
43. B. L. Jacobs and C. E. Samuel, *Virology* **143,** 63 (1985).
44. H. Ernest and A. J. Shatkin, *PNAS* **82,** 48 (1985).
45. G. Sarkar, J. Pelletier, R. Bassel-Duby, A. Jayasuriya, B. N. Fields and N. Sonenberg, *J. Virol.* **54,** 720 (1985).
46. B. L. Jacobs, J. A. Atwater, S. M. Munemitsu and C. E. Samuel, *Virology* **147,** 9 (1985).
47. M. W. Shaw and P. W. Choppin, *Virology* **139,** 178 (1984).
48. M. W. Shaw, P. W. Choppin and R. A. Lamb, *PNAS* **80,** 4879 (1983).
49. J. L. Leibowitz, S. Perlman, G. Weinstock, J. R. DeVries, C. Budzilowicz, J. M. Weissemann and S. R. Weiss, *Virology* **164,** 156 (1988).
50. C. J. Budzilowicz and S. R. Weiss, *Virology* **157,** 509 (1987).
51. M. A. Skinner, D. Ebner and S. G. Siddell, *J. Gen. Virol.* **66,** 581 (1985).
52. S. M. Munemitsu and C. E. Samuel, *Virology* **163,** 643 (1988).
53. R. A. Katz, B. R. Cullen, R. Malavarca and A. M. Skalka, *MCBiol* **6,** 372 (1986).
54. C.-C. Liu, C. C. Simonsen and A. D. Levinson, *Nature* **309,** 82 (1984).
55. R. C. Herman, *J. Virol.* **58,** 797 (1986).
56. D. Hassin, R. Korn and M. S. Horwitz, *Virology* **155,** 214 (1986).
57. A. Srivastava, E. W. Lusby and K. I. Berns, *J. Virol.* **45,** 555 (1983).
58. M. R. Green and R. G. Roeder, *J. Virol.* **36,** 79 (1980).
59. C. A. Laughlin, H. Westphal and B. J. Carter, *PNAS* **76,** 5567 (1979).
60. K. C. Gupta and D. W. Kingsbury, *BBRC* **131,** 91 (1985).
61. T. Jacks and H. Varmus, *Science* **230,** 1237 (1985).
62. K. Shimotohno, Y. Takahashi, N. Shimizu, T. Gojobori, D. W. Golde, I. S. Y. Chen, M. Miwa and T. Sugimura, *PNAS* **82,** 3101 (1985).
63. Y. Yoshinaka, I. Katoh, T. D. Copeland, G. W. Smythers and S. Oroszlan, *J. Virol.* **57,** 826 (1986).
64. T. Jacks, M. D. Power, F. R. Masiarz, P. A. Luciw, P. J. Barr and H. E. Varmus, *Nature* **331,** 280 (1988).
65. R. Moore, M. Dixon, R. Smith, G. Peters and C. Dickson, *J. Virol.* **61,** 480 (1987).
66. T. Jacks, K. Townsley, H. E. Varmus and J. Majors, *PNAS* **84,** 4298 (1987).
67. Y. Yoshinaka, I. Katoh, T. D. Copeland and S. Oroszlan, *PNAS* **82,** 1618 (1985).
68. Y. Yoshinaka, I. Katoh, T. D. Copeland and S. Oroszlan, *J. Virol.* **55,** 870 (1985).

69. K. Khalili, J. Brady and G. Khoury, *Cell* **48,** 639 (1987).
70. K. Ozawa, J. Ayub and N. Young, *JBC* **262,** 10922 (1988).
71. K. Ozawa J. Ayub, H. Yu-Shu, G. Kurtzman, T. Shimada and N. Young, *J. Virol.* **61,** 2395 (1987).
72. S. Hughes, K. Mellstrom, E. Kosik, F. Tamanoi and J. Brugge, *MCBiol* **4,** 1738 (1984).
73. R. J. Kaufman, P. Murtha and M. V. Davies, *EMBO J.* **6,** 187 (1987).
74. D. S. Peabody and P. Berg, *MCBiol* **6,** 2695 (1986).
75. D. S. Peabody, S. Subramani and P. Berg, *MCBiol* **6,** 2704 (1986).
76. C.-C. Liu, C. C. Simonsen and A. D. Levinson, *Nature* **309,** 82 (1984).
77. M. Kozak, *MCBiol* **7,** 3438 (1987).
78. G. Jay, S. Nomura, C. W. Anderson and G. Khoury, *Nature* **291,** 346 (1981).
79. A. P. Geballe and E. S. Mocarski, *J. Virol.* **62,** 3334 (1988).
80. R. M. L. Buller and J. A. Rose, *J. Virol.* **25,** 331 (1978).
81. M. Kozak, *NARes* **9,** 5233 (1981).
82. M. Kozak, *Nature* **308,** 241 (1984).
83. M. Kozak, *Cell* **44,** 283 (1986).
84. A. J. Shatkin, *Cell* **9,** 645 (1976).
85. K. H. Levin and C. E. Samuel, *Virology* **77,** 245 (1977).
86. T. G. Lawson, M. H. Chadaras, B. K. Ray, K. A. Lee, R. D. Abramson, W. C. Merrick and R. E. Thach, *JBC* **263,** 7266 (1988).
87. M. Kozak, *Nature* **280,** 82 (1979).
88. M. Konarska, W. Filipowicz, H. Domdey and H. J. Gross, *EJB* **114,** 221 (1981).
89. M. Kozak and A. J. Shatkin, *JBC* **253,** 6568 (1978).
90. M. Kozak, *Cell* **22,** 459 (1980).
91. M. Kozak, *Cell* **19,** 79 (1980).
92. J. Pelletier and N. Sonenberg, *Cell* **40,** 515 (1985).
93. M. Kozak, *PNAS* **83,** 2850 (1986).
94. D. Chevrier, C. Vézina, J. Bastille, C. Linard, N. Sonenberg and G. Boileau, *JBC* **263,** 902 (1988).
95. L. P. van Duijn, S. Holsappel, M. Kasperaitis, H. Bunschoten, D. Konings and H. O. Voorma, *EJB* **172,** 59 (1988).
96. M. Kozak, *NARes* **12,** 3873 (1984).
97. C.-C. Liu, C. C. Simonsen and A. D. Levinson, *Nature* **309,** 82 (1984).
98. A. T. Panganiban, *J. Virol.* **62,** 3574 (1988).
99. R. J. Kaufman, P. Murtha and M. V. Davies, *EMBO J.* **6,** 187 (1987).
100. G. Sarkar, I. Edery, R. Gallo and N. Sonenberg, *BBA* **783,** 122 (1984).
101. J. Pelletier and N. Sonenberg, *Nature* **334,** 320 (1988).
102. S. K. Jang, H.-G. Kräusslich, M. J. H. Nicklin, G. M. Duke, A. C. Palmenberg and E. Wimmer, *J. Virol.* **62,** 2636 (1988).
103. M. R. Roner, R. K. Gaillard and W. K. Joklik, *Virology* **168,** 292 (1989).
104. A. J. Dorner, B. L. Semler, R. J. Jackson, R. Hanecak, E. Duprey and E. Wimmer, *J. Virol.* **50,** 507 (1984).
105. D. S. Shih, I.-W. Park, C. L. Evans, J. M. Jaynes and A. C. Palmenberg, *J. Virol.* **61,** 2033 (1987).
106. N. Kitamura, B. Semler, P. G. Rothberg, G. R. Larsen, C. J. Adler, A. J. Dorner, E. Aemini, R. Hanecak, J. J. Lee, S. Werf, C. W. Anderson and E. Wimmer, *Nature* **291,** 547 (1981).
107. V. R. Racaniello and D. Baltimore, *PNAS* **78,** 4887 (1981).
108. H. Toyoda, M. Kohara, Y. Kataoka, T. Suganuma, T. Omata, N. Imura and A. Nomoto, *JMB* **174,** 561 (1984).

109. A. J. Dorner, L. F. Dorner, G. R. Larsen, E. Wimmer and C. W. Anderson, *J. Virol.* **42,** 1017 (1982).
110. N. Sonenberg, *Adv. Virus Res.* **33,** 175 (1987).
111. D. Etchison, S. Milburn, I. Edery, N. Sonenberg and J. W. B. Hershey, *JBC* **257,** 14806 (1982).
112. J. Pelletier, G. Kaplan, V. R. Racaniello and N. Sonenberg, *MCBiol* **8,** 1103 (1988).
113. D. Trono, J. Pelletier, N. Sonenberg and D. Baltimore, *Science* **241,** 445 (1988).
114. D. Trono, R. Andino and D. Baltimore, *J. Virol.* **62,** 2291 (1988).
115. K. Bienkowska-Szewczyk and E. Ehrenfeld, *J. Virol.* **62,** 3068 (1988).
116. R. Y. Walder and J. A. Walder, *PNAS* **85,** 5011 (1988).
117. V. M. Rivera, J. D. Welsh and J. V. Maizel, Jr., *Virology* **165,** 42 (1988).
118. R. D. Abramson, T. E. Dever and W. C. Merrick, *JBC* **263,** 6016 (1988).
119. R. D. Abramson, T. E. Dever, T. G. Lawson, B. K. Ray, R. E. Thach and W. C. Merrick, *JBC* **262,** 3826 (1987).
120. K. Moldave, *ARB* **54,** 1109 (1985).
121. B. K. Ray, T. G. Lawson, J. C. Kramer, M. H. Cladaras, J. A. Grifo, R. D. Abramson, W. C. Merrick and R. E. Thach, *JBC* **260,** 7651 (1985).
122. D. S. Peabody, *JBC* **262,** 11847 (1987).
123. G. Ramsay, G. I. Evan and J. M. Bishop, *PNAS* **81,** 7742 (1984).
124. S. R. Hann and R. N. Eisenman, *MCBiol* **4,** 2486 (1984).
125. W. J. Craigen and C. T. Caskey, *Cell* **50,** 1 (1987).
126. R. Weiss, N. Teich, H. Varmus and J. Coffin, "Molecular Biology of Tumor Viruses: TNA Tumor Viruses." CSH Laboratory, Cold Spring Harbor, New York, 1985.
127. J. E. Celis and J. D. Smith, "Nonsense Mutations and tRNA Suppressors." Academic Press, London, 1979.
128. F. A. Laski, R. Belagaje, U. L. RajBhandary and P. A. Sharp, *PNAS* **79,** 5813 (1982).
129. J. F. Young, M. R. Capecchi, F. A. Laski, U. L. RajBhandary, P. A. Sharp and P. Palese, *Science* **221,** 873 (1983).
130. J. M. Sedivy, J. P. Capone, U. L. RajBhandary and P. A. Sharp, *Cell* **50,** 379 (1987).
131. M. L. Celma and E. Ehrenfeld, *JMB* **98,** 761 (1975).
132. J. Tooze, "Molecular Biology of Tumor Viruses: DNA Tumor Viruses." CSH Laboratory Cold Spring Harbor, New York, 1980.
133. Y. Kuchino, H. Beier, N. Akita and S. Nishimura, *PNAS* **84,** 2669 (1987).
134. Y. Kuchino, S. Nishimura, H. C. Schröder, M. Rottmann and W. E. G. Müller, *Virology* **165,** 518 (1988).
135. A. Barkan and J. E. Mertz, *MCBiol* **4,** 813 (1984).
136. J. Pelletier, M. E. Flynn, G. Kaplan, V. Racaniello and N. Sonenberg, *J. Virol.* **62,** 4486 (1988).
137. T. Jacks, H. D. Madhani, F. R. Masiarz and H. E. Varmus, *Cell* **55,** 447 (1988).
138. M. A. Williams and R. A. Lamb, *J. Virol.* **63,** 28 (1989).
139. J. Pelletier and N. Sonenberg, *J. Virol.* **63,** 441 (1989).
140. P. J. Dillon and K. C. Gupta, *J. Virol.* **63,** 974 (1989).
141. L-J. Chang, P. Pryciak, D. Ganem and H. E. Varmus, *Nature* **337,** 364 (1989).

Mammalian β-Glucuronidase: Genetics, Molecular Biology, and Cell Biology

Kenneth Paigen*

University of California
Berkeley, California 94720

* Present address: The Jackson Laboratory, Bar Harbor, Maine 04609.

Progress in Nucleic Acid Research
and Molecular Biology, Vol. 37

The β-glucuronidase system has a considerable intellectual history. The animal enzyme was originally described by Masamune in 1934 (*1*); it was the the first gene for a mammalian enzyme to be identified (*2*); and the androgen induction of β-glucuronidase in mouse kidney was one of the earliest known enzyme responses to a hormone (*3*). More recently, the study of β-glucuronidase has led to the discovery of progressive gene activation by a steroid hormone (Section IV,E) and to a novel type of tissue-specific regulation involving the yield of mature enzyme molecules synthesized from each mRNA molecule (Section VI). β-Glucuronidase is also a very ancient protein, and its early evolutionary origin is evidenced by the appreciable sequence identity between the *Escherichia coli* and mammalian enzymes (*4, 5, 5a*).

Growing out of these earlier studies, β-glucuronidase has become a model system for the study of genetic regulatory polymorphism, mechanisms of androgen induction, lysosomal enzyme processing, and mechanisms of intracellular enzyme localization. As a result, it is now one of the best understood mammalian proteins from the combined standpoint of genetics, molecular biology, cell biology, and regulation. Some of the genetic resources created in the course of this work have provided experimental tools useful in other areas of research as diverse as cell lineage studies in mammalian development and the evolution of genetic regulatory mechanisms.

Like other lysosomal enzymes β-glucuronidase is present at highest levels in macrophages, at relatively high levels in liver, spleen, large intestine, and kidney, at lower levels in other internal organs, and at still lower levels in muscle, but is absent from erythrocytes. Its biological function is to participate in the degradation of the polysaccharides that contain glucuronide residues, such as chrondroitin sulfate; deficiency of the enzyme results in a severe mucopolysaccharidosis. The recent discovery (*6*) of a null enzyme mutant exhibiting a typical mucopolysaccharidosis syndrome in an inbred laboratory strain of mice may provide an exceptional model for studying the pathophysiology and gene therapy of the corresponding human disease (*7*).

Several reviews of β-glucuronidase research have appeared.

These describe the substrate specificities of the enzyme, its distribution in nature, and its relation to several human diseases (*8*); androgen induction of the enzyme in kidney (*9*); intracellular localization mechanisms (*10, 11*); genetic regulation (*12, 13*); and clinical aspects of human enzyme deficiency (*14*). This review focuses on more recent findings, presenting briefer outlines of the research described in these earlier papers.

I. Genetic Variants of β-Glucuronidase[1]

The mouse β-glucuronidase gene was first detected as a marked difference in enzyme activity between the C3H and A strains of mice (*15*). This difference is determined by a single Mendelian gene (*2*), which was later mapped to the distal end of mouse chromosome 5 (*16–19*). Other experiments established that the locus determining the difference in activity, [*Gus*] (which was originally called (*G*), is closely linked to the structural gene for the enzyme (*20, 21*). These original mouse strains, C3H and A, carry the $[Gus]^H$ and $[Gus]^A$ haplotypes, respectively.[1]

The human β-glucuronidase gene is located on chromosome 7 (*22–26*) in a chromosomal region homologous to mouse chromosome 5 (*27*). This agrees with estimates that the human and mouse chromosome sets can be derived from each other by a limited number of chromosome rearrangements, whose number is estimated at 150–200 (*28*). The rat gene has not yet been mapped.

A. The [*Gus*] Gene Complex

Following the established conventions of mouse genetic nomenclature (*29*), the chromosomal region determining the collective structural and regulatory features of β-glucuronidase is called the β-glucuronidase gene complex, [*Gus*].[1] Within this DNA domain, the components determining specific phenotypic aspects are given separate designations. Thus, *Gus-s* determines the amino acid sequence of the enzyme; *Gus-r*, the response to androgen induction; *Gus-t*, the temporal switches in enzyme expression occurring during development and differentiation that lead to tissue differences in β-glucuronidase expression; and *Gus-u*, aspects of the systemic regulation of the enzyme common to all cells in the organism. Each component is defined genetically by the occurrence of mutants with

[1] Gene complexes are denoted by brackets according to genetics nomenclature rules. Symbols of genes or gene complexes that show dominant or codominant inheritance are capitalized. Genes that are only known by the existence of recessive mutations are not capitalized. All gene names and symbols are italicized.

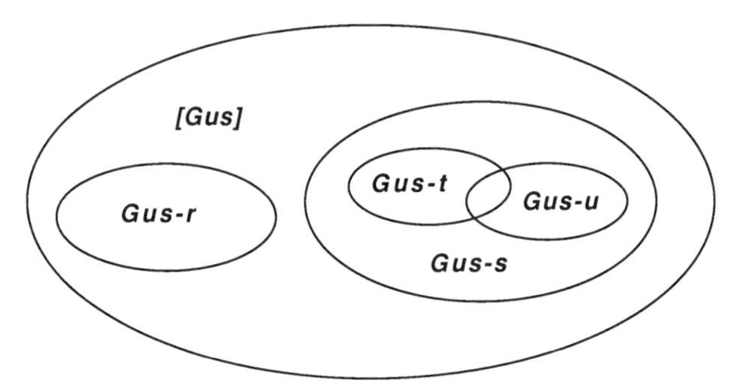

FIG. 1. A Venn diagram indicating the probable relationships among the sets of DNA sequences determining the components of the [*Gus*] gene complex.

altered phenotype. The mutations not only identify phenotypes determined by sequences within the gene complex, but they also show that the various regulatory and structured phenotypes can vary independently of each other. Alternate versions of the complex are referred to as haplotypes with given alleles at each separate component. Thus, the [*Gus*]^{A} haplotype has the constitution *Gus-s*^{a}, $-r^{a}$, $-t^{a}$, $-u^{b}$, indicating that it carries the *a* allele at the first three separate components and the *b* allele at the last.[2]

Although this nomenclature has considerable operational utility and is still the most convenient menas of designating the multiple phenotypic properties of the [*Gus*] complex, it does require some reconciliation with our current understanding of the molecular details of both gene structure and the mechanisms of regulatory processes. The amino-acid sequence of the protein is determined by sets of DNA sequences that are not contiguous along the DNA helix, and this interrupted arrangement is almost certainly true for regulatory sequences as well. Moreover, the set of sequences determining one phenotype may be interspersed among, partially overlap, or even be nested within other sets of sequences. Indeed, the most recent evidence suggests that the *Gus-u* and *Gus-t* sequence sets may overlap, and that both may be subsets of the *Gus-s* sequences; i.e., *Gus-u* and *Gus-t* may be nested within *Gus-s*. In contrast, *Gus-r* sequences almost certainly lie outside *Gus-s* and do not overlap, although they may be interspersed with *Gus-s* sequences along the DNA helix. The situation is described by a Venn diagram (Fig. 1), and

[2] The two original mouse strains, C3H and A, whose differences initiated the study of β-glucuronidase genetics, carry the [*Gus*]^{A} and [*Gus*]^{H} haplotypes, respectively. Other haplotypes that have been studied include [*Gus*]^{B}, which is extensively used as a reference for comparison, [*Gus*]^{CL}, [*Gus*]^{CS}, and [*Gus*]^{N}.

one of the experimental challenges is to determine the identity, location, and function of the sequences within each set, and how the sets of sequences relate to each other. Although the various sets of sequences comprising the gene complex are not distinct "loci" in the classic genetic sense of being separate genes arranged in a linear order on the chromosome, the notation system treating them as distinct components of [*Gus*] nevertheless remains very useful as long as it is recognized that the [*Gus*] complex includes the entire DNA domain, and the components are sets of DNA sequences within this domain.

1. Gus-s

Eight structural alleles of β-glucuronidase have been reported. Their products are distinguished by differences in thermolability was well as by charge differences detected as changes in electrophoretic mobility or isoelectric focusing (Table I). Four of these alleles (*a*,*b*,*h*,

TABLE I
β-GLUCORONIDASE ALLOZYMES[a]

Relative thermostability[c]	Charge[b]		
	−1	0	+1
2.0	*CS*[f]		
1.0	*A*[g]	*B*	*N*[d,h], *S*[i]
0.4	*CL*[f]		
0.1		*H*[j], *Ho*[e,i]	

[a] For a general description of allozyme properties, see refs. *12* and *35*.

[b] Charge on enzyme subunits relative to the *B* allozyme.

[c] Stability relative to the *B* allozyme at 71°C. There are also differences among allozymes in the activation energy of thermal denaturation, leading to differences in relative stability at different temperatures (*41a*).

[d] The *N* allozyme is about 3 kDa smaller than the other allozymes.

[e] The *H* and *Ho* allozymes have the same charge and thermal stability; they are assumed to be separate allozymes only on the basis of their occurrence in distinct species of *Mus*.

[f] Ref. *41a*.

[g] Refs. *18* and *19*.

[h] Refs. *18* and *35*; also L. T. Bracey, personal communication.

[i] Ref. *35*.

[j] Ref. *20*.

and *n*) are derived from inbred strains of laboratory mice. Other alleles are derived from interfertile species of *Mus*, including *cl* and *cs* from *M. castaneus*, *s* from *M. spretus*, and *ho* from *M. hortulanus*.[3]

The marked thermolability of GUS-H protein,[3] coded by *Gus-s*^{h}, compared with enzymes coded by other alleles from inbred mouse strains, has been exploited as a cell-specific marker; by heating tissue sections before applying histochemical stains for enzyme activity, individual cells carrying *Gus-s*^{h} and *Gus-s*^{b} can be distinguished from each other in chimeric mice (*30*). This technique has been applied to neurological mutants, for example, to analyze numbers of stem cells and cells lineages of Purkinje cells (*31–32b*).

2. Gus-r

The *a* and *b* alleles of the *Gus-r* regulatory locus controlling the magnitude of androgen induction of β-glucuronidase in kidney were initally described by Swank *et al.* (*18*); the *cl*, *cs*, and *n* alleles, differing in magnitude and kinetics of induction, were described subsequently (*33, 34*), as was the *ho* allele that fails to induce at all (*35*). The detailed properties of the various alleles are described in Section IV,C.

3. Gus-t

The locus controlling the temporal expression of β-glucuronidase was initially defined by developmental and organ-specific differences in β-glucuronidase expression between mouse strains C3H and DBA/LiHa (*21*). Within liver, *Gus-t* regulates β-glucuronidase expression in hepatocytes but not non-hepatocytic cells (*35a*). The mechanism of *Gus-t* action involves developmental changes in the translational yield of β-glucuronidase mRNA (see Section VI,B). Several reports suggested that *Gus-t* might act *trans* rather than *cis* (*36, 37*), and this was confirmed when Lusis *et al.* (*38*) showed that the [*Gus*]^H haplotype present in C3H mice exhibits two regulatory differences in β-glucuronidase expression, both genetically linked to the [*Gus*] complex. One, *Gus-t*, is a *trans*-acting factor that determines the presence of tissue-specific developmental switches in β-glucuronidase expression. The other is a *cis*-acting factor that

[3] Again following the conventions of genetics, a separate nomenclature is used to designate β-glucuronidase protein and its distinct allozymes (i.e., isozymes resulting from mutations in the structural gene). GUS is the symbol for the generic protein, β-glucuronidase, and GUS-A, GUS-B, etc., are the allozymes coded by the *Gus-s*^{a}, *Gus-s*^{b}, etc., alleles of *Gus-s*.

behaves like an allele of the systemic regulator *Gus-u*, affecting all cells more or less uniformly (Section I,A,4). *Gus-t* is the only [*Gus*] component that has been shown to act *trans*. It is not known whether the *trans* action of *Gus-t* results from interaction between β-glucuronidase polypeptides themselves during synthesis of the enzyme, or whether it results from the action of an independent gene product coded by DNA located within or near [*Gus*].

4. Gus-u

Several haplotypes differ from the standard *B* haplotype $[Gus]^B$ in showing a reduced enzyme activity in all tissues. This effect was initially described for the *CL* and *CS* haplotypes (*39*); it is also present in the *N* haplotype (*34*) and in the *H* haplotype (*38*). Like *Gus-t*, *Gus-u* determines changes in translational yield (see Section VI,B); *Gus-t* and *Gus-u* are distinguished by their *trans* versus *cis* expression. The unusual mechanism of regulation shared by *Gus-t* and *Gus-u*, involving translational yield, suggests that some relationship exists between these two regulatory loci. However, determining the exact nature of this relationship will depend upon clarifying the details of translational yield control and the exact DNA sequences determining *Gus-u* and *Gus-t*.

B. Haplotypes and Congenic Strains

Many inbred strains of mice have been examined for the *Gus-s*, *-r*, *-t*, and *-u* alleles they carry; these results have been summarized by Chapman *et al.* (*40*). The original haplotypes were discovered in surveys of inbred laboratory strains of mice; later, haplotypes were identified among wild-trapped mice and their descendants, and within sibling species of mice that are interfertile with laboratory mice. To date, nine distinct haplotypes of [*Gus*] have been identified, based on the varying combinations of alleles they carry for *Gus-s*, *Gus-r*, *Gus-t*, and *Gus-u*. The genetic constitutions of those haplotypes presently available in laboratory strains of mice are listed in Table II. Nearly all inbred strains of laboratory mice carry one of three common haplotypes: *A*, *B*, or *H*. These differ in multiple regulatory aspects as well as in the structural alleles they carry, and the differences in structural alleles provide a simple means of typing strains. The remaining haplotypes are rare. This limited repertoire of haplotypes among inbred strains probably reflects the fact that the common inbred strains are derived from a somewhat limited stock of progenitor mice (*41*). Several of the haplotypes identified among wild-trapped *M. m. domesticus* and *M. m. musculus* mice (e.g.,

TABLE II
CHARACTERISTICS OF β-GLUCURONIDASE HAPLOTYPES[a]

Haplotype	Origin	-s	-r	-t	-u
B	C57BL/6 (v.i.s.)[b]	*b*	*b*	*b*	*b*
H	C3H/HeJ (v.i.s.)[b]	*h*	*b*[c]	*h*	*h*
A	A/J (v.i.s.)[b]	*a*	*a*	*b*	*b*
W16, W17, W18	*M. m. domesticus*	*a*	*a*	—	—
W12	*M. m. domesticus*	*a*	*b*[d]	—	—
W26	*M. m. domesticus*	*a*	*b*[d]	—	—
N	*M. m. domesticus*	*n*	*n*	*n*	*n*
CS	*M. castaneus*	*cs*	*cs*	*b*	*cs*
CL	*M. castaneus*	*cl*	*cl*	*b*	*cl*
S	*M. spretus*	*s*	*s*	—	—
Ho	*M. hortulanus*	*ho*	*ho*	—	—

[a] Detailed descriptions of the phenotypes determined by alleles of the loci making up the [*Gus*] complex are provided in their respective sections. For additional genetic background information, see refs. *2*, *35*, and *41a*.

[b] v.i.s., Various inbred strains. For a detailed listing, see ref. *40*.

[c] The *H* haplotype was thought for a long time to carry a different allele of *Gus-r*. More recent evidence using mRNA measurements in congenic strains shows it to be identical to $Gus\text{-}r^b$ (see Section IV,C for details).

[d] These haplotypes show the $Gus\text{-}r^b$ phenotype when assayed for β-glucuronidase enzyme activity, but have not yet been analyzed for mRNA levels, induction kinetics, or transcription rates.

haplotypes W-12 and W-26) appear to be recombinants between $[Gus]^A$ and $[Gus]^B$, but these have yet to be confirmed in molecular detail.

Through the efforts of Verne Chapman at Roswell Park Memorial Institute, nearly all of these haplotypes have been transferred into a constant genetic background. This has generated a series of so-called congenic strains of mice that are genetically identical except that each carries a different version of the chromosome-5 segment containing [*Gus*]. The congenic strains were constructed by initially crossing each donor stock with the C57BL/6J (B6) strain of mice, and then repeatedly backcrossing the progeny to the B6 strain for 7–14 generations. At each generation, the progeny carrying the new haplotype were chosen for continued backcrossing, but other donor strain genes were allowed to segregate away. Depending on the number of generations of backcrossing, the average genetic length of the chromosome segment transferred in this way is 10–20 centiMorgans. This corresponds to about 20–40 $\times$ 10^6 bases of DNA and comprises 0.6–1.2% of the mouse genome. Even though the transferred piece is small genetically, it is still large enough to contain 200–500 distinct genes.

These congenic strains have proven extremely useful in studying the regulatory properties of various [*Gus*] haplotypes. Their use allows haplotypes to be compared in the absence of possible confounding effects by other genetic differences between strains or species (*41a*). Congenic strains can also be used as a means of identifying new *trans* regulatory genes affecting β-glucuronidase. When the *H* haplotype was moved to strain B6 from its C3H strain of origin, it acquired a new regulatory phenotype for the androgen induction of β-glucuronidase, pointing to the modifying action of an unlinked regulatory gene with different alleles in these two strains (see Section IV,G). Congenic strains have also been used to construct new combinations of genes affecting β-glucuronidase expression; for example, mutant alleles of the *Eg* (Section V,B) and *pearl* (Section VII,B) genes have been combined with various [*Gus*] haplotypes, all on the B6 genetic background (*41b*).

II. Molecular Biology of the β-Glucuronidase Gene

β-Glucuronidase was first cloned as a cDNA from the two most abundant sources of its mRNA—androgen-stimulated mouse kidney and rat preputial gland (*42–44*)—and the complete mRNA sequences have been reported from the human (*5*), rat (*4*), and mouse (*45, 46*). A mouse cDNA probe was used by M. Meisler to obtain the first genomic clone containing [*Gus*] and its flanking DNA. This mouse clone provided the material for a restriction map of the [*Gus*] region (*47*) and the identification of several restriction-fragment-length polymorphisms that distinguish various haplotypes (*47, 48*). Comparisons of the restriction map of the cloned gene with genomic DNA indicates that only a single β-glucuronidase gene is present, and no duplicated genes or pseudogenes were detected (*47*). The detailed organization of the mouse gene has been determined (*45, 49*) and the sequencing of the entire gene has been completed (*49*).

A. cDNA Clones and mRNA

Several criteria have been used to establish the identity of the mouse cDNA clones. These included: homology with the sequences of several rat β-glucuronidase peptides (*50*); hybridization with an mRNA that directs the synthesis of β-glucuronidase in an *in vitro* translation system; the ability of the clones to detect an mRNA with the regularity properties predicted for β-glucuronidase; and the fact that the clones also detect genomic DNA polymorphisms mapping to the chromosomal region of [*Gus*]. The rat clones were identified by their hybridization with an mRNA coding for β-glucuronidase protein. Isolation of another rat cDNA clone has also been reported (*51*), but the identity of this clone has not been fully established. A mouse clone provided the initial probe used for obtaining a human cDNA sequence (*52*).

The full-length cDNA coding sequences have now been reported for β-glucuronidase from rat preputial gland (*4*), rat liver (*53*), human placenta (*5*), and mouse kidney (*45, 48*). The two rat sequences are not identical, suggesting that genetic polymorphism for β-glucuronidase structure may be present in this species as well as in mice. The identity of the human clones was confirmed by the fact that they code for the correct N-terminal amino-acid sequence of human β-glucuronidase, and by their ability to code for catalytically active human enzyme when transfected into COS cells using a simian virus 40 (SV40) expression vector. For reasons that remain obscure, it has been difficult to obtain full-length mouse cDNA clones (none have yet been reported), and the full-length coding sequence in mice was derived from a combination of cDNA and genomic DNA clones.

The 5′ end of the message has been identified in mice by primer extension of induced kidney mRNA (*45, 48*); it has a rather short 5′ leader sequence of only 12 nucleotides. The 5′ end of the message has not been identified in humans or rats, but the cDNA clones obtained for human β-glucuronidase show that at least 26 bases are present on the 5′ side of the initiation codon. In all three species, the 3′ untranslated region is approximately 500 nucleotides long. In both the rat and the mouse, polyadenylation begins with a genomic poly(A) tract. In the rat, two tandemly arrayed AATAAA polyadenylation signals are present, separated by a few nucleotides; 18 nucleotides downstream from the second of these is a sequence of 18 A's, and polyadenylation is thought to begin somewhere within this A_{18} sequence. In the mouse, the 3′ untranslated region contains a single AATAAA signal, followed 16 nucleotides downstream by an A_{14} sequence (*45, 49*), and polyadenylation begins somewhere within the A_{14} sequence (*46*). The genomic sequence immediately 3′ to the poly(A) tract does not hybridize to mouse β-glucuronidase mRNA.

B. Organization of the β-Glucuronidase Gene

The mouse β-glucuronidase gene (Fig. 2) contains 12 exons and 11 introns (*45, 49*); it is 14 kb long and contains a number of B1 and B2

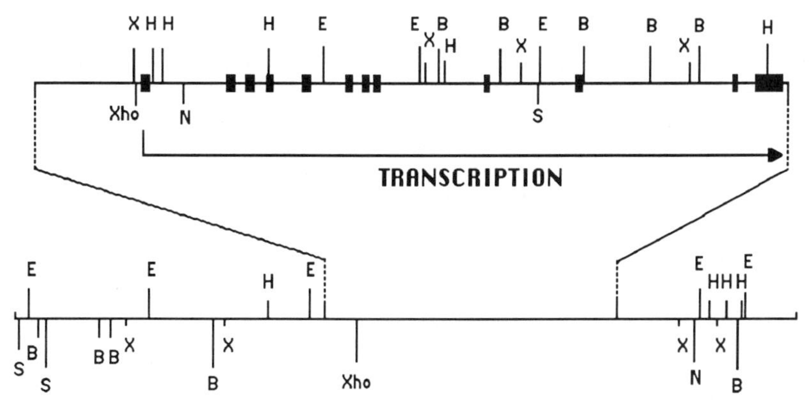

FIG. 2. A restriction map of the chromosomal region containing [*Gus*]. The transcription unit is 14.0 kb long and contains 12 exons.

repetitive elements (*49*). The TATA box is represented by a TAAAAA sequence 28 nucleotides upstream from the cap site, but there are no other recognizable promoter elements in the immediate 5′ region (*49*). Some haplotypes do not even contain a recognizable TATA box, but have the sequence AAAAAA instead; however, this substitution is not correlated with any obvious change in the regulation of β-glucuronidase (D. Schott, R. Jaussi, and K. Paigen, unpublished). The promoter regions of the rat and human genes have not yet been identified.

Nearly all of the mouse [*Gus*] haplotypes are distinguished by restriction-fragment-length differences within the gene itself or in the immediately adjacent flanking regions (*47*) (Fig. 3). Several of these polymorphisms involve the gain or loss of restriction sites, and two of them involve small insertions of several hundred base-pairs. The insertion present in the *A* haplotype has been identified as a B2 repetitive element (*48*); the same insertion is invariably present in the genomic DNA of mouse strains carrying the *A* haplotype and absent from strains carrying the *B* and *H* haplotypes.

C. Multiple Human mRNAs

There is definitive evidence that two distinct β-glucuronidase mRNAs are present in human tissue (*5*). There is a larger, complete mRNA, which comprises most of the mRNA and codes for an enzymatically active protein product. There is also a smaller mRNA, missing an internal 153-nucleotide segment, that does not code for a catalytically active β-glucuronidase. The two types of clones were

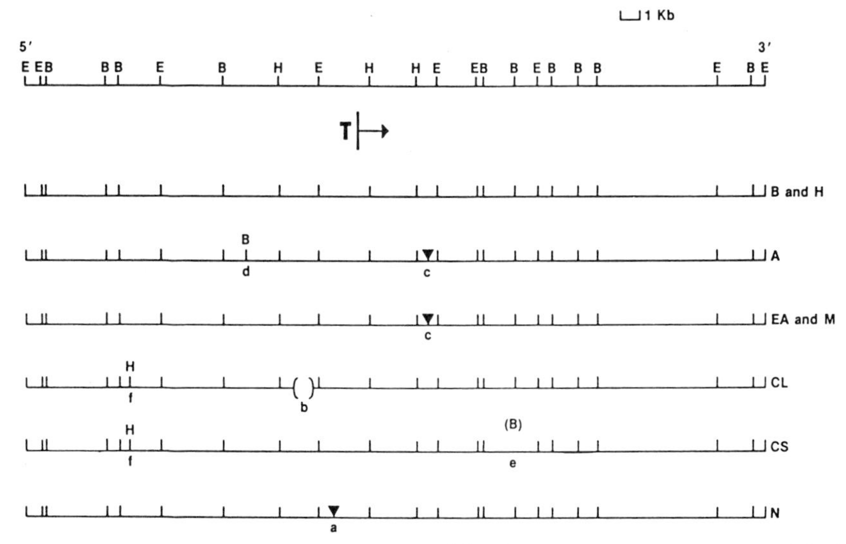

FIG. 3. A comparative restriction map of the chromosome region containing [*Gus*] for eight haplotypes, indicating the known polymorphisms within this region. The transcription unit is indicated above, along with all of the *Bam*HI (B) and *Eco*RI (E) sites. However, only those *Hin*dIII (H) sites used in defining the polymorphisms are shown. For additional details of the restriction map, see Fig. 2. The *B* haplotype is used here as in the standard type. Transcription initiates at the site marked "T." The lowercase letters refer to individual polymorphisms in which the various haplotypes differ from *B* as follows: (a) a 200 to 250-bp insertion in *N*; (b) a 400-bp deletion in *CL*; (c) a 200-bp B2 element insertion in *A*, *EA*, and *M*; (d) an additional *Bam*HI site in *A*; (e) a missing *Bam*HI site in *CS*; (f) an additional *Hin*dIII site in *CS* and *CL*. No DNA differences between the *B* and *H* haplotypes have yet been identified (Adapted from ref. *47*.)

discovered when sequencing and restriction mapping of cDNAs obtained from placental mRNA showed that one type lacked the internal fragment. Transfection experiments indicate that only the larger clone can code for enzymatically active β-glucuronidase. S1 nuclease mapping experiments with placental mRNA established that both types of RNA are present *in vivo*, demonstrating that the presence of two classes of clones was not a cloning artifact. Later studies of genomic clones demonstrated that the 153-nucleotide segment comprises an entire exon coding for 51 amino acids (*54*). Thus, the shorter mRNA probably arises by a splicing error skipping over this exon, although the possible presence of a duplicated gene carrying a deletion of the missing exon has not been eliminated. The human 153-bp exon begins with an extra 9-bp sequence (three amino acids) that is missing in the mouse and the rat (*45*). It was suggested that the presence of this additional sequence at the 5′ splice site of exon 6 may be conducive to alternative splicing and the production of two human mRNAs. As yet, there is no indication that two mRNAs occur in the mouse or the rat, despite extensive cDNA cloning and mRNA analyses, but the possibility of a second, aberrant mRNA is a significant consideration in experiments demonstrating differences in translational yield of mouse β-glucuronidase (see Section VI).

D. Coding Sequences and Evolutionary Relationships

No significant sequence similarities have been found between the nucleotide or deduced amino-acid sequences of β-glucuronidase and those of other lysosomal enzymes, including β-hexosaminidase, glucocerebrosidase, α-fucosidase, cathepsins B, D, and H, and α-galactosidase A. There is also no similarity to the microsomal UDP glucuronosyltransferase (EC 2.4.1.17) (*4, 5*). β-Glucuronidase does show a statistically significant amino-acid sequence similarity with *E. coli* β-galactosidase, and the rodent β-glucuronidase sequence between amino acids 320 and 457 is 39% similar to positions 330–472 of bacterial β-galactosidase. An even greater similarity is observed with *E. coli* β-glucuronidase. In this case, the overall nucleotide sequences of the two coding regions are 53% similar and the overall peptide sequences are 39% similar. There is an internal segment with even greater similarity; the regions between mouse amino-acid residues 197 and 430 and bacterial residues 154–380 are 50% similar. The greater similarity between rodent and bacterial sequences at the nucleotide level than at the peptide level is curious.

As would be expected, mammalian β-glucuronidases show even greater similarity with each other. The magnitude of these similarities are in accord with the established phylogenetic relationships of the three species, in which the primate–rodent divergence occurred approximately 75 million years ago and the rat–mouse divergence occurred 15–20 million years ago. Overall, there is 77% amino-acid homology between the deduced rat and human β-glucuronidase protein sequences (*55*), 75% between mouse and human (*45*), and 88% between mouse and rat (*45*). The rodent and human cDNA sequences differ by two small insertion/deletions (*55*). The 3′ regions of mouse, rat, and human are too divergent to allow accurate alignment, and the 5′ regions are too short to provide much comparison. As is customary, sequence divergence was greatest in the third base position of codons and least in the second (*55a*). Only seven codons constitute 30% of the entire human sequence: CUG (Leu), GUG (Val), CAG (Gln), GAG

(Glu), GAC (Asp), UAC (Tyr), and UUC (Phe); these codons are more conserved than the rest of the sequence, suggesting a significant selective advantage in evolution for the use of these codons in preference to their cognates (*55*).

III. Structure and Properties of the Protein

Mature, enzymatically active β-glucuronidase is a tetrameric glycoprotein with four identical subunits. The enzyme has been purified from a number of sources including cow (*56*), rat (*57–62*), rabbit (*63, 64*), and mouse (*65–68*). The richest sources of the mammalian enzyme are the rat preputial gland, where it accounts for several percent of the total protein (*57*), and the urine of androgen-stimulated mice (*67, 68*).

A. Amino-acid Sequence

The complete amino-acid sequence of the primary translation product has been deduced from cDNA clones for the human, rat, and mouse enzymes (*45, 54*). Four potential N-linked glycosylation sites exist in each sequence: the first three are located at identical positions in all species, but the most C-terminal is at a different position in humans and rodents. The human protein is three amino acids longer than the rodent enzyme, corresponding to a 9-bp insertion at the beginning of exon 6.

B. Protein Properties, Catalytic Activity, and Assay

β-Glucuronidase has an isoelectric point between 5.5 and 6.5, depending on the source of the enzyme and the particular isoform tested. It is a glycoprotein and binds to appropriate lectin columns; this property has been exploited for the rapid purification of the enzyme (*39*). The most unusual feature of β-glucuronidase is its stability to both thermal denaturation and proteolytic attack. Although there are differences between β-glucuronidase of different species, and even the enzymes coded by alleles within species, β-glucuronidase typically is stable up to temperatures of 70°C and can be assayed at temperatures approaching this. It is also extremely resistant to attack by most proteases.

The natural substrates of the enzyme are thought to be β-glucuronide linkages in polysaccharides such as chrondroitin sulfate. Since the enzyme has an absolute requirement for an unmodified glucuronic acid residue, it probably functions exclusively as an exoglycosidase. Tests with many potential substrates show that it will cleave almost any aglycone in a β linkage to glucuronic acid (*8*).

A number of convenient synthetic substrates have been developed for the assay of β-glucuronidase, taking advantage of this lack of structural specificity for the aglycone. The most useful of these is the 4-methyl umbelliferyl (4MU) derivative, which liberates a fluorescent product and provides a sensitive assay (*69, 70*). With this substrate, Labarca and Paigen (*71*) detected the enzyme formed when mouse RNA preparations were injected into *Xenopus* oocytes, and they developed this into a quantitative assay for β-glucuronidase mRNA activity. Using the same substrate, Wudl and Paigen (*72*) assayed the enzyme in single cells, using a microscope equipped with a photomultiplier. Histochemical staining for the enzyme in tissue sections is carried out using the naphthol AS-BI derivative (*73*); the same stain works well for electrophoresis gels.

Several interesting inhibitors of the enzyme have been developed. 1,5-Saccharolactone, a glucuronic acid analogue in which the pyranose ring oxygen is replaced by an ester linkage, is a commonly used and powerful inhibitor. The most potent inhibitor is a glucuronic acid analogue of the antibiotic kojirimycin, in which the ring oxygen is replaced by a nitrogen atom (*74*). Activity is also inhibited by such polycarboxylic acids as citrate and tartrate, presumably because they bind to a catalytic site adapted to recognize the carboxyl residue in the substrate.

IV. Androgen Induction

The epithelial cells of the kidney proximal tubule are responsive to androgen. When androgens are administered to castrated male mice or to intact female mice, these cells hypertrophy and the synthesis of a number of mRNAs and enzymes is induced (*75, 76*). The induction of β-glucuronidase is one of the more dramatic responses; among the various [*Gus*] haplotypes, there is a 10–50-fold increase in enzyme activity and a 20–80-fold increase in both the rate of enzyme synthesis (*18, 33*) and mRNA levels (*43, 44, 77–79*). A pulse-labeling method to assay rates of synthesis of specific mRNAs showed that induction results from a marked increase in the rate of synthesis of β-glucuronidase mRNA, presumably from transcriptional activation of the gene, along with some stabilization of the mRNA (*80*).

Much of the induced enzyme is secreted from the proximal tubule cells into the glomerular filtrate, from which it finds its way into urine (*3*). This androgen-stimulated secretion is the reason that tissue levels of the enzyme do not increase in proportion to the increased enzyme synthesis. It has been suggested that the urinary enzyme may function in the pheromone signaling system that mice use to control reproductive behavior, and that this may be the biological function of the induction response (*81*).

A variety of mutations have been identified affecting the induction process; these have made the β-glucuronidase system a model for studying androgen function and the mechanisms of androgen induction. One outcome of these studies was the discovery that induction of the [*Gus*] gene by androgen is a slow, progressive process over a period of many days, a phenomenon not previously observed in other steroid response systems. Two explanations have been advanced for this effect. One is that β-glucuronidase is not actually induced by androgen receptor, but, instead, is induced secondarily by a novel regulatory protein that is itself androgen-induced. The other is that [*Gus*] can bind more than one molecule of androgen receptor protein; a minimum number of receptors must bind before transcription can increase, and, beyond this point, transcription increase in proportion to the number of additional receptors bound. The present weight of evidence favors the latter explanation (Section IV,D).

β-Glucuronidase is also induced in the submaxillary gland by androgen, but this occurs by a very different mechanism. There is an increase in enzyme synthesis, but no increase in mRNA levels, suggesting an androgen-stimulated change in translational yield (Section VI).

A. Receptor and Hormonal Requirements

Induction of kidney β-glucuronidase by androgen depends on the availability of androgen receptor protein, and there is no hormonal response in the *Tfm* mouse mutant (*82, 83*), which is deficient in receptor protein (*84–86*). The hormonal specificity for androgens and their structural analogues has been examined in some detail (*87*). That evidence, plus the relative lack in kidney of the enzyme testosterone 5 α-reductase, which is responsible for converting testosterone to dihydrotestosterone, suggests that testosterone is the physiological effector in this tissue (*75*). One curious observation is that the anti-androgen cyproterone acetate potentiates induction when administrated along with small amounts of testosterone, but it antagonizes induction when administered along with larger amounts (*87*). β-Glucuronidase is not induced by progesterone or glucocorticoid, but the induction is partially antagonized by estrogen (*3*). Although progesterone itself is not an inducer, some progestin analogs can function as inducers, notably medroxyprogesterone acetate. This compound does not act through the progestin receptor, but instead acts as a weak effector for androgen receptor protein and stabilizes less than maximal concentrations of active receptor at saturation binding (*87b*). This partial response was used to demonstrate that the altered induction phenotype of some [*Gus*] haplotypes is equivalent to a reduced responsiveness to androgen receptor protein (*88*).

Maximal induction of β-glucuronidase also requires a normal, pulsatile supply of growth hormone (*89*). The induction response is only about one-fourth of normal in hypophysectomized mice and in the *dw* (*dwarf*) mouse mutant, which is deficient in virtually all anterior pituitary functions (*9*). Several observations indicate that the critical pituitary factor is growth hormone: (i) induction is also reduced in the *lit* (*little*) mutant (*9*), which appears to be specifically deficient in growth hormone and prolactin; (ii) much of the induction response can be restored by the injection of growth hormone but not prolactin (*9*); and (iii) continuous infusion of large amounts of growth hormone produces a hypophysectomy-like refractoriness to induction, presumably by down-regulating surface receptors for growth hormone (*41b*). The last phenomenon had been observed previously in the growth-hormone-dependent androgen induction of major urinary protein (MUP) in liver (*90*). It is not known whether growth hormone acts directly on kidney epithelial cells, or if the effect is mediated through the production of insulin-like growth factor (IGF) 1 or 2.

The stimulation of β-glucuronidase induction by growth hormone is gene-specific; the induction of eight other androgen-responsive proteins and mRNAs tested in mouse kidney is not stimulated by growth hormone (*90a*). The growth hormone effect is also limited to the induction process itself, and hypophysectomy does not affect basal levels of the enzyme in the absence of androgen stimulation (*89*).

In addition to being induced in kidney epithelial cells by androgen, β-glucuronidase is also induced in kidney medullary cells by potassium depletion, although the magnitude of the induction is much less. This induction is not affected by *Gus-r* regulatory mutations (*91*) (Section VI,C).

B. Molecular Biology of Induction

β-Glucuronidase induction was first described as an increase in enzyme activity (*3*), then as an increase in the rate of β-glucuronidase synthesis (*18*), later as an increase in mRNA levels (*42–44, 77*), and finally as an increase in the rate of mRNA synthesis (*80*). Thus, the major mechanism of induction appears to be transcriptional, although androgen administration also induces some stabilization of β-glucuronidase mRNA as well as other androgen-inducible mRNAs in kidney (*80*). Stabilization of specific mRNAs by steroids has been observed in other systems, notably the stabilization of vitellogenin mRNA by estrogen (*91a*).

The kinetics of β-glucuronidase mRNA induction are unusual (*79*). Following androgen administration, there is a lag period of 18 hours or more without any discernible increase in mRNA levels, even though androgen receptor protein reaches maximal levels in the nucleus within 30 minutes after hormone administration (*92*), and even though some other inducible mRNAs begin to increase almost immediately (*90a*). The lag period is followed by a slow rise of β-glucuronidase mRNA to a new plateau level over a period of 1–2 weeks. This slow, progressive induction of β-glucuronidase mRNA results from a slow activation of transcription, as shown by measurements of rates of β-glucuronidase mRNA synthesis at various times after androgen administration (*90a*). There is a lag of at least 16 hours before the rate of mRNA synthesis begins to increase; after the lag period, the rate of mRNA synthesis slowly rises to a maximum 7–14 days later. [*Gus*] haplotypes vary in both the duration of the lag period and the time course of the subsequent response (See Section IV,C). Other androgen-responsive genes in mouse kidney vary considerably in the time at which they reach maximum mRNA synthesis, and [*Gus*] is not unique in being very slow (*90a*).

The time course of mRNA accumulation after androgen administration approximates the equation

$$dR/dt = k_1 - k_2\,(R) \tag{1}$$

where t is time, R is the concentration of β-glucuronidase mRNA at time t, R_0 is the initial RNA concentration, and k_1 and k_2 are the zero and first-order rate constants for accumulating and losing the androgen-induced capacity to synthesize β-glucuronidase mRNA (*79, 93*). This will be the situation when the tissue level of activated [*Gus*] chromatin (i.e., its rate of transcription) is in a dynamic steady state, with chromatin being activated at rate k_1 and deactivated at rate k_2; and when the half-life of β-glucuronidase mRNA is much shorter than the half-time for induction, so that the concentration of β-glucuronidase mRNA at any moment in time is a fairly accurate reflection of its rate of synthesis. Experimentally, k_1, but not k_2, proved to be dependent on androgen receptor concentration when an androgen analogue was used to produce a reduced concentration of active androgen receptor (*88*).

In this equation, k_1 determines the initial slope of the induction curve, k_2 determines the half-time for induction (a measure of the length of time required to reach the induction plateau), and the ratio of the two rate constants, k_1/k_2, determines the final plateau level of induction. Mutations affecting k_1, then, will affect the initial slope and the final plateau level, but not the time required to reach maximum induction; whereas mutations affecting k_2 will not affect the initial slope, but will affect the half-time for induction and the final plateau level (Fig. 4). This distinction between the consequences of changing k_1 and k_2 is observed experimentally, and the various [*Gus*] haplotypes have proved to be altered in one or another of these two rate constants.

A more extensive kinetic analysis of the induction process has been developed (*93*) that is based on multiple receptor binding and explicitly includes the rate of mRNA turnover (see Section V,E for details).

C. *Gus-r* Regulatory Mutations

Many of the [*Gus*] haplotypes vary in their response to androgen induction. Such differences were first observed when Swank *et al.* (*18*) surveyed a number of inbred strains of mice for their androgen responsiveness and found that they fall into two major classes. High

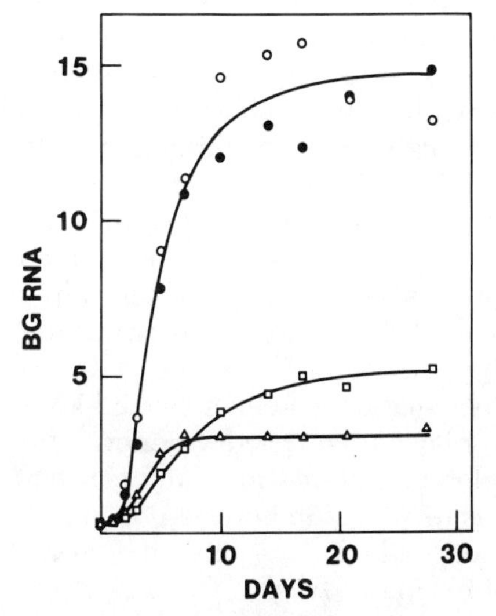

FIG. 4. Androgen induction of β-glucuronidase mRNA (BG RNA) in kidney. Results are shown for the *B* haplotype in B6 mice (□), the *CS* haplotype in B6.CS congenic mice (Δ), and the *A* haplotype in both B6.A congenic mice (○) and in strain A/J mice (●). β-Glucuronidase (BG) mRNA accumulation is expressed in micrograms per gram of total kidney RNA in the days following administration of testosterone to female mice. The salient points are the presence of a significant lag period before mRNA accumulation begins, the slow time course of accumulation, and the differences among haplotypes in the initial rate, duration, and final extent of induction. (From ref. *33*.)

inducers, typified by mice of the A/J strain, carrying the [*Gus*]^A haplotype, are several-fold more inducible than low inducers, typified by mice of the B6 strain carrying the *B* haplotype. The difference is determined by a regulatory locus, designated *Gus-r*, that maps within the [*Gus*] complex and controls the rate and magnitude of the induction response. *Gus-r* affects only androgen induction and does not influence basal levels of β-glucuronidase. When analyzed kinetically (*94*), mice carrying the *Gus-r*^{a} allele proved to have a higher k_1 and a shorter lag period.

Gus-r is closely linked genetically to *Gus-s*, the structural gene, and acts *cis*, so that $s^a,r^a/s^b,r^b$ heterozygotes produce a preponderance of GUS-A rather than GUS-B subunits (*18*, *95*). Initially, it was suspected that recombination between *Gus-s* and *Gus-r* could occur in laboratory crosses, but later, more extensive studies showed that this is not the case (*9*). However, haplotypes with the properties expected of recombinants between *Gus-s* and *Gus-r* have been identified in wild populations of mice (*35*).

Additional haplotypes with altered inducibility have been analyzed kinetically. In each case, the time course of induction of β-glucuronidase mRNA follows the kinetic equation given above (*33*, *34*) (Fig. 4), and each haplotype can be characterized by its induction rate constants. Using the *B* haplotype present in B6 strain mice as a reference, some haplotypes, such as *A* and *N*, differ primarily in k_1, while others, such as *CL* and *CS*, differ primarily in k_2.

Since k_1, but not k_2, is dependent on androgen receptor concentration (*88*), *Gus-r* mutations affecting k_1 probably alter the ability to attach androgen receptor, either by changing the number of potential binding sites for androgen receptor or the physical accessibility of chromatin to receptor. If so, then k_2 describes the detachment rate of receptor from the [*Gus*] gene.

The haplotypes that have been analyzed kinetically also differ in the length of their lag period. The lag period is inversely proportional to k_1 among haplotypes altered in k_1 and constant among haplotypes altered in k_2. This relationship between k_1 and the duration of the lag period is also seen when k_1 is altered by changing the effective concentration of receptor (*88*).

D. Progressive Induction of Transcription and Multiple Receptor Binding

1. Alternative Models

Compared with other steroid systems, the androgen induction of β-glucuronidase is exceptional in both the presence of a significant lag period before transcription begins to increase and the long time (7–10 days) required before transcription finally reaches its maximum. There are a number of mechanisms conceivable for this slow, progressive induction following a lag period. Several of these can be eliminated, from the data already available. First, progressive induction is not a consequence of the unique growth hormone requirement for β-glucuronidase induction; at least one other gene in mouse kidney, *mark I*, shows an identical slow, progressive induction following a lag period, but its induction is not growth hormone dependent (E. A. Allen and G. Watson, personal communication). Second, progressive induction is not due to slow entry of androgen receptor into the nucleus, as entry is maximal within 30 minutes (*92*) and there are other androgen-induced mRNAs in kidney whose induction does not exhibit a lag and that are induced appreciably faster than β-glucuronidase (*90a, 90*). Third, there is no autoinduction of the receptor by androgen, and receptor levels do not increase during induction (*97*). And finally, there is not a slow recruitment of newly responding cells (*98*). The number of responding cells does not change with time; instead, there is a progressively more intense response within a fixed population of cells.

Two remaining possibilities can be suggested. The first is a process of sequential induction in which androgen induces a novel regulatory protein that in turn induces [*Gus*]. As this protein accumulates, it induces [*Gus*] to higher levels of expression. In this case, the lag period would be the time required before the regulatory protein begins to increase in concentration, and progressive induction describes the accumulation of this substance. Differences between haplotypes in k_1 would reflect their relative sensitivities to the new transcription factor. However, this model makes several additional predictions that are not borne out experimentally. One is that the half-time for accumulation of the hypothetical transcription factor, and hence the half-time to reach maximal induction would depend on the half-lives of this novel protein and its mRNA. These should be genetically determined by the sequences of the gene coding for the transcription factor. Instead, the half-time for induction, which is a function of k_2, is genetically determined by the [*Gus*] gene complex itself. Another prediction is that the duration of the lag period should depend on k_2, the time required to accumulate the new factor, but this is not the case. Finally, this model predicts that changing the concentration of androgen receptor should alter k_1 by altering the rate of accumulation of regulatory protein. This should not affect the duration of the lag period, which is dependent on the half-lives of the new regulatory protein and its mRNA; however, changing the receptor concentration does alter the duration of the lag period.

The second possibility is that the [*Gus*] complex can bind multiple molecules of androgen receptor protein. In this case, a minimum number must bind before transcription can increase (the lag period), and transcription would increase progressively as additional molecules of the receptor bind to the [*Gus*] complex. Changes in the number of potential binding sites would produce haplotypes with altered k_1, corresponding to $[Gus]^A$ and $[Gus]^N$ (*94*), and changes in the binding affinity of the receptor for these sequences would produce haplotypes with altered k_2, corresponding to $[Gus]^{CS}$ and $[Gus]^{CL}$ (*33*). This model also correctly predicts that changing either the concentration of active receptor molecules or the total number of potential binding sites will affect the time required to attach the minimum number of receptors required for activation, and hence the duration for the lag period (*79, 88, 94*). Since the experimental data agree much better with the multiple-binding model than with the sequential-induction model, a mechanism involving multiple receptor binding appears to be the most likely explanation for progressive induction.

2. Multiple Receptor Binding

Several observations in other systems suggest that the multiple-receptor-binding model may apply elsewhere. The metallothionein gene resembles β-glucuronidase in both the requirement for a minimum number of binding sites and the proportionality of transcription to the additional number of sites occupied (*100*). Gene constructions tested in a cell culture system did not induce when only one response sequence was present, and when more than one response sequence was present, inducibility was proportional to the number of additional sequences present. In the case of the glucocorticoid induction of mouse mammary tumor virus (*100a*), response is proportional to the number of glucocorticoid response sequences adjacent to the gene, but there is no minimum number requirement.

Progressive induction by multiple binding has several interesting biological correlates. Modulation of gene activity is quantitative, rather than all-or-none, and the level of gene expression depends on the concentration of hormone. As a consequence, an individual cell is not restricted to an on–off response, but can be regulated over a range of response as the intensity of the biological signal varies. The minimum number requirement and lag period also function as noise-suppression devices, blocking response to transient pulses of hormone and effectively preventing a gene from responding until a sufficiently intense hormonal signal has been received for a sufficient duration of time.

In the multiple-binding mechanism, the slow response time of the system is the consequence of a very slow off-rate for dissociation of the androgen receptor from [*Gus*] DNA. These slow dissociation rates and response times may provide a biological advantage selected for in the evolution of androgen responses. Androgens primarily regulate growth and differentiation in contrast to steroids such as glucocorticoids, which are primarily rapid metabolic regulators, and the lag period and slow responses of androgen-inducible genes may reflect this biological difference in hormone function. However, when dissociation is slow, it is also necessary to reduce the rate at which steroid receptor attaches to DNA in order to keep the binding equilibrium within a reasonable range. This may be one reason why androgen receptors are characteristically present at relatively low concentrations compared to other steroid receptors. Assaying whole mouse kidneys, Isomaa *et al.* (92) estimated that there are about 1500 molecules of androgen receptor per nucleus. Even after adjusting for the fact that only 25–30% of kidney cells are androgen-responsive, this means that only 5,000–6,000 molecules of receptor are present per nucleus in the responding cell population.

3. Physical Basis of Kinetic Parameters

Almagor and Paigen (93) developed a detailed kinetic treatment of the multiple-binding model that defines the limitations of the kinetic treatment used earlier [Eq. (1) and Section IV,B] and provides a physical explanation for the several kinetic parameters of induction. The treatment relates rates of transcription, mRNA concentration, and enzyme concentration to the potential number of binding sites, the on and off rates for receptor binding, and a transcription rate factor describing the increase in transcription for each additional receptor molecule bound. The minimum number requirement was treated as a special case of a more general model. The essential rate equation for mRNA accumulation after the lag period is given by:

$$dR/dt = (\alpha N K^{+}/K^{0})(1 - e^{-K^{0}t}) - k_{r}(R) \qquad (2)$$

where R is β-glucuronidase mRNA concentration, t is time, α is the increase in transcription rate for each additional receptor molecule bound to the gene, N is the potential number of binding sites, K^{+} is the attachment rate constant for receptor binding to DNA, K^{-} is the detachment rate constant, K^{0} is the relaxation time constant of the induction process (and is equal to the sum of K^+ and K^-), and k_r is the is the first-order rate constant for mRNA degradation in the process of intracellular mRNA turnover.

The relationship between the parameters of this equation and those of Eq. (1) is derived by integrating both equations and solving for the final plateau of mRNA, $R(\infty)$, (93). From Eq. (1):

$$R\,(\infty) = R_0 + k_1/k_2 \tag{3}$$

and from Eq. (2):

$$R(\infty) = R_0 + \alpha NK^+/K^0 \cdot k_r \tag{4}$$

From this, it can be seen that k_1 is proportional to αNK^+, and the constant k_2 is proportional to the transcription relaxation constant K^0.

The shape and half-time of the induction curve for mRNA accumulation depend upon both k_r and K^0. However, in the approach to a new steady state, the observed shape of the curve is dominated by the slower of these constants, and this becomes increasingly true with time. K^0 must be appreciably slower than k_r, since the half-time for induction of β-glucuronidase mRNA, which depends upon both K^0 and k_r, is not much longer than the half-time for induction of transcription, which depends only upon K^0 (*90a*). It is for this last reason that the time course of β-glucuronidase mRNA induction closely approximates Eq. (1), which was used in earlier work.

In conceptual terms, the more detailed analysis suggests that the initial slope of the induction curve depends on the number of receptor binding sites in [*Gus*] and the rate of receptor attachment to each site, which together describe the overall rate of receptor attachment, along with the increase in transcription as each successive receptor molecule binds. The half-time for induction is primarily determined by K^0 when K^0 is much slower than k_r, as appears to be the case. It is because K^0 is so slow that progressive induction is observed on an experimental time scale.

The final plateau level of induced mRNA is determined by the ratio of the four parameters affecting the rate of mRNA synthesis, $\alpha NK^+/K^0$, and the rate of mRNA degradation, k_r. Thus, the degree of induction of β-glucuronidase mRNA, $R(\infty)/R_0$, reflects changes in both the rate of synthesis and the rate of degradation of the message.

Although this kinetic treatment was developed for a multiple-binding model, it is mathematically applicable to any induction process that depends upon two sequential exponential processes, and can be used to describe sequential induction mechanisms as well as those involving multiple binding (for details, see 93).

E. Developmental Appearance of Inducibility

The developmental appearances of both β-glucuronidase inducibility and androgen receptor protein in mouse kidney have been examined (*97*). Inducibility appears synchronously with a rise in receptor concentration beginning about 20 days after birth, indicating that all other factors, including the differentiation of the gene into a responsive state, have already occurred. Interestingly, a modest level of receptor is already present earlier, but this is not adequate to cause induction. A crude estimate of the quantitative relationships between receptor concentration and inducibility can be obtained from the developmental data, and it suggests a relationship similar to that predicted by the multiple-binding model.

F. *trans* Regulation of Induction

The androgen responsiveness of the $[Gus]^H$ haplotype was originally described using rates of β-glucuronidase protein synthesis to analyze the kinetics of induction (*94*). These experiments suggested that the *H* haplotype has a unique inducibility phenotype; however, subsequent work has shown that, because of reduced translational yield in the *H* haplotype (Section VI,B), measurements of β-glucuronidase protein synthesis do not provide a valid comparative measure of β-glucuronidase mRNA concentrations in $[Gus]^H$ mice. Induction of β-glucuronidase mRNA in the *H* haplotype has now been characterized using hybridization probes (*90a*). These measurements show that the *H* haplotype induces identically with the *B* haplotype when the congenic strain B6.H is compared with strain B6 itself, indicating that $[Gus]^H$ and $[Gus]^B$ are probably identical in the *Gus-r* regulator controlling androgen induction. However, when C3H strain mice, carrying $[Gus]^H$, are compared with B6.H mice, also carrying $[Gus]^H$, there is a significant difference in the induction process (*90a*). The lag period is longer and k_1 is slower in C3H mice, suggesting the existence of a *trans* regulatory gene affecting β-glucuronidase induction that differs between the C3H and B6 strains of mice. The putative *trans* regulatory gene has no effect on basal levels of β-glucuronidase mRNA, indicating that it only affects the induction process. It is also gene-specific in that the induction of several other androgen-responsive mRNAs in kidney does not differ between strains B6 and C3H.

G. Androgen Induction in the Submaxillary Gland

Among mammals, saliva and sweat are the principal routes of pheromone release in addition to urine. It is interesting, then, for its possible relationship to pheromone function (Section IV,H), that β-glucuronidase is also androgen-inducible in submaxillary gland (*101*), and there is evidence that some of the induced enzyme may be secreted into saliva.

Induction occurs by a different mechanism than in the kidney. In the submaxillary gland, there is a marked increase in the rate of synthesis of β-glucuronidase protein with very little increase in β-glucuronidase mRNA. Thus, the mechanism of induction is an increase in translational yield (Section VI) and not transcriptional activation of [*Gus*] as in the kidney. Induction is genetically variable, and haplotypes differ in their responsiveness, with the *N* haplotype showing the maximum response. The means of achieving this induction of translational yield is unknown, but presumably it is related to the developmental and genetic regulation of translational yield seen under other circumstances (see Section VI).

H. Species Variation and the Evolution of Induction

Surveys of *Mus* species for β-glucuronidase inducibility (summarized in *81*) show that the ability to induce is not present in all species (Fig. 5), and appears to have been gained and/or lost several times in the 10 million years of *Mus* evolution. The various subspecies of *M. m. musculus*, including *M. m. domesticus* (the primary genetic source of inbred laboratory mice), *M. m. musculus*, and *M. m. castaneus*, are all inducible, but vary quantitatively in their responses. *M. hortulanus*, a species very closely related to *M. m. domesticus*, cannot induce, and this is a result of a *cis*-acting mutation in the close vicinity of the [*Gus*] gene (*35*, *101a*). However, a more distant species, *M. spretus*, can induce; the still more distant species *M. caroli* and *M. cervicolor* cannot induce, but the most distant *Mus* species that has been tested, *M. saxicola*, is inducible. Among other mammals, rats do not induce (E. Allen, personal communication), and humans probably do not induce kidney β-glucuronidase, as males do not excrete higher urinary levels of enzyme than females (*101b*).

The ability to induce the mRNA known as RP-2 also shows a sporadic distribution among the same *Mus* species (*101c*). The abilities to induce β-glucuronidase and RP-2 appear to have evolved independently of each other, as there is little congruence among species in the inducibility of these two mRNAs. Some species can induce one, but not the other.

The androgen response system in kidney appears to be more widespread among mammals than is β-glucuronidase induction itself. Humans possess androgen receptor protein in kidney (*102*), and

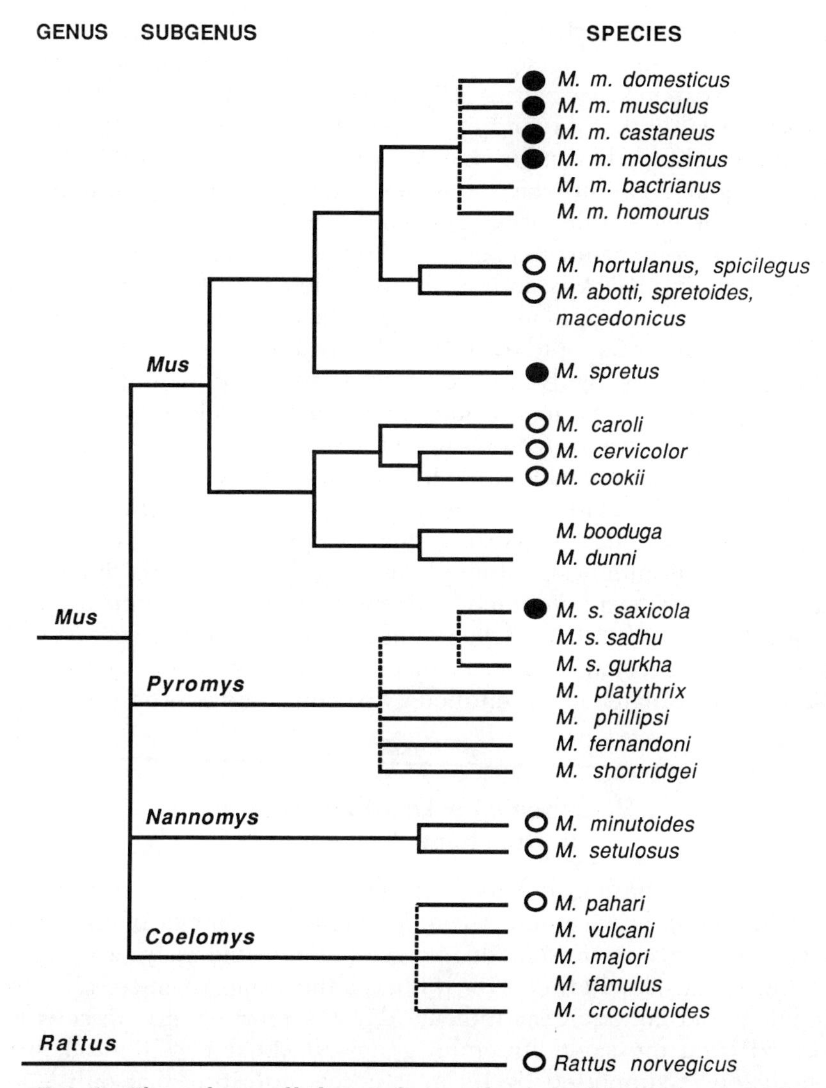

FIG. 5. The evolution of kidney β-glucuronidase inducibility in the genus *Mus*. A consensus phylogeny of the genus *Mus* is used, with vertical lines indicating cases in which the order of separation of species is not known. The one exception is *M. molossinus*, which is thought to have arisen by hybridization of mice from the species *M. m. musculus* and *M. m. castaneus*. Species are marked as to whether they are inducible (●) or noninducible (○); untested species are left blank. Given this distribution of inducibility among species, it is apparent that the inducibility trait has appeared and/or disappeared several times during the approximately 10 million years of *Mus* evolution.

several mRNAs other than β-glucuronidase are androgen-inducible in rat kidney (E. Allen, personal communication). The function of the androgen response system is still a matter of speculation. It is, of course, possible that the androgen induction system actually has no physiological function and is simply undergoing genetic drift because it is not under evolutionary selection. However, this is difficult to accept in the face of the occurrence of the androgen response system among divergent mammalian groups and the fact that as much as 3% of mouse kidney mRNA is androgen-inducible (*90a*). It has been suggested (*81*) that the system functions in pheromone signaling, which could account for its rapid evolution. All mammals release a complex set of pheromones involved in regulating estrus and mating behavior (*105*). In mice, the primary route of pheromone release is through urine. Much of the androgen-induced β-glucuronidase is secreted into the lumen of the proximal tubule by extrusion of lysosomes across the brush border of the kidney epithelial cells. From there, it finds its way into urine, where it is enzymatically active. A functional role for this activity in pheromone signaling is suggested by the observation (*106*) that an aggression pheromone is present in urine as a glucuronide conjugate that is split by β-glucuronidase to release the active substance. This substance appears to be distinct from the aggression pheromones isolated and identified by Novotny and co-workers (*107*).

V. Subcellular Localization and Intracellular Processing

β-Glucuronidase is unusual among acid hydrolases in exhibiting a dual intracellular location. An appreciable percentage of the total enzyme activity is present in microsomes as well as lysosomes of several organs, including liver, kidney, and lung. In mice, all biochemical and genetic tests indicate that the enzyme at both sites is derived from the same structural gene (*19, 20, 109, 110*), and this conclusion is supported by recent studies indicating that only one β-glucuronidase structural gene is present in mice (*42*). This dual enzyme localization, plus the existence of mutants with altered subcellular localization of the enzyme, has encouraged research on the posttranslational processing pathways involved, the mechanisms by which this dual intracellular localization is achieved, and its possible physiological significance. The major finding has been the involvement of a second protein, egasyn, in the localization mechanism.

A. β-Glucuronidase Isoforms

β-Glucuronidase protein is a tetramer of four identical subunits; however, the subunits present in the microsomal and lysosomal forms of the enzyme differ from each other in both size and charge as a result of posttranslational processing events. The microsomal enzyme, called the **X** form of β-glucuronidase, has subunits that are a few thousand daltons heavier and have a higher isoelectric point than the subunits of the enzyme found in lysosomes (*110, 111, 112*). The **X**-form tetramers present in microsomal particles are associated with an accessory protein, egasyn (*110,111*), and four β-glucuronidase–egasyn complexes, called **M** forms, are found (**M1–M4**). These contain one to four molecules of egasyn bound to each β-glucuronidase tetramer. The complexes are free within the lumen of microsomal vesicles or, at best, only weakly bound to the inside of the membrane vesicles (*113*). The lysosomal enzyme actually occurs in two forms, **La** and **Lb**, whose subunits also differ from each other in molecular weight and isoelectric point (*114, 115*). Altogether, then, seven isoforms of β-glucuronidase are found: free **X**, and four possible complexes of the **X** form with egasyn (**M1–M4**), and the two lysosomal forms **La** and **Lb**.

B. Egasyn Protein and the Eg^0 Mutation

When purified (*69*), egasyn proved to be a 64-kDa, amphipathic glycoprotein lacking β-glucuronidase activity. The two proteins do not cross-react immunologically and do not appear to possess any structural relationship. The isolation of egasyn allowed the development of a radioimmune assay for it that was used to determine its biological distribution among tissues and to confirm its physical association with β-glucuronidase in microsomes (*69, 117*).

Understanding the properties and function of egasyn followed the discovery of a mutation, Eg^0, in mice that results in loss of microsomal β-glucuronidase (*109*) and that later proved also to lack egasyn (*117*). The *Eg* (*endoplasmic glucuronidase*) gene is not linked to the structural gene for β-glucuronidase itself. Instead, it is located on chromosome 8 in the mouse, very close to the gene coding for (carboxyl) esterase-1,[4] *Es-1* (*118*), a location that eventually provided an important clue in understanding egasyn function.

Several lines of evidence indicate that the formation of β-glucuronidase–egasyn complexes is required for the localization of β-glu-

[4] Carboxylesterase (EC 3.1.1.1).

curonidase in endoplasmic reticulum, from which the microsomal particles are derived during tissue homogenization. Eg^0 mutant mice lack both egasyn and microsomal β-glucuronidase, even though nearly normal levels of lysosomal β-glucuronidase are present (*10, 109, 111*). Additionally, in liver, the developmental appearance of microsomal β-glucuronidase **M**-form complexes is correlated with the developmental appearance of egasyn, and in mature animals, only tissues that express egasyn also contain microsomal β-glucuronidase (*120*).

The ability to interact with egasyn and be localized in microsomal membranes depends, of course, upon the structure of β-glucuronidase protein, and the enzyme coded by the $Gus\text{-}s^h$ and $Gus\text{-}s^n$ alleles exhibits a much altered affinity for egasyn (*121, 121a*).

The earlier studies of the dual intracellular localization of β-glucuronidase, and the discovery and properties of the Eg^0 mutation and egasyn have been reviewed in detail (*10*).

C. Egasyn Is an Esterase

1. Mouse

The *Eg* gene is located within a cluster of esterase genes on chromosome 8 (*118, 122*), and egasyn resembles microsomal esterases in several properties (*123, 124*). These facts led Medda and Swank (*125*) to the discovery that egasyn has carboxylesterase activity. Anti-egasyn antibody precipitates a protein with esterase activity; this esterase is complexed with β-glucuronidase in a manner identical to egasyn; and the esterase is missing in mice carrying the Eg^0 mutation. Using the enzymatic assay for egasyn, it now appears that β-glucuronidase is the only protein associated with egasyn (*125*), demonstrating the highly specific nature of the egasyn · β-glucuronidase interaction and the role of egasyn in determining the intracellular localization of β-glucuronidase, and correcting earlier suggestions that egasyn may be associated with proteins other than β-glucuronidase.

A variety of criteria (*126*) identify egasyn as the previously described esterase-22[4] of mice (*127*). Egasyn and esterase-22 are virtually identical in their molecular weights and isoelectric points, substrate specificities, and sensitivities to inhibitors. They also have the same tissue distribution and appear synchronously during liver development. Genetically, a number of allelic forms of esterase-22 differing in their electrophoretic mobility had previously been described, and Medda *et al.* (*126*) demonstrated that egasyn displays an indentical strain distribution of electrophoretic variation. They did so by using anti-β-glucuronidase antibody to isolate β-glucuronidase/egasyn complexes, and then determining the electrophoretic mobility of the egasyn extracted from the complexes. Finally, genetic mapping placed the genes coding for egasyn and esterase-22 at the same location on chromosome 8 of the mouse. Wassmer *et al.* (*128*) had located the esterase-22 gene, *Es-22*, within cluster 1 of the esterase gene region, distal to the gene *Es-6* and very close to the gene *Es-9*. More detailed mapping of the *Eg* gene showed that it is also distal to *Es-6* and close to *Es-9* (*126*).

Although egasyn forms β-glucuronidase complexes containing up to four molecules of egasyn, free egasyn is only present as a monomer. When strains carrying different egasyn/esterase-22 alleles were crossed to produce F1 heterozygotes, these progeny expressed only the two parental esterase bands; no heteromer bands indicative of subunit interaction were seen.

These comparative studies of egasyn and esterase-22 also showed that esterase-22 activity is induced 2 to 3 fold by androgen in mouse kidney. This is in contrast to the earlier report (*120*), based on radioimmune assays measuring amounts of egasyn protein, that egasyn is not androgen-inducible. It is possible that androgen alters the activity, but not the amount of egasyn protein. Along with other microsomal proteins, egasyn is also induced in liver by phenobarbital (*70*), causing a corresponding shift in the localization of β-glucuronidase from lysosomes to microsomes.

A total of eight allelic forms of the *Eg/Es-22* gene coding for different electrophoretic forms of egasyn have so far been distinguished (Table III). These alleles also show considerable regulatory variation. The alleles present in *Mus spretus* and *Mus molossinus* produce very high levels of egasyn; the allele present in common inbred mouse strains produces a moderate amount of egasyn; the allele present in the Peru/1 line of inbred mice produces a very low level of egasyn; and the Eg^0 mutant allele produces no egasyn at all. Together, these alleles provide the potential for a systematic quantitative study of the role of egasyn in the microsomal localization of β-glucuronidase. Additionally, since variant structural forms of β-glucuronidase differ in their ability to interact with egasyn, the possibility also exists for a study of the interaction between alternate structural forms of both egasyn and β-glucuronidase.

2. Other Species

The presence of microsomal β-glucuronidase as a result of complex formation between β-glucuronidase and egasyn appears to be a general mammalian characteristic. In addition to mouse, micro-

TABLE III
ALLELIC FORMS OF EGASYN[a]

Phenotype	Electrophoretic mobility	Egasyn–esterase activity	Species and strains
A	3	++	*M. m. molossinus,* MOL/35, *M. m. castaneus* (Buffalo)
B	4	?	*M. m. castaneus* (Freiburg)
C	5	+++	IS/cam
D	7	+	C57BL/6J, v.i.s.[b]
E	6	+/−	Peru/1Fre, other Peru Coppock lines
F	1	+++	*M. spretus*
G	2	+++	*M. hortulanus*
H or O	—	none	YBR, STS/A, LTS/A, LIS/A, C57BL/6.Eg^0 [c]

[a] Adapted from Medda *et al.* (*126*).
[b] v.i.s., Various inbred strains.
[c] C57BL/6.Eg^0 is a congenic strain with the Eg^0 mutation of strain YBr transferred onto a C57BL/6J genetic background.

somal β-glucuronidase or **M** forms have been observed in rat (*132, 133*), rabbit (*63, 64*), bovine (*56*), human (A. J. Lusis, personal communication), and even tadpole liver (*134*). In rabbit, the β-glucuronidase–egasyn **M**-form complexes are particularly stable and easily discerned (*63, 64*).

Rat liver egasyn also possesses carboxylesterase activity (*121*). Isoelectric focusing gels show the presence of microsomal **M** forms of β-glucuronidase. When precipitated with anti-β-glucuronidase antibody, these contain a 61-kDa accessory protein that has carboxylesterase activity and is precipitated by anti-mouse egasyn antibody. Identity of this esterase with the previously described rat esterase-3[4] was suggested by similarities in size, isoelectric point, and substrate specificities (*121*). The two proteins cross-react immunologically, and both are precipitated by antibody to either purified rat esterase-3 or mouse egasyn. The identification of rat esterase-3 as egasyn was confirmed genetically. Rat egasyn and esterase-3 show identical genetic polymorphism, with three different alleles of the common egasyn/esterase-3 gene, each allele coding for a different electrophoretic form of the protein. The equivalent of the mouse Eg^0 mutation is also found in rats. Out of 103 inbred rat strains, 90 possess both rat egasyn and esterase-3, and 13 strains are negative for both, and the presence or absence of both activities cosegregated in a genetic cross between strains of the two types.

Rat β-glucuronidase/egasyn complexes are appreciably less stable than are the mouse complexes. They dissociate at lower temperatures and are more easily disrupted by the reagent bis(*p*-nitrophenyl) phosphate (see Section V,F). The fact that the rat complex eluded detection in earlier studies (*135*) may reflect this relative lability; it is also possible that the Eg^0 mutation is segregating among commercial rats.

Rat egasyn is probably not identical with the rat β-glucuronidase binding protein isolated by Strawser and Touster (*136*). This latter binding protein is an intrinsic membrane protein, can only be extracted at much higher detergent concentrations than are required to extract egasyn, and does not cross-react with anti-egasyn antibody.

D. Metabolic Relationships among β-Glucuronidase Isoforms

The physical and metabolic relationships among the multiple intracellular forms of β-glucuronidase have been examined in some detail (*114, 115*), establishing that the subunit of the microsomal form of β-glucuronidase, **X**, is 75 kDa, the subunit of the lysosomal form **La** is 73 kDa, and the subunit of the lysosomal form **Lb** is 71.5 kDa, and that tissues containing egasyn have predominantly the **Lb** form in their lysosomes, while tissues lacking egasyn (either because they do not normally make it or because the Eg^0 mutation is present) have predominantly the **La** form.

Pulse-labeling studies in porcine kidney have shown that β-glucuronidase is synthesized there as a 75-kDa precursor protein. This precursor is converted to the predominant 72-kDa mature form by carboxy-terminal proteolytic processing (*136a*). In mouse kidney, pulse labeling studies show that the precursor has the same physical properties as **X**; it is also 75 kDa, and located in microsomes (*137*). This precursor is then converted to a 71-kDa lysosomal product that is presumably **Lb** (*138*). In macrophages, which lack egasyn, the precursor is converted to a 73-kDa lysosomal product that is presumably **La**. Brown and Swank (*139*) used Percoll gradients to separate the various intracellular components in macrophages and showed that the conversion of the 75kDa product to its 73-kDa product was coincident with transport to lysosomes. Combining the evidence from both species, the most direct assumption is that the β-glucuronidase precursor loses a carboxy-terminal peptide when it enters the lysosomal space, probably by the action of one of the catheptic proteases abundant there. There is less evidence regarding the processing of the rat precursor, which has been reported to be only 72 kDa (*140*). Since the rat and mouse cDNA sequences predict precursor polypeptides of identical length, this apparent lower molecular weight of the precursor may reflect some species difference in the timing or extent of glycosylation.

Pfister *et al.* (*142*) have used pulse-chase techniques in primary hepatocyte cultures of normal and egasyn-deficient mice to establish in detail the metabolic relationships of these multiple intracellular forms (Fig. 6). It appears that when **X** enters the lysosomal space, its fate depends upon whether or not it is complexed with egasyn. When it is complexed, it is converted to **Lb**; and when it is not complexed, it is converted to **La**. This accounts for the correlation between the occurrence of egasyn in tissues and whether **La** or **Lb** is the predominant lysosomal form of β-glucuronidase. The transport of **X** to lysosomes is much retarded by egasyn, explaining why the presence of egasyn produces an appreciable accumulation of microsomal β-glu-

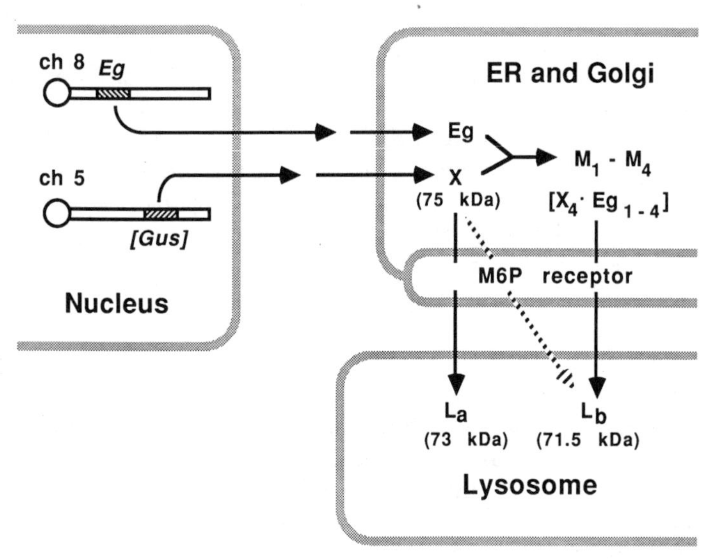

FIG. 6. The intracellular flow of β-glucuronidase isoforms and of egasyn in liver. Translation of β-glucuronidase mRNA with concurrent removal of the signal peptide gives rise to the **X** form of β-glucuronidase. Complex formation between the **X** form and egasyn gives rise to the **M** forms, which contain from one to four molecules of egasyn bound to each β-glucuronidase tetramer. When **M** forms of β-glucuronidase are translocated to the lysosomal space, they give rise to the **Lb** form of the enzyme by proteolytic cleavage. When **X** tetramers lacking bound egasyn enter the lysosomal space, they give rise predominantly to the **La** form, although a small amount of **Lb** also arises. In kidney, **X** tetramers give rise to **Lb** whether they enter as free tetramers or bound to egasyn. Molecular weights of the various β-glucuronidase isoforms are indicated in parentheses. (See text and ref. *142* for further details.)

curonidase and why tissues lacking egasyn also lack microsomal β-glucuronidase. Egasyn has an additional indirect effect on the cellular content of β-glucuronidase because the half-life of **La** is somewhat shorter than that of **Lb**. The net result is that egasyn-deficient cells have less β-glucuronidase than their normal counterparts, and nearly all of the enzyme they have is in lysosomes. These results are in accord with earlier data (*137*) comparing precursor conversion in normal and egasyn-deficient kidney and suggesting that the presence of egasyn slows conversion of the microsomal form, thus allowing it to accumulate. Smith and Ganschow (*142a*) obtained somewhat different results for mouse liver, probably because they used longer labeling times and their work was carried out before the distinction between the **La** and **Lb** forms of β-glucuronidase was appreciated.

It is now possible to suggest how these conversions come about and what the role of egasyn is by using the structural data now available for mouse β-glucuronidase. The **X**, **La**, and **Lb** forms are 75, 73, and 71.5 kDa, respectively, and their respective isoelectric points are 6.9, 5.4, and 5.9. From examination of isoelectric focusing gels (*115*), it appears that **Lb** is 7–9 charges per tetramer, or 2 charges per subunit, more negative than **X**, and **La** is 1–2 charges per subunit more negative than **Lb**. **La** also appears to contain more complex polysaccharide with some negative charge than does **Lb**; the precise nature of this charge is uncertain. The formation of **La** and **Lb** from **X** presumably involves a carboxy-terminal proteolytic cleavage when β-glucuronidase enters the lysosomal space. Figure 7 shows the C-terminal sequence of β-glucuronidase, including the location of positive and negative charges, and the putative N-linked glycosylation site in this region. Indicated on the sequence are two potential

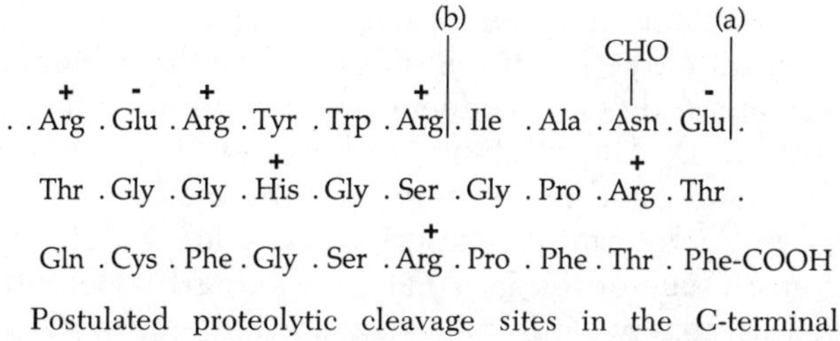

(b) (a)
CHO
\+ - + + -
. . Arg . Glu . Arg . Tyr . Trp . Arg | . Ile . Ala . Asn . Glu | .
\+ +
Thr . Gly . Gly . His . Gly . Ser . Gly . Pro . Arg . Thr .
\+
Gln . Cys . Phe . Gly . Ser . Arg . Pro . Phe . Thr . Phe-COOH

FIG. 7. Postulated proteolytic cleavage sites in the C-terminal region of β-glucuronidase protein. Cleavage occurs upon entry of β-glucuronidase into the lysosomal space. When free tetramers enter, cleavage occurs predominantly at site (a), giving rise to the **La** form. The presence of egasyn complexed to β-glucuronidase shifts the cleavage point to site (b), giving rise to the **Lb** form, which is about 1.5 kDa smaller and one charge less negative than **La**.

proteolytic cleavage sites, marked **a** and **b**. Much of what is known about β-glucuronidase processing would be explained if the choice of cleavage sites is dependent upon whether or not β-glucuronidase is complexed to egasyn. Cleavage at site **a** alone would occur if no egasyn was complexed. This would generate a subunit with the properties of **La**; it would be 2kDa smaller than **X** and 3 charges more negative. Cleavage at site **b** would occur when egasyn is present. This would generate a subunit with the properties of **Lb**; it would be 1–2 kDa still smaller, but only two charges more negative than **X** because a negatively charged glutamate residue would also be lost. It is not possible to say whether egasyn inhibits cleavage at site **a** in order to promote cleavage at site **b**, or whether both sites **a** and **b** are cleaved in the presence of egasyn; **La** would be produced in either case.

An interesting question is what happens when the **M** forms $\mathbf{M}_1$, $\mathbf{M}_2$, and $\mathbf{M}_3$ enter the lysosomal space. These β-glucuronidase–egasyn complexes do not have an egasyn molecule attached to every β-glucuronidase subunit. Is the presence of one egasyn molcule on a β-glucuronidase tetramer enough to alter the cleavage pattern of the entire tetramer, or does the presence of each egasyn molecule only affect the cleavage site of the β-glucuronidase polypeptide subunit to which it is attached?

E. [*Gus*] Haplotypes Vary in the Intracellular Location of β-Glucuronidase

At least two [*Gus*] haplotypes, *H* and *N*, produce enzyme with a markedly altered distribution between lysosomes and microsomes, albeit by apparently different mechanisms.

Compared to the *A* and *B* haplotypes, which have slightly more than half their enzyme in lysosomes, *H* has a much greater predominance of lysosomal enzyme in liver (*121*), almost certainly as a secondary result of the structural difference in the enzyme. Quantitatively, the magnitude of the effect depends upon the total amount of enzyme produced. The relationship between the proportion of enzyme in lysosomes and microsomes and the total amount of enzyme produced can be determined experimentally for any given haplotype by taking advantage of the fact that even inbred mice with identical genotypes vary somewhat in their enzyme content. In $[Gus]^A$ mice, at the lowest levels of enzyme encountered, there is an equal amount of β-glucuronidase in lysosomes and microsomes. As the total level of enzyme activity increases, about 75% of the increment goes to lysosomes and about 25% to microsomes. $[Gus]^H$ mice show a very different distribution. The lysosomal compartment always has a fixed number of enzyme units more than the microsomal compartment, regardless of the total amount of enzyme present.

The $[Gus]^N$ haplotype behaves in a very different manner (*121a*), showing a predominance of microsomal enzyme in liver. $[Gus]^N$ mice contain about 60–65% of their β-glucuronidase in microsomes, compared to an average value of 35–45% in $[Gus]^B$ animals. The effect is actually a deficiency of lysosomal enzyme; $[Gus]^N$ animals have the same absolute amount of microsomal enzyme as $[Gus]^B$ mice, but have lost an appreciable part of their lysosomal enzyme. This effect would be accounted for if the retention of β-glucuronidase in microsomes and its subsequent transport to lysosomes was normal, but the half-life of the enzyme was shorter once it entered the lysosomal space.

F. The Active Site of Egasyn and Complex Formation

Complex formation between egasyn and β-glucuronidase almost certainly involves the carboxylesterase active site of egasyn. When egasyn is complexed with β-glucuronidase, its active site is inaccessible to both esterase substrates and irreversible inhibitors (*143*). These compounds can interact with free egasyn after it has been released from its complex with β-glucuronidase by exposure to elevated temperatures. Additionally, the β-glucuronidase–egasyn complex can be specifically dissociated *in vitro* by the carboxylesterase inhibitor bis(*p*-nitrophenyl) phosphate but not by a variety of naturally occurring mono- and diphospho compounds that have been tested (*143*).

In contrast to the egasyn active site, the active site of β-glucuronidase is almost certainly not involved in complex formation. β-Glucuronidase is fully active as an enzyme in β-glucuronidase–egasyn complexes, and the stability of complexes is not affected by exposure to saccharolactone, a potent competitive inhibitor of β-glucuronidase. Although bisphosphates do interact with the active site of egasyn, it is not likely that attachment to β-glucuronidase is at the phosphodiester linkages that arise in lysosomal oligosaccharide side-chains during synthesis of the mannose-6-phosphate side-group. Low-molecular-weight analogs of this side group do not dissociate the complex (*143*), and other lysosomal enzymes sharing the same structure do not form egasyn complexes. If, as suggested in Section V,D, the presence of egasyn attached to β-glucuronidase affects proteolytic cleavage of β-glucuronidase at the carboxyl end of the molecule, this would seem to be the most likely region involved in complex formation.

Organophosphorus Insecticides

In both mouse and rat strains possessing egasyn, 30–50% of total hepatic β-glucuronidase is typically bound to microsomal membranes. These complexes dissociate *in vivo*, as well as *in vitro*, when animals are exposed to the carboxylesterase inhibitor bis(*p*-nitrophenyl) phosphate (BPNP). They also dissociate *in vivo* when animals are exposed to organophosphorus insecticides (parathion and paraoxon) and serine protease inhibitors (diisopropyl fluorophosphate and *p*-toluenesulfonyl fluoride). Some of the dissociated β-glucuronidase is secreted into blood, causing a dramatic rise in plasma β-glucuronidase, which is otherwise very low. The elevation in blood β-glucuronidase is one of the most rapid and sensitive indicators of exposure to these classes of compounds (*143*).

The biological response is specific to β-glucuronidase; other lysosomal enzymes do not react, probably because they do not form egasyn complexes. Dissociation precedes secretion when animals are exposed to active compounds, and only the freed β-glucuronidase is secreted; the released egasyn remains bound to microsomal membranes. The mechanisms of dissociation *in vivo* appear to be the same as those *in vitro*. Mouse β-glucuronidase/egasyn complexes are much more resistant to dissociation by BPNP *in vitro* than are rat complexes, and correspondingly higher concentrations of BPNP are required to induce dissociation *in vivo* (*143*).

G. Physiological Significance of Egasyn · Glucuronidase Complexes

The association between egasyn and β-glucuronidase almost certainly existed prior to the separation of the various mammalian lineages some 75 million years ago, as rabbits, humans, rats, and mice all possess β-glucuronidase/egasyn complexes. The association may well have existed much earlier even than that since amphibians also appear to possess **M** form complexes (*134*). The fact that contemporary rats and mice are both genetically polymorphic for the presence or absence of these complexes suggests that, despite their ancient origin, some antagonistic or cyclic selection for and against the presence of complexes may be occurring at the present time.

What the physiological role of β-glucuronidase · egasyn complexes might be, and the possible nature of the selective forces in evolution, is uncertain. Complex formation is not required for the intracellular conversion of the precursor form of β-glucuronidase to a lysosomal form. This conversion occurs in tissues that lack egasyn in normal mice and in all tissues of Eg^0 mutant mice. Additionally, although the possibility has not been examined in detail, there is no reason to suspect a difference in the catalytic activities of the **Lb** and **La** forms of β-glucuronidase that arise in the presence and absence of egasyn. It is also unlikely that egasyn functions to hydrolyze linkages in the carbohydrate sidechains of β-glucuronidase, as egasyn does not bind to the phosphodiester linkages present in the oligosaccharide side chains of newly synthesized enzyme molecules (*143*).

Whittaker *et al.* (*144*) have suggested that microsomal β-glucuronidase functions in the β-glucuronide conjugation–deconjugation cycle involved in the formation of glucuronosyl conjugates of xenobiotic metabolites. Compounds such as the carcinogen benzo[*a*]pyrene are oxidized by the microsomal mixed function oxidase system to phenols, as well as other products, and these phenols are in turn conjugated with glucuronic acid by microsomal UDP-glucuronosyltransferase (EC 2.4.1.17). The relative levels of free and conjugated phenols depend on the balance between this conjugation reaction and the hydrolysis of the conjugates by β-glucuronidase (*145–147*). Calcium ion, which activates microsomal glucuronidase by altering its affinity for substrate (*144, 147*), alters this reaction balance *in vivo* and inhibits the production of glucuronide conjugates from *p*-nitrophenol by the perfused liver (*146*).

It is quite possible that the function of β-glucuronidase/egasyn complexes is to regulate esterase rather than β-glucuronidase activity; particularly since complex formation affects the function of the active site of egasyn but not that of β-glucuronidase. Rat esterase-3, which is the egasyn of this species, has strong long-chain fatty acid esterase activity against several naturally occurring substrates, including palmityl carnitine (*148*) as well as monoglycerides and lysophospholipids (*149*), and formation of β-glucuronidase complexes could function to modulate long-chain fatty acid metabolism.

VI. Translational Yield

Translational yield is defined as the rate at which mature enzyme molecules are synthesized from each mRNA molecule. Differences among tissues in translational yield are a major factor in the tissue specificity of [*Gus*] expression. Translational yield is the product of the rate of synthesis of nascent chains per mRNA molecule and the fraction of nascent chains that are successfully converted to mature enzyme molecules. The presently available evidence suggests that it is the latter step, the conversion of nascent chains to mature enzyme molecules, that is the site of regulation. In addition to being a major factor in the tissue specificity of β-glucuronidase expression, changes in translational yield underlie all of the systemic and temporal mutations known to affect β-glucuronidase expression, and androgen administration can induce a marked increase in the translational yield of β-glucuronidase in the submaxillary gland. The first suggestion that translational yield might play an important role in the expression of β-glucuronidase came during the characterization of the *CS* and *CL* haplotypes of [*Gus*] when a significant reduction was observed in the rate of enzyme synthesis in these mice without any corresponding reduction in mRNA levels (*39*).

β-Glucuronidase is probably not unique in being regulated by translational yield. Levels of β-glucuronidase and several other lysosomal enzymes change coordinately during development (*151, 152, 152a,*), and if the developmental regulation of β-glucuronidase is achieved through translational yield, it seems likely that these other enzymes are regulated in a similar manner.

A. Tissue Specificity and Inducibility of Translational Yield

There is at least a 20-fold range in levels of β-glucuronidase activity among mammalian tissues, even after exceptional tissues such as the androgen-induced mouse kidney and rat preputial gland are excluded. To determine the source of these differences, Bracey and Paigen (*150*) estimated numbers of enzyme molecules per cell, rates of enzyme synthesis and degradation, and β-glucuronidase mRNA concentration in six representative tissues. Nearly all differences in tissue expression resulted from differences in rates of enzyme synthesis. Differences in the relative half-life of the enzyme were not a significant factor except for brain, which showed a tripling in enzyme turnover. The differences in enzyme synthesis between tissues were not accounted for by changes in β-glucuronidase mRNA concentrations; instead, they resulted from differences in the rate of β-glucuronidase synthesis per mRNA molecule, i.e., translational yield.

As described in Section IV,G, androgen administration induces a marked increase in translational yield of β-glucuronidase in the submaxillary gland (*101*). The extent of this induction varies among [*Gus*] haplotypes, with the highest induction seen in mice carrying the $[Gus]^N$ haplotype.

B. Haplotype Differences in *Gus-u* and *Gus-t*

Haplotype differences in translational yield were first observed in studies of *Gus-u* mutations in the *CS* and *CL* haplotypes (*39*). These haplotypes show reduced rates of enzyme synthesis but no reduction in β-glucuronidase mRNA levels. This is also true of the *N* haplotype (*34*).

The *H* haplotype, which was the original [*Gus*] mutant described, differs in both its systemic and developmental regulation of β-glucuronidase. As described earlier (Section I,A), the general reduction in β-glucuronidase activity seen in *H* is the result of two regulatory changes. One is a *cis*-acting regulatory mutation at the *Gus-u* locus that acts systemically. The other is a *trans*-acting regulatory mutation in the *Gus-t* locus that determines developmental switches in the rate of enzyme synthesis. These switches occur in selected tissues at times specific to each tissue. Despite its *trans* action, *Gus-t* is also linked genetically to [*Gus*]. mRNA measurements at different developmental stages in [*Gus*]^H and [*Gus*]^B mice show that both *cis*-acting and *trans*-acting regulatory changes in [*Gus*]^H result from changes in translational yield (*152c*). There are no measurable changes in β-glucuronidase mRNA levels in [*Gus*]^H mice despite the fact that these mice have one-tenth or less the normal rate of enzyme synthesis in some tissues.

All of the *Gus-u* and *Gus-t* mutations so far tested have proved to be changes in translational yield. The fact that both *Gus-u* and *Gus-t* mutations involve changes in translational yield suggests that they may act through a common mechanism, despite the fact that one locus acts *cis* and the other *trans*.

C. Mechanism of Translational Yield

The molecular basis of translational yield has been tested by comparing β-glucuronidase synthesis in mice of the B6.H congenic strain carrying the *H* haplotype with that in B6 mice carrying the *B* haplotype (*152b*). These experiments were carried out using androgen-induced kidney where the translational yield of β-glucuronidase in B6.H mice is one-half that of B6 animals. The lower yield of enzyme molecules in *H* was not accounted for by the presence of a significant pool of untranslated β-glucuronidase mRNA, or by a failure to translate β-glucuronidase mRNA on membranes; virtually all of the β-glucuronidase mRNA was present in membrane-bound polysomes in both B6.H and B6 animals. There was also no evidence for a reduced rate of ribosome initiation of translation, which would be expected to decrease the average size of β-glucuronidase polysomes, or a reduced rate of termination, which would be expected to increase the average size of β-glucuronidase polysomes, since the size distribution of β-glucuronidase polysomes was identical in B6.H and B6 mice.

Reduced translational yield also does not result from a diminished capacity to bind signal recognition particle to the β-glucuronidase signal peptide. β-Glucuronidase is translated on membrane-bound polysomes, and the recently initiated polypeptide chain must bind a signal recognition particle[5] to enter the lumen of the endoplasmic reticulum and be processed properly. Failure to bind a signal recognition particle would lead to extra-lumenal synthesis and release of the polypeptide chain to the cytosol, where it could not be appropriately glycosylated and processed and would presumably be rapidly degraded. However, sequencing of the signal peptide in the *A*, *B*, CS, and *H* haplotypes showed that they are identical, eliminating this mechanism. The most likely remaining explanation is that only some nascent polypeptide chains are successfully matured to active β-glucuronidase, and that the fraction of newly synthesized chains that do become enzyme molecules can change during development and can be hormonally affected in the submaxillary gland.

At first glance, translational yield seems an unlikely biological mechanism in view of the energy waste involved in destroying newly synthesized molecules of β-glucuronidase. However, this may not be as inefficient as it seems (*150*). β-Glucuronidase is present in cells in relatively small amounts, typically only thousands of molecules per cell, and only a few molecules of β-glucuronidase mRNA are present per cell. It may be even more costly to establish and maintain a separate and specific regulatory system to control tissue-specific expression of this gene than it is to throw some of the molecules away.

VII. Deficiency Mutants

Human mutations resulting in β-glucuronidase deficiency were first reported by Sly *et al.* (*7*). Homozygous-affected individuals typically have a severe mucopolysaccharidosis (MPS VII) resulting from defective glycosaminoglycan degradation, but there is considerable heterogeneity in expression (*14, 153*). Typical manifestations of the disease include lysosome accumulation, the presence of metachromatic granules in leukocytes, urinary excretion of glycosaminoglycans, hepatosplenomegaly, and poor immunologic response. [For further details of the clinical aspects of this disease, see recent

[5] A paper on "Structure and Function of Signal Recognition Particle RNA" by C. Zwieb appears in this volume [Eds.].

reviews (*14, 153*)]. A similar mutation has been reported in dogs (*154*) and another has recently been discovered in the mouse (*6*). This latter mutant supplements other animal models of lysosomal storage diseases by providing an opportunity to analyze the pathophysiology and possible means of gene therapy of a typical MPS deficiency disease in a convenient and well-defined genetic system.

VIII. Additional Genes Modulating β-Glucuronidase Expression

A. Pigmentation Mutants

There is a group of mouse pigmentation mutants that exhibits a common syndrome of abnormalities involving defective processing of intracellular organelles. The affected organelles include melanosomes in melanogenesis, resulting in hypopigmentation; exocytosis of lysosomes, resulting in decreased secretion of induced β-glucuronidase; and abnormal dense granule function in platelets, leading to prolonged bleeding times. Initially, Brandt *et al.* (*155, 155a*) tested one of these mutants, the *beige* (*bg*) mouse, for androgen induction and found that it accumulates excess amounts of β-glucuronidase in kidney. They were prompted to this experiment by the fact that *beige* is the homolog of the human mutation causing Chediak–Higashi syndrome, which is characterized by the presence of abnormally large lysosomes in addition to hypopigmentation, immunodeficiency, and prolonged bleeding times. Subsequently, Meisler (*156*) observed a similar effect in the *light ears* (*le*) mouse mutation. Following up these observations, Novak and Swank (*157*) carried out a systematic screen of 34 mouse pigmentation mutants and discovered that a number of additional mutants (*pale ears*, *ruby-eye*, *maroon*, *pallid*, and *pearl*) show increased levels of kidney β-glucuronidase following induction.

Further studies showed that all but one of the mutants accumulate excess β-glucuronidase because secretion is defective. Rates of induced enzyme synthesis were normal, but very little of the induced enzyme was secreted into urine, resulting in excess accumulation in the kidney. The exceptional mutant, *pearl*, instead of being secretion-defective, is actually induced to higher levels of enzyme synthesis; this is discussed in the following Section (VIII,B). Other lysosomal enzymes are affected as well. All of the mutants also exhibit prolonged bleeding times resulting from the presence of abnormal dense granules in platelets.

The mechanism of the secretion-defect of the *pale ear* (*ep*) mutant in macrophages has been characterized (*158*). Based on the effects of ammonia on secretion of precursor and mature enzyme forms in normal and mutant mice, it was concluded that two secretion pathways normally exist in these cells. One leads from endoplasmic reticulum through Golgi directly to the plasma membrane. The other leads from Golgi to lysosomes with enzyme release in via exocytosis of lysosomal contents. It is the latter that is apparently defective in the *pale ear* mutant.

Further analysis of these mutants has focused on the syndrome of effects, and the results have been reviewed (*159*).

B. *pearl*, a Class Regulator of Several Lysosomal Enzymes

Among the pigmenation mutations affecting β-glucuronidase induction, *pearl* is exceptional in showing an elevated level of β-glucuronidase synthesis following hormone treatment (*157*). β-Glucuronidase mRNA levels are increased in *pearl* mice to the same extent that enzyme synthesis is, showing that translational yield is not altered in the *pearl* mutant (*41b*). Mice carrying the *pearl* mutation also have increased basal enzyme levels in the absence of androgen stimulation, indicating that *pearl* action is not limited to induction but involves some more general aspect of gene regulation (*157*). Tests of other lysosomal glycosidases have shown that all are equally affected by the *pearl* mutation in mouse kidney, although the others are not androgen-responsive (*41b, 157*). The action of *pearl* shows considerable tissue specificity and is not manifest in all tissues.

In genetic crosses, *pearl* exhibits additive inheritance, and levels of β-glucuronidase mRNA in pe/pe^{+} heterozygotes are intermediate between the two homogygous parents. The additive inheritance of β-glucuronidase mRNA levels parallels observations of other *trans*-acting regulatory genes. This is in contrast to *trans*-acting factors that act posttranslationally, where heterozygotes typically have the phenotype of one or the other parent (*12*). The *pearl* gene is located on chromosome 13 and is not close to [*Gus*] or any other known lysosomal enzyme genes.

The properties of the *pearl* gene suggest that it may be a novel type of regulatory gene, regulating mRNA levels in a tissue-specific manner for a class of genes that are related by the intracellular location of their gene products—the lysosomal hydrolases. Whether *pearl* does this by modulating the transcription of these genes or by affecting the stability of their mRNAs is not known.

In considering mechanisms of *pearl* action, it is interesting that the *pearl* mutation is probably the result of a transposon insertion, since the mutant allele is genetically unstable in some genetic backgrounds and readily reverts to wild-type (*160*). Thus, it is possible that *pearl* is not a simple inactivation mutant and may represent an overproduction of the normal gene product.

IX. Conclusions

Beyond the specifics of β-glucuronidase, the properties and functions of the gene, and its enzyme product, there are several general conclusions that can be derived from the study of this system.

One is the utility of investigating problems of gene expression and gene regulation in systems with varied biological properties. Much of what is generally known about the mechanisms involved in regulating gene expression is derived from the study of a relatively small number of intensively investigated model systems. This is appropriate, but most of these systems involve special function genes that are expressed at very high levels in one or only a few tissues, and there is relatively much less known about housekeeping genes (*161*). It is probable that somewhat different mechanistic solutions have evolved to cope with the distinct biological problems involved in controlling the expression of genes whose products are found in many cell types, but in only very small amounts. There is a real need to extend the repertoire of model systems to encompass the varieties of gene expression that are observed in nature. In the case of β-glucuronidase, a typical housekeeping gene, the study of its expression has introduced us to two previously unsuspected mechanisms of gene regulation, progressive induction and translational yield, as well as to a novel aspect of enzyme localization involving complex formation between β-glucuronidase and a second protein, egasyn.

The β-glucuronidase system also demonstrates the feasibility and utility of bringing together the tools and concepts of genetics, cell biology, and molecular biology in a mammalian system. Each of these approaches has reinforced the others, and our understanding of β-glucuronidase would be much diminished if this combined approach had not been possible. Such combined approaches have been traditional in lower organisms; it is now apparent that advances in mammalian genetics have reached the stage where this is also quite practicable in an experimental mammal.

Finally, β-glucuronidase provides an exceptional example of a gene as an operating system, with a multiplicity of functions and components that interact with each other in the same sense that a cell or an organism is an operating system. Even a cell has too many components for us to grasp the system in its entirety; we must describe the overall properties in terms of general principles, and are limited to examining only specific aspects in detail. However, a gene as an operating system is on a small enough scale that its details are conceptually within our grasp. It is also an exceedingly attractive system to understand, serving as it does as the ultimate determinant of the structure and function of the organism, in addition to being the product of an evolutionary history.

Viewed as a system, enough is already known of β-glucuronidase expression to realize that it expresses one unexpected and surprising feature. Every genetic regulatory mutation that has been analyzed so far is altered in the same step in the transcription/translation pathway that is normally modulated by the organism (*81*). All variation in kidney androgen inducibility is transcriptional, whereas variation in submaxillary androgen inducibility involves translational yield, as does variation in systemic and developmental expression. This pattern is the more marked when contrasted with β-globin, a mammalian gene in which the majority of mutations affecting levels of gene expression involve RNA splicing and function (*162*). In principle, regulatory mutations can affect any step in the transcription/translation pathway, and mutations affecting nearly every step in the pathway have been described for one or another gene. The restricted patterns of polymorphism for β-glucuronidase and β-globin suggest that evolutionary selection can act to choose between alternative mechanistic changes for achieving the same alteration in final levels of gene product, as well as acting on the final level of product itself.

A question for the future is whether there will be equally interesting and surprising aspects to the manner in which the DNA sequences responsible for various aspects of β-glucuronidase regulation are organized relative to each other and how they interact in chromatin.

Acknowledgments

We are all indebted to the many workers who have contributed to our current understanding of β-glucuronidase, many of whom have generously shared their thoughts over the years and provided advanced texts of current publications. It is impossible to mention them all, but there are several whose thoughts have been particularly valuable, and I would like to express my personal appreciation to Verne Chapman, Roger Ganschow, A. J. Lusis, Kurt Pfister, Richard Swank, Shiro Tomino, and especially Gordon Watson. It would not have been the same without them. This work was supported by NIH Grant GM 31656.

References

1. I. H. Masamune, *Jpn. J. Biochem.* **19**, 353 (1934).
2. L. W. Law, A. G. Morrow and E. M. Greenspan, *JNCI* **12**, 909 (1952).
3. W. H. Fishman, *Ann. N.Y. Acad. Sci.* **54**, 548 (1951).
4. Y. Nishimura, M. G. Rosenfeld, G. Kreibich, U. Gibler, D. D. Sabatini, M. Adesnik and R. Andy, *PNAS* **83**, 7292 (1986).
5. A. J. Oshima, W. Kyle, R. D. Miller, J. W. Hoffmann, P. P. Powell, J. H. Grubb, W. S. Sly, M. Tropak, K. S. Guise and R. A. Gravel, *PNAS* **84**, 685 (1987).
5a. R. A. Jefferson, S. M. Burgess and D. Hirsh, *PNAS* **83**, 8447 (1986).
6. E. Birkenmeier. *Cell* in press (1988).
7. W. S. Sly, B. A. Quinton, W. H. McAlister and D. L. Rimoin, *J. Pediatr.* **82**, 249 (1973).
8. M. Wakabayashi, *in* "Metabolic Conjugation and Metabolic Hydrolisis" (W. H. Fishman, ed.), p. 519. Academic Press, New York, 1970.
9. R. T. Swank, K. Paigen, R. Davey, V. Chapman, C. Labarca, G. Watson, R. Ganschow, E. J. Brandt and E. K. Novak, *Recent Prog. Horm. Res.* **34**, 401 (1978).
10. A. Lusis and K. Paigen, *Curr. Top. Biol. Med. Res.* **2**, 63 (1977).
11. M. D. Skudlarek, E. K. Novak and R. T. Swank, *in* "Lysosomes in Biology and Pathology (J. T. Dingle, R. T. Dean and W. S. Sly, eds.), Vol. 7, p. 17. Elsevier, Amsterdam, 1984.
12. K. Paigen, *ARGen* **13**, 417 (1979).
13. A. J. Lusis, *in* "Glycosaminoglycans and Proteoglycans" (R. S. Varina and R. Varma, eds.), p. 55. Karger, Basel, Switzerland, 1982.
14. J. E. Sheets Lee, R. E. Falk, W. G. Ng and G. N. Donnell, *Am. J. Dis. Child.* **139**, 57 (1985).
15. A. G. Morrow, E. M. Greenspan and D. M. Carroll, *JNCI* **10**, 657 (1949).
16. K. Paigen and W. K. Noell, *Nature* **190**, 148 (1961).
17. R. L. Sidman and M. C. Green, *J. Hered.* **56**, 23 (1965).
18. R. T. Swank, K. Paigen and R. Ganschow, *JMB* **81**, 225 (1973).
19. P. A. Lalley and T. B. Shows, *Science* **185**, 442 (1974).
20. K. Paigen, *Exp. Cell Res.* **25**, 286 (1961).
21. K. Paigen, *PNAS* **47**, 1641 (1961).
22. C. J. Chern and C. M. Croce, *Am. J. Hum. Genet.* **28**, 350 (1976).
23. U. Francke, *Am. J. Hum. Genet.* **28**, 357 (1976).
24. P. A. Lalley, J. A. Brown, R. L. Eddy, L. L. Haley and T. B. Shows, *Cytogenet. Cell Genet.* **16**, 184 (1976).
25. K.-H. Grzeschik, *Cytogenet. Cell. Genet.* **16**, 142 (1976).
26. K.-H. Grzeschik, *Somatic Cell Genet.* **2**, 401 (1976).
27. "Human Gene Mapping 9 (1987): Ninth International Workshop on Human Gene Mapping," *Cytogenet. Cell Genet.* **46**, 1 (1987).
28. J. Nadeau and B. Taylor, *PNAS* **81**, 814 (1984).
29. M. Lyon, *Biochem. Genet.* **9**, 369 (1973).
30. H. J. Condamine, R. P. Custer and B. Mintz, *PNAS* **68**, 2032 (1971).
31. R. J. Mullen and K. Herrup, *in* "Neurogenetics: Genetic Approaches to the Nervous System" (X. O. Breakfield, ed.), p. 173. Elsevier/North-Holland, Amsterdam, 1979.
32. R. Wetts and K. Herrup, *J. Neurosci.* **2**, 1494 (1982).
32a. K. Herrup and K. Sunter, *Dev. Biol.* **117**, 417 (1986).
32b. K. Herrup and K. Sunter, *J. Neurosci.* **7**, 829 (1987).

33. K. Pfister, G. Watson, V. Chapman and K. Paigen, *JBC* **259,** 5816 (1984).
34. L. T. Bracey and K. Paigen, *Biochem. Genet.* in press.
35. V. M. Chapman, D. R. Miller, E. Novak and R. W. Elliott, *Curr. Top. Microbiol. Immun.* **127,** 114 (1986).
35a. K. Paigen and A. F. Jakubowski, *Dev. Genet.* **5,** 83 (1985).
36. K. Herrup and R. J. Mullen, *Biochem. Genet.* **15,** 641 (1977).
37. S. A. Meredith and R. E. Ganschow, *Genetics* **90,** 725 (1978).
38. A. J. Lusis, V. M. Chapman, R. W. Wangenstein and K. Paigen, *PNAS* **80,** 4398 (1983).
39. K. Pfister, V. Chapman, G. Watson and K. Paigen, *JBC* **260,** 11588 (1985).
40. V. M. Chapman, K. Paigen, L. Siracusa and J. Womack, *in* "FASEB Biological Handbook III. Inbred and genetically defined strains of laboratory animals. Part I: Mouse and Rat" (P. Altman and D. Katz, eds.), p. 77. FASEB, Bethesda, Maryland, 1979.
41. M. C. Green, "Genetic Variants and Strains of the Laboratory Mouse." Fischer, Stuttgart, FRG, 1981.
41a. K. Pfister, K. Paigen, G. Watson and V. Chapman, *Biochem. Genet.* **20,** 519 (1982).
41b. K. Paigen and D. Tabron, unpublished.
42. J. F. Catterall and S. L. Leary, *Bchem* **22,** 6049 (1983).
43. R. Palmer, P. M. Gallagher, W. L. Boyko and R. E. Ganschow, *PNAS* **80,** 7596 (1983).
44. G. Watson, M. Felder, L. Rabinow, K. Moore, C. Labarca, C. Tietze, G. Vander Molen, L. Bracey, M. Brabant, J. Cai and K. Paigen, *Gene* **36,** 15 (1985).
45. B. Funkenstein, S. L. Leary, J. C. Stein and J. F. Catterall, *MCBiol* **8,** 1160 (1988).
46. P. M. Gallagher, M. A. D'Amore, S. D. Lund and R. E. Ganschow, *Genomics* **2,** 215 (1988).
47. K. J. Moore and K. Paigen, *Genomics* **2,** 25 (1988).
48. P. M. Gallagher, M. A. D'Amore, S. D. Lund, R. W. Elliott, J. Pazik, C. Hohman, T. R. Korfhagen and R. E. Ganschow, *Genomics* **1,** 145 (1987).
49. M. A. D'Amore, P. M. Gallagher, T. R. Korfhagen and R. E. Ganschow, *Bchem* **27,** 7131 (1988).
50. P. H. Leighton, W. K. Fisher, K. E. Moon and E. Thompson, *Aust. J. Biol. Sci.* **33,** 513 (1980).
51. V. C. Hieber, *BBRC* **104,** 1271 (1982).
52. K. S. Guise, R. G. Korneluk, J. Wayne, A.-M. Lamhonwah, F. Quan, R. Palmer, R. E. Ganschow, W. S. Sly and R. A. Gravel, *Gene* **34,** 105 (1985).
53. P. P. Powell, J. W. Kyle, R. D. Miller, J. Pantanno, J. H. Grubb and W. S. Sly, *BJ* **250,** 547 (1988).
54. J. W. Hoffman, P. P. Powell, R. D. Miller, A. Oshima, J. W. Kyle and W. S. Sly, *Am. J. Hum. Genet.* **41,** A219 (1987).
55. R. D. Miller, P. P. Powell, A. Oshima, J. W. Kyle and W. S. Sly, *Genetics* **116,** S21 (1987).
55a. T. Maruyama, T. Gojobori, S. Aota, and T. Ikemura, *NARes* **14** (Suppl.), 4151 (1986).
56. B. V. Plapp and R. D. Cole, *ABB* **116,** 193 (1966).
57. K. Ohtsuka and M. Wakabayashi, *Enzymologia* **39,** 109 (1970).
58. P. D. Stahl and O. Touster, *JBC* **246,** 5398 (1971).
59. D. R. P. Tulsiani, R. K. Keller and O. Touster, *JBC* **250,** 4770 (1975).
60. R. K. Keller and O. Touster, *JBC* **250,** 4765 (1975).

61. M. Himeno, H. Ohkara, Y. Arakawa and K. Kato, *Jpn. J. Biochem.* **77**, 427 (1975).
62. M. Himeno, Y. Nishimura, H. Tsuji and K. Kato, *Jpn. J. Biochem.* **70**, 349 (1976).
63. R. T. Dean, *BJ* **138**, 395 (1974).
64. R. T. Dean, *BJ* **138**, 407 (1974).
65. S. Tomino, K. Paigen, D. R. P. Tulsiani and O. Touster, *JBC* **250**, 8503 (1975).
66. C.-W. Lin, M. L. Orcutt and W. H. Fishman, *JBC* **250**, 4737 (1975).
67. N. C. Mills, C. Gupta and C. W. Bardin, *ABB* **185**, 100 (1978).
68. A. J. Lusis and K. Paigen, *JBC* **253**, 7336 (1978).
69. A. J. Lusis, S. Tomino and K. Paigen, *JBC* **251**, 7753 (1976).
70. D. Owerbach and A. J. Lusis, *BBRC* **69**, 628 (1976).
71. C. Labarca and K. Paigen, *Anal. Biochem.* **102**, 344 (1980).
72. L. Wudl and K. Paigen, *Science* **184**, 992 (1974).
73. M. Hayashi, Y. Nakijima and W. H. Fishman, *J. Histochem. Cytochem.* **12**, 293 (1964).
74. T. Niwa, T. Tsuruoka, S. Inouye, Y. Naito, T. Koeda and T. Niida, *Jpn. J. Biochem.* **72**, 207 (1972).
75. W. Bardin and J. F. Catterall, *Science* **211**, 1285 (1981).
76. J. F. Catterall, K. K. Kontula, C. S. Watson, P. J. Seppanen, B. Funkenstein, E. Melanitou, N. J. Hickok, C. W. Bardin and O. A. Janne, *Recent Prog. Horm. Res.* **42**, 71 (1986).
77. K. Paigen, C. Labarca and G. Watson, *Science* **203**, 554 (1979).
78. C. S. Watson and J. F. Catterall, *Endocrinology* **118**, 1081 (1986).
79. G. Watson and K. Paigen, *MCBiol* **7**, 1085 (1987).
80. G. Watson and K. Paigen, *MCBiol* **8**, 2117 (1988).
81. K. Paigen, *Am. Nat.* in press (1989).
82. R. Dofuku, V. Tettenborn and S. Ohno, *Nature NB* **232**, 5 (1971).
83. R. Dofuku, V. Tettenborn and S. Ohno, *Nature NB* **234**, 259 (1971).
84. L. P. Bullock and C. W. Bardin, *Ann. N.Y. Acad. Sci.* **286**, 321 (1977).
85. V. Gehring and G. M. Tomkins, *Cell* **3**, 59 (1974).
86. B. Attardi and S. Ohno, *Cell* **2**, 205 (1974).
87. W. Fishman, *Methods Horm. Res* **4**, 273 (1965).
87a. I. Mowszowicz, D. E. Bieser, K. W. Chung, L. P. Bullock and C. W. Bardin, *Endocrinology* **95**, 1589 (1974).
87b. L. P. Bullock, C. W. Bardin and M. R. Sherman, *Endocrinology* **103**, 1768 (1978).
88. L. Bullock, G. Watson and K. Paigen, *Mol. Cell. Endocrinol.* **41**, 179 (1985).
89. R. T. Swank, R. Davey, L. Joyce, P. Reid and M. R. Macey, *Endocrinology* **100**, 473 (1977).
90. G. Norstedt and R. Palmiter, *Cell* **36**, 805 (1984).
90a. G. Watson and K. Paigen, unpublished.
91. C. E. Cleveland and R. T. Swank, *BJ* **170**, 249 (1978).
91a. D. J. Shapiro and M. L. Brock, *in* "Biochemical Action of Hormones" (G. Litwack, ed.), Vol. 12, p. 139. Academic Press, Orlando, Florida, 1985.
92. Y. Isomaa, A. E. I. Pajunen, C. W. Bardin and O. A. Janne, *Endocrinology* **111**, 833 (1982).
93. H. Almagor and K. Paigen, *Bchem* **27**, 2094 (1988).
94. G. Watson, R. A. Davey, C. Labarca and K. Paigen, *JBC* **256**, 3005 (1981).
95. A. J. Lusis, V. M. Chapman, C. Herbstman and K. Paigen, *JBC* **255**, 8959 (1980).
96. F. G. Berger, D. Loose, H. Meisner and G. Watson, *Bchem* **25**, 1170 (1986).
97. K. Paigen and J. Peterson, *Dev. Genet.* **2**, 269 (1981).

98. K. Paigen and A. Jakubowski, *Biochem. Genet.* **20,** 875 (1982).
100. P. F. Searle, G. W. Stuart and R. D. Palmiter, *MCBiol* **5,** 1480 (1985).
100a. M. G. Toohey, K. L. Morley and D. O. Peterson, *MCBiol* **7,** 1085 (1986).
101. L. T. Bracey and K. Paigen, *Mol. Endocrinol.* **2,** 701 (1988).
101a. S. D. Lund, D. Miller, V. Chapman and R. E. Ganschow, *Genetics* **119,** 151 (1988).
101b. K. Paigen and J. Peterson, *J. Clin. Invest.* **61,** 751 (1978).
101c. J. Tseng-Crank and F. G. Berger, *Genetics* **116,** 593 (1987).
102. J. J. Corrales, I. Pastor, L. C. Garcia, J. M. Gonzales, J. Montero, S. DeCastro and J. M. Miralles, *Urol. Int.* **40,** 307 (1985).
105. J. G. Vandenburgh (ed.), "Pheromones and Reproduction in Mammals." Academic Press, New York, 1983.
106. D. W. Ingersoll, G. Bobotas, C.-T. Lee and A. Lukton, *Physiol. Behav.* **29,** 789 (1982).
107. M. Novotny, S. Harvey, B. Jemiolo and J. Alberts, *PNAS* **82,** 2059 (1985).
109. R. Ganschow and K. Paigen, *PNAS* **58,** 938 (1967).
110. S. Tomino and K. Paigen, *JBC* **250,** 1146 (1975).
111. R. T. Swank and K. Paigen, *JMB* **77,** 371 (1973).
112. R. E. Ganschow and B. G. Bunker, *Biochem. Genet.* **4,** 127 (1970).
113. J. Brown, E. K. Novak, K. Takeuchi, K. Moore, S. Medda and R. T. Swank, *J. Cell Biol.* **105,** 1571 (1987).
114. P. Beltramini-Guarini, R. Gitzelmann and K. Pfister, *Eur. J. Cell Biol.* **34,** 165 (1984).
115. R. T. Swank, K. Pfister, D. Miller and V. Chapman, *BJ* **240,** 445 (1986).
117. A. J. Lusis, S. Tomino and K. Paigen, *Biochem. Genet.* **15,** 115 (1977).
118. T. R. Karl and V. M. Chapman, *Biochem. Genet.* **11,** 367 (1974).
119. S. Medda, K. Takeuchi, D. Devore-Carter, O. von Deimling, E. Heymann and R. T. Swank, *JBC* **262,** 7248 (1987).
120. A. J. Lusis and K. Paigen, *J. Cell Biol.* **73,** 728 (1977).
121. K. Paigen and R. Ganschow, *Brookhaven Symp. Biol.* **18,** 99 (1965).
121a. R. T. Swank, K. Moore and V. M. Chapman, *Biochem. Genet.* **25,** 161 (1987).
122. J. Peters, *Biochem Genet.* **20,** 585 (1982).
123. E. Heymann, *Enzym. Basis Detox.* **2,** 291 (1980).
124. M. Robbi and H. Beaufay, *EJB* **137,** 293 (1983).
125. S. Medda and R. T. Swank, *JBC* **260,** 14802 (1985).
126. S. Medda, O. von Deimling and R. T. Swank, *Biochem. Genet.* **24,** 229 (1986).
127. E. Eisenhardt and O. von Deimling, *Comp. Biochem. Physiol. B* **73,** 719 (1982).
128. B. Wassmer, S. M. deLooze and O. von Deimling, *Biochem. Genet.* **23,** 759 (1985).
132. C. B. C. DeDuve, R. Pressman, P. Gianetto, R. Wattiaux and F. Appelmans, *BJ* **60,** 604 (1955).
133. J. W. Owens, K. L. Grammen and P. D. Stahl, *ABB* **166,** 158 (1975).
134. A. T. Varute, *Acta Histochem.* **38,** 227 (1970).
135. J. W. Owens and P. Stahl, *BBA* **438,** 474 (1976).
136. L. D. Strawser and O. Touster, *JBC* **254,** 3716 (1979).
136a. A. H. Erickson and G. Blobel, *Bchem* **22,** 5201 (1983).
137. J. A. Brown, G. P. Jahreis and R. T. Swank, *BBRC* **99,** 691 (1981).
138. M. D. Skudlarek and R. T. Swank, *JBC* **256,** 10137 (1981).
139. J. A. Brown and R. T. Swank, *JBC* **258,** 15323 (1983).
140. M. G. Rosenfeld, G. Kreibich, D. Popov, K. Kato and D. D. Sabatini, *J. Cell Biol.* **93,** 135 (1982).

142. K. Pfister, N. Bosshard, M. Zopfi and R. Gitzelmann, *BJ* **255**, 825 (1988).
142a. K. Smith and R. E. Ganschow, *JBC* **253**, 5437 (1978).
143. S. Medda, A. M. Stevens and R. T. Swank, *Cell* **50**, 301 (1987).
144. M. Whittaker, P. K. Sokolove, R. G. Thurman and F. C. Kauffman, *Cancer Lett.* **26**, 145 (1985).
145. I. Schellhammer, D. S. Pell and M. H. Bickel, *Enzymes* **20**, 269 (1975).
146. S. A. Belinsky, F. C. Kauffman, P. M. Sokolove, T. Tsukuda and R. G. Thurman, *JBC* **259**, 7705 (1984).
147. P. M. Sokolove, M. A. Wilcox, R. G. Thurman and F. C. Kauffman, *BBRC* **121**, 897 (1984).
148. R. Mentlein, M. Suttorp and E. Heymann, *ABB* **228**, 230 (1984).
149. R. Mentlein, G. Reuter and E. Heymann, *ABB* **240**, 801 (1985).
150. L. T. Bracey and K. Paigen, *PNAS* **84**, 9020 (1987).
151. M. H. Meisler and K. Paigen, *Science* **177**, 894 (1972).
152. W. L. Daniel, *Genetics* **82**, 477 (1976).
152a. A. J. Lusis and K. Paigen, *Cell* **6**, 371 (1975).
152b. K. Denich, G. Watson, P. Scheller, K. Pfister, D. Schott, C. Voliva, N. Bosshard and K. Paigen, unpublished.
153. D. Irani, H. S. Kim, H. Ep-Hibri, R. V. Dutton, A. Beaudet and D. Armstrong, *Ann. Neurol.* **14**, 486 (1983).
154. M. E. Haskins, R. J. Desnick, N. DiFerrante, P. F. Jezyk and D. F. Patterson, *Pediatr. Res.* **18**, 980 (1984).
155. E. J. Brandt and R. T. Swank, *Am. J. Pathol.* **82**, 573 (1976).
155a. E. J. Brandt, R. W. Elliott and R. T. Swank, *J. Cell Biol.* **67**, 774 (1975).
156. M. H. Meisler, *JBC* **253**, 3129 (1978).
157. E. K. Novak and R. T. Swank, *Genetics* **92**, 189 (1979).
158. J. A. Brown, E. K. Novak and R. T. Swank, *J. Cell Biol.* **100**, 1894 (1985).
159. M. Reddington, E. K. Novak, E. Hurley, C. Medda, M. P. McGarry and R. T. Swank, *Blood* **69**, 1300 (1987).
160. L. B. Russell, *in* "The Role of Chromosomes in Development" (M. Locke, ed.), p. 153. Academic Press, New York, 1964.
161. J.-H. Park, H. V. Hershey and M. W. Taylor, *in* "Molecular Genetics of Mammalian Cells" (G. M. Malacinski, ed.) p. 79. Macmillan, New York, 1986.
162. H. H. Kazazian, Jr., S. E. Antonarakis, H. Youssoufian, C. E. Dowling, D. G. Phillips, C. Wong and C. D. Boehm, *CSHSQB* **51**, 371 (1986).

Structure and Function of Signal Recognition Particle RNA

CHRISTIAN ZWIEB

Laboratory of Molecular Biology
National Cancer Institute
National Institutes of Health
Bethesda, Maryland 20892

This article concentrates on the structure and function of the RNA component of the signal recognition particle (SRP). It starts with a brief description of the function of SRP in the process of translocation of secretory proteins across the membrane of the endoplasmic reticulum. Then, all known SRP-RNA sequences are compared to form a solid basis for a common secondary structure. This is followed by a discussion on intramolecular interactions that might be responsible for the folding of the SRP-RNA into a three-dimensional molecule and on the dynamic property of the RNA. Finally, an attempt is made to bring the two partners of the SRP into the picture: the translating ribosome, and the membrane of the endoplasmic reticulum.

SRP-RNA was previously named 7-SL RNA because of its sedimentation value, and to distinguish it from the unrelated 7-SK RNA. SRP-RNA was first detected in avian and murine ocornavirus particles (*1, 2*). It is also a stable component in the cytoplasm of cells not infected by virus, and is associated with membrane fractions of HeLa cells (*3*). There are two reports of the presence of SRP-RNA in polysomes (*2, 4*).

Progress in Nucleic Acid Research
and Molecular Biology, Vol. 37

The process of the functional involvement of SRP-RNA remained unknown for some time. Prior to the discovery of SRP-RNA in an 11-S RNP (the SRP), it had been shown that this particle promotes translocation of secretory proteins across the membrane of the endoplasmic reticulum (*12–14*). Only when it was shown that SRP-RNA is an integral component of a purified preparation of SRP (*15*) did the search for a function of SRP-RNA come to an end.

Portions of the SRP-RNA are similar to sequences of the dominant *Alu* family of middle repetitive DNA sequences of the human genome (*5, 6*). A more detailed analysis shows that, of the 300 bases of the SRP-RNA, only 100 at the 5′ and 40 at the 3′ end are related to those from *Alu*, while the remaining central portion contains unique SRP-RNA sequences (*7, 8*). The arrangement of one block of SRP-specific sequences being inserted into *Alu*-DNA can be explained by the structure of the SRP-RNA itself: three RNA fragments are obtained by mild digestion of the SRP-RNA with micrococcal nuclease at the boundaries of the *Alu* sequences (*9*). It is very likely that *Alu*-DNA originated from SRP-RNA by an excision of the central SRP-RNA-specific fragment, followed by reverse transcription and integration into new chromosomal sites (*10*).

Since this article discusses the structure of the SRP-RNA itself, the interplay between RNA structure and genomic organization cannot be covered in detail. Some aspects of the transcription of SRP-RNA by polymerase III (*11*) and its assembly into an RNP are also not part of this presentation.

I. Function of the Signal Recognition Particle

The SRP combines with ribosomes that are involved in the translocation of proteins across the membrane of the endoplasmic reticulum (ER). The association is temporary; once a functional translocation complex has been formed, SRP leaves the ribosome. Therefore, it cannot be considered as a third ribosomal subunit. A protein that is an integral part of the membrane, called "SRP receptor" or "docking protein," is essential for the attachment of the SRP to the membrane. The SRP recognizes the membrane receptor as well as a ribosome engaged in the synthesis of a secretory protein *via* the leader- or signal sequence of the protein. The order in which these interactions actually occur *in vivo* is not clear. Purified SRP affects the synthesis of a secretory protein *in vitro* by causing a transient inhibition of translation, which is released by binding of the SRP–ribosome complex to the ER membrane (*14, 16–23*).

7-SL RNAs have been isolated from a wide variety of sources, emphasizing their important and essential role in the process of protein targeting. The term SRP-RNA is used here, because RNAs of slightly different sizes have been found in organisms as varied as mammals (*7, 8, 24, 25*), *Drosophila* (*10, 26*), *Xenopus* (*10*), plants (*27–29a*), fungi (*30–32*), and halobacteria (*34, 35*).

So far, no SRP-RNA has been clearly identified in *Saccharomyces cereviseae* nor in eubacteria. For a while, 6-S RNA was suspected to be the eubacterial equivalent of SRP-RNA (*36–38*), but it has been shown that 6-S RNA is not essential for cell growth (*39*). An interesting homology between *Escherichia coli* 4.5-S RNA and the conserved central domain of SRP-RNA, as well as the centrally located 100 nucleotides of *Bacillus subtilis* small cytoplasmic (sc) RNA, has been recognized (*39a,c,d*). It is possible that eubacterial 4.5-S RNA carries out functions similar or identical to the ones identified in the SRP-RNA of eukaryotes.

From comparisons of the sizes [calculated from the molecular weights assuming globular protein structures (*38*)] of the six proteins of the SRP (*9, 14, 19, 54, 64, and 72kDa*), with the size of the SRP-RNA, it is suggested that the RNA determines the structural organization of the SRP to a large extent (*38*). Shape and dimensions of the RNA alone, derived from physical and computer modeling studies assuming A-form RNA, agree well with the shape and the dimensions of SRP and the SRP-RNA as seen in the electron microscope (*40–42*). The SRP appears as an elongated rod with terminal small and large globular domains. This bonelike structure is consistent with the finding that at least two functions can be attributed to SRP: the effect on translation (which resides in the small domain containing the *Alu* parts of the RNA, and the 9-kDa/14-kDa protein dimer) and the effect on protein translocation itself (which resides in the large domain containing centrally located sequences of the RNA and four of the remaining proteins of the SRP) (*9, 12, 43*). Five SRP proteins bind to SRP-RNA, the exception being the 54-kDa protein (*36, 44, 45*).

One obvious function of the RNA is to ensure the proper spatial arrangement of its proteins. Secondly, SRP-RNA might be involved directly in some of the steps necessary for the regulation of translation and targeting of secretory proteins. Both aspects will be discussed in the following in the light of the structure of the SRP-RNA. Many different, but clearly related, SRP-RNA primary sequences have become available recently (*29–33*). A comparison with this database can be used in the future to determine the membership of a given RNA in the club of SRP-RNAs, and is essential for the identification of the characteristic features of SRP-RNA and the particle.

```
        1          11         21         31         41         51         61
zmA     NCCGAGCUCU GUAGCGAGAG CUUGUAACCC GAGCGGGGGC AUUAAGGU-G GUGCGGAUUC UUUGCGAUGG
zmB                      .... .......... .......... ........-. .......... ..........
zmC                            ......... .......... ........-. .......... ..........
zmD                                    . .......... ........-. .C......G. ..........
zmE                                                               U.A..G. ..........
zmF                      .... .........U A......... ........-. ...U....GU ..CCU..C..
zmG                                                              .U....G. ..G.UACC.-
zmH                        U. .......... .......... .A......-A ...U...--- .G.CAAU..U
wg      .........A ..U...G... .....C.... AU.U...... ...G...C-G ...------- ---UG....C
tom     GG......UA ...A..U.G. .......U.. A..U..A.A. ..C..A..-. .------..G AA.AUUGG.C
cin     A.......UA ...A..CU.. UC.A....U. A..U.AG..U ..C..A..-. .G-----..G AACUUUGA.C
humA    G...G..G.G ..G...C.U. .C....GU.. C...UACUCG GGAGGCUGA. .------C.G GAG.AUCGCU
humB    G...G..G.G ..G...C.U. .C....GU.. C...UACUCG GGAGGCUGA. .------.GG GAG.AUCGCU
rat     G...G..G.G ..G...CAC. .C....GU.. C...UACUCG GGAGGCUGA. .-----.CAG GAG.AUCGCU
xen     G...G..G.. ..G.....U. .C.....U-. C...UACU-U GGAGGCU.G. .------C.G .CG.AUCGCU
dro      GACU.GAAG ..U.GC..CU UC.....U-. AC..UUCU.U -GAGG-UCU. A------..G .GG.AUGGCC
pom     ---------- ---------. .-.....U-- ---------- ---------- ----..C..G G.C.AAGU.U
lipol    --------- ---------. .-.....UGG C.UUUU.U-- ---------- ---....G.G G.AAAUCGAC
hbact   GGACUAGG.C .GGCG.UUU. GC.CCGC..G ACA.CC.U.A GAC.GUCA-U CA....GGG. CGAA.ACC..


        71         81         91         101        111        121        131
zmA     CUUUC-UGGG C-CC-GGGCU -CG--CUAUG UGCC-UUUGG CCGGCCUGCC CGUCCCAAGU UG-GUAGUGG
zmB     .....-.... .-..-..... -..--..... ....-..... .......... .......... ..-.......
zmC     .....-.... .-..-..... -..--..... ....-..... .......... .......... ..-.......
zmD     -....-.... .-.U-..... -..----U.. ..A.-.C.A. .....U.... .A........ ..-......U
zmE     -....-.... .-..U..... -..----U.. ..A.-AC... .....U.... .A........ ..-......U
zmF     A..---CU.. G-..U..... -U.-G..... ....-AC... .......... ...U...... ..-..U....
zmG     -.G..-.UU. .-.U-...UC -.A--UGG.. ....-...-. UU........ ...U...... ..-.......
zmH     GG.G.-AU.U G-..-U..G. -.U--AGUGC .A.A-AAC-U G...A.CA.. .A.U...... ..C.......
wg      U.GGUGC..U UUGUU..C.. GG.CUUGUGA ..UAACC.U. .U........ ...U...... ..-.......
tom     U..A.-CA.A A-GG-UU.GG -.U--UGG.. G..U-.AGUU .U........ .UGU.....C AC-AG...U.
cin     U.GG.U.A.C U-AG-UU.GG --UC-U-G.A .A.ACCGCAC .U...U.... ...U...... A.-AG.....
humA    UGAGUCCA.. A-GUUCU.GG -.UG-UAG.. C..UA.GCC. AUC.GG..U. ..CA.U.... .C--GGCAUC
humB    UGAG.CCA.. A-GUUCU.GG -.UG-UAG.. C..UA.GCC. AUC.GG..U. ..CA.U.... .C--GGCAUC
rat     UGAGUCCA.. A-GUUCU.GG -.UG-UAG.. C..UA.GCC. AUC.GG..U. ..CA.U.... .C--GGCAUC
xen     UGAGUCCAU. A-GUUCU.GG -.UG-UAC.. A..UA.G.C. AUC.GG..U. ..CA.U.... .C--GGUUAC
dro     UGAGGC.... AU.UACU..G -UAG-.GGAC CAG.UCA..U U...AAC.U. ..CA.U...C .U--GCCAUC
pom     U.AGUACUCC .-AAUA.UGC -AUG-U.CG. ..GUC.CG.. UUC.AG.CU. GC.UU.G.UC CC--.C.AUC
lipol   U.C.UG...U GCGUUC.AGU CUUGG..C.. CA.UUGGCCA UUU..UU.U. .U.U..G.A. .C-UGC.GUU
hbact   GCGCGUCC.A .-.GCC.CGG -UCG-GCCCC G.AAGCCAAC GU..AAGC.U .....GUC.G GG--AC.GC.


        141        151        161        171        181        191        201
zmA     CUGGCGGAGG CUUU--AGCG GAAGCUUU-G GUC-UCUCCA GAC-CUGAAG UGGCAGGAAU GGCGUGAGGC
zmB     .......... ....--.... ........-. ...-...... ...-...... .......... ..........
zmC     .......... ....--.... ........-. ...-...... ...-...... .......... ..........
zmD     ...A...-.. ..C.--.... A.......G. ...-...G.. ...-..AC.. C......... .....A....
zmE     ....U..G.. ..C.--.... A.......G. ...-...G.. ...-...G.. C......... .....A....
zmF     ....U..-.. ..CG--G... A.G..C.G-. .C.-...UU. ...-...... C......C.. ..........
zmG     .C.UU...A. ...G--G... U.G..CC.-. .G.U..CUUG ...-..AU.. .........C ..........
zmH     G.UU....U. .C..--G... A.......-. --.-AUC... ACA-..A... ....G.AC.. U.........
wg      GCU.G.UG.C UGGG---... A......C-. --.C....AG .C--..U... .......C.C CA...U....
tom     GAUCAC.... .CCA--.AU. A..AUA.GG. C.--...UG. UU.-..A... ...GGC.G.C C..A......
cin     G.UCAC.UUU .C.G--.... A.G...CGG. UGA-.G.GAG CUU-UAG..U C..GG..C.A U....U....
humA    AAUAU..U.A .C.C--CCG. ..GCGGGG-. AC.-A.CAGG UUG-..A.G. A..GGU...C C.GCCC...U
humB    AAUAU..U.A .C.C--CCG. ..GCGGGG-. AC.-A.CAGG UUG-..A.G. A..GGU...C C.GCCC...U
rat     AAUAU..U.A .C.C--CCG. ..GCGGGG-. AC.-A.CAGG UUG-..A.G. A..GGU...C C.GCCC...U
xen     AAUAU..UUU UCC.--G.G. ..GC..CG-. A..-A.CAGG UUGU..A.G. A..GGU...C C.GCCU...U
dro     AAUAU..GU. .CA.--G.A. ..GU.CG.-. .CA-.UCAGG UUGGCUA.G. A..G.U...C C.G.CC...G
pom     UGCCAC.UCU G..CGA..A. U.GU..UCGU GG.-AACUGG C.GUU.A..C C.UGUA.U.C C.AUG....U
lipol   GAU.G.C.UC UCGGUCU.A. U..U.GGCUU UGAGAU.U.C .UUCUAAG.U .AA.U..G.A ACUUC-..UG
hbact   G.CCGC.GC. UGCG..CC.. C.G.GGCG-U UCC-GUCGUG .UU-.GACG. ....---..C C-..CC....
```

```
      211        221        231        241        251        261        271
zmA   UGGCUUCACA GAGCAGCGAU -CACU---CG CCCGCUUCCA ACGGUGGGAG GAUAACG-GG CCGCUGCACU
zmB   .......... .......... -....---.. .......... .......... .......-A. ..........
zmC   .......... .......... -....----. .......... .......... .......-.. ..........
zmD   .......... .......... -....---.. ...A...... .......... ......A-A. ..........
zmE   .......... .......... -....----. ...A..C... .......... .......AA. ..........
zmF   .......... .......... -....---.. .......... .......... ......A-.. ........U.
zmG   ..U....... .......... -....----. ......AUU. .......A.. ......A-.. ..A.....GC
zmH   ........U. .........G -U..----.. .....CA... G......A.. .....U.-A. U....AUG..
wg    ...U...... ........U. -....----. G......... .......A.. ......A-.. ........-G
tom   .......... ......U..- --..G---.U ......CUGU G.A....A.. .....U.-.. U..G..UCU.
cin   ...U...... ....U....A -A..----.U .UAC..CU.G ..U....A.. ...U..A-A. ...GA.UGUC
humA  C..--AA..G ......GUCA --.AA---.U ....UGCUG. U.A..A.UG. ...--..-C. ..UG..-.A.
humB  C..--AA..G ......GUCA --.AA---.U ....UGCUG. U.A..A.UG. ...--..-C. ..UG..-.A.
rat   C..--AA..G ......GUCA --.AA---.U ....UGCUG. U.A..A.UG. ...--..-C. ..UG..-.A.
xen   C..--AA..G ......GUCA --.AA---.C ....UGC.G. U.A..A.UG. ...--..-C. ..UG..-.A.
dro   GU.--AA.AC C....A.C.A --GAG---UU ....UGGUAG G.A..A.UG. ...--A.-C. UAC.G.-.G.
pom   ...--AA... AU...CAUCA --CUA----- ..-.-GGU.U UG..CA.UGC ...--A.-C. -AUGG.-.U.
lipol A-..--A.UC C.....A... -.CAG---UU G...UGGGU. UG.CG.U.G- ...-C-.-CA A.CAA.UGG.
hbact AC.--GA.GU ........GA C...CGAA.. ....UCG.UC GAC.G.UCGC .GGGUG.A.A AG..GA.CGG
```

```
      281        291        301        311        321
zmA   UCGAGCCCAA CUCAGGCCCA GA--GCCUCA CUAAGCAGAC CACCAUCUUU
zmB   .......... .......... ..--...... .......... .........
zmC   .......... .......... ..--...... .......... ..........
zmD   .U........ .A...A.... ..--U....U ..U....A.. ..........
zmE   .U.....U.. .......U.. ..--A..... .......A.. ..........
zmF   G.A....... ...-..U... ..--...... U...A..... ..........
zmG   AU.G..U.GC U.A.....U. ..CUU.A.AC UA........ .......U..
zmH   AUAG...... U
wg    CAUG.G..CG ...U....UC CUACC..G.C A.A-...
tom   AUC.AGUUC. G.A.C...U. AUG-.GU.GC UCC.AU.A.. A....C...
cin   .UAG..U..U A.A.C-.AGC CUG-..UAGC .C-.A...GA ....CAU..
humA  AGCCA.UGC. ...CA...UG UG---.AA.. UAGC.AGAC. .CGUC.....
humB  AGCCA.UGC. ...CA...UG AG---.AA.. UAGC.AGAC. .CGUC.....
rat   AGCCA.UGC. ...CA...UG UG---.AA.. UAGC.AGAC. .CGUC.....
xen   AGCC.GUGC. GAACA...UG A.---.AA.. .AGC.AGACA ..GU-.....
dro   GGACGG..GU UAUCA..... AC---.GAU. UGGUUGGAC. -..A......
pom   CACCUU.GC. GGAUGUG.AU .GAA-GUAU. AAC.CA.CGG UCGUU
lipol AUU..UUAUG GAAGAUAUUU AC---GA... .G.------- ------....
hbact CUACA...GG .CGG.AA.GC CGG-..UA.C .CG.CUGUC. ..
```

FIG. 1. Alignment of primary-sequence data of SRP-RNAs. One of the maize endosperm sequences (zmA) is shown on top of each panel, to which all other SRP-RNAs are related. Bases identical to the zmA sequence are indicated by dots; dashes indicate deletions. The sequence of *H. halobium* (hbact) is included in this figure, although its membership is justified on the basis of its secondary structure.

II. SRP-RNA Sequences

A family of related SRPs was discovered in maize endosperm (*28*, *29*). Some of these contain SRP-RNAs that are very similar, while others show a homology of less than 70%. All can be aligned unambiguously as shown in Fig. 1. One of the maize endosperm sequences (zmA) is shown on top of each panel, to which all other SRP-RNAs were related. The presence of these RNAs in particles has not been confirmed experimentally in all cases. All exhibit a particular pattern of variable and conserved regions, which can be considered as being typical for SRP-RNAs. The SRP-RNA of *Halobacterium halobium* (*34*) is far removed from the other SRP-RNAs when compared on the primary sequence level, but is clearly an SRP-RNA by the analysis of its secondary-structure elements, discussed below.

Bases at the immediate 5′ end of the SRP-RNAs are generally quite conserved, although sometimes typical secondary-structure features are missing. Another region where erosion occurs, even between the different maize RNAs, is located around position 55 (Fig. 1). Changes in the number of nucleotides in this particular area, as well as deletions and insertions at the immediate 3′ end, are mainly responsible for differences between the SRP-RNAs. Extensive sequence variation exists around positions 60–110, after which a short stretch of 10–20 relatively conserved nucleotides is encountered. The following 50–60 nucleotides are again quite variable; nucleotides after position 190 are conserved, while the rest of the sequence toward the 3′ end is variable. In agreement with this pattern, conserved nucleotides are found at the following positions (Fig. 1): 22 (Y), 23 (Y), 25 (Y), 28 (Y), 114 (G), 119 (Y), 120 (Y), 125 (C), 160 (G), 162 (A), 163 (R), 185 (Y), 189 (R), 192 (R), 207 (A), 208 (G), 218 (A), 223 (G), 224 (C), 256 (R), 259 (R), 261 (R), and 270 (R).

A curious sequence homology was detected previously between SRP-RNA and 5S rRNA in human, *Xenopus*, and *Drosophila* (*46, 47*). This homology is not present in the other SRP-RNAs. It might be of structural and functional significance only in some species, but clearly is not a feature common to all SRP-RNAs.

III. Secondary Structure of SRP-RNA

Investigation of the secondary structure of RNA by the use of single-strand- or double-strand-specific nucleases cannot give direct information about the pattern of base-paired RNA regions, because different secondary structures can be deduced from one set of digestion data. Nevertheless, this approach can be used to confirm or disprove a suspected secondary structure (*26*). A more direct approach involves the analysis of interacting fragments after a mild digestion of the RNA and electrophoresis in a nondenaturing polyacrylamide gel. The base-paired fragments can then be separated under denaturing conditions and identified by fingerprinting. The structural analysis of SRP-RNA shows that a distinct 5′ domain consisting of about 50 nucleotides exists in the SRP. Virtually all regions of the RNA are involved in strong base-pair interactions (*46*).

The most successful approach for arriving at reliable secondary-structure models is one involving little experimentation. By comparison of phylogenetically related primary sequences, a pattern of compensatory base changes emerges that can be used to confirm or disprove particular intramolecular interactions (*48*). Only three SRP-RNA sequences have been available previously for the construction of preliminary secondary-structure models using this approach. These structures were still rather "loose" and were not confirmed in the very conserved regions of the molecule (*46*). With more and more sequence information becoming available, additional base-pair interactions can be established. A universal scheme for the secondary structure of SRP-RNA has emerged (Figs. 2–11).

Secondary structures of SRP-RNAs have been derived from the alignment of the primary-sequence data shown in Fig. 1. It is useful to distinguish four domains: the 5′ domain; the "adaptor"; the central domain; and the "finger" (see Fig. 12, top). Typically, the 5′ domain consists of 40–50 nucleotides, but it can shrink considerably as in *S. pombe* and *Y. lipolytica* (see Fig. 10); it is largest in the *H. halobium* SRP-RNA (see Fig. 11). The adaptor includes the 3′ end of the RNA. A shortening of the 3′ end is apparently linked to an erosion of the region around position 50. No sequence conservation is found in the adaptor, which seems to have the function of simply varying the distance between the 5′ and the central domains. The latter consists of most of the conserved nucleotides of the SRP-RNA. It excludes the finger, a structural element very variable in sequence but unusually conserved in the maintenance of a stem–loop. A peculiar conservation (G-N-A-R) is found in the unpaired four bases of the loop. Since the 19-kDa protein of the SRP protects these bases from digestion by α-sarcin (*45*), it has been suggested that they are directly required for the binding of the 19-kDa protein (*29*).

IV. Tertiary Structure of SRP-RNA

By comparing simple models built from tubes (representing the double-helical segments of the RNA) and wire (representing the single-stranded parts of the molecule) and models derived from physical studies of SRP-RNA using computer graphics (*40*), with the known size of the SRP (*41*, *42*), I conclude that the RNA must be folded in three dimensions in order to fit into a cylinder of 50-Å diameter and 240-Å length (see Fig. 12). Since the SRP is not at present easily accessible to crystallization and X-ray structural analysis, other methods to investigate the higher-order structure of this particle must be used. A number of tertiary base-pair interactions have been suggested (*40, 49*), and an RNA–RNA cross-link has been located (*40*). Another approach, which limits the number of possible three-dimensional models considerably, is described in the following section.

A. "Limbs" and "Joints"

Similar to the analysis of compensatory base changes for the determination of the RNA secondary structure, the approach of the identification of "limbs" (defined as stretches of continuously stacked RNA double-helices) and "joints" (stretches of single-stranded RNA between limbs) takes advantage of the availability of a large number of related secondary-structure models in order to define common morphological features of the RNA. The secondary-structure models are based on the phylogenetic comparison of primary-sequence data. Using an analogous approach, the existence of double-helical segments can be confirmed and their sizes can be estimated by comparing the secondary-structure features of all SRP-RNAs. As a consequence, the identification of the location and the number of limbs and joints can provide valuable clues for the construction of sensible tertiary-structure models. The medium-sized SRP-RNA provides a prime example for the usefulness of this approach.

The finger (the structural feature around position 150 in most SRP-RNAs) is likely to be defined as one single limb of about two double-helical turns with precisely four unpaired bases at the top of the stem–loop. Apparently, this structure has been realized only once in the *Drosophila* SRP-RNA (see Fig. 7, bottom) where it includes one possible A–G interaction. Nevertheless, the construction principles of the finger in the SRP-RNA from other organisms makes it likely that this morphological feature is present as one single stem–loop in all SRP-RNAs. Their comparison shows that not more than one or two nucleotides have been inserted in one strand of the helix. Insertions of one nucleotide are not likely to disturb the continuous stacking of the double-helix (*50–52*). When an insertion of two nucleotides occurs in one strand, another one or two pyrimidine bases are added to the opposite strand. Also in these cases the direction of the double-helix is likely not to change at the nucleotide insertion point. In one single example (the tomato SRP-RNA, Fig. 7, top) two adenosines are juxtaposed near the tip of the finger, expecting to cause also very little, if any, change in the direction.

The stem–loop of the central domain (the structural feature around position 200 in most SRP-RNAs and pointing "downward" in the secondary-structure models (see Fig. 2–12) has a size of roughly three double-double-helical turns and, like the finger, possesses precisely four unpaired bases at the top. In contrast to the situation in the finger, in none of the SRP-RNAs has a single continuously stacked double-helix been realized. This part consists of three limbs connected by two joints. The joints usually are composed of adenosines and guanosines. Frequently, one or two A–G interactions are suggested to be located close to the top of the stem–loop. The conservation of the joint structures in all SRP-RNAs supports the idea that changes in the direction at these two locations of the central domain are essential for maintaining the proper three-dimensional structure of the RNA.

No distinct pattern of limbs and joints in the remaining part of the central domain can be easily recognized. The largest stretch of potentially continuous stacked base-pairs is found in the human and rat SRP-RNAs at the borderline between the adaptor and the central domain (Fig. 8, top and bottom; and Fig. 9, top). It is composed of almost three turns and contains insertions of single pyrimidine nucleotides. Somewhat shorter double-helical segments can be found in this region in most, but not all, SRP-RNAs. As expected, not only the size but also the composition of the adaptor vary considerably.

B. A tRNA-Like Structure

The 5′ domain of SRP-RNA resembles tRNA (*45, 49, 53*). In *S. pombe* and *Y. lipolytica*, the 5′ domain is reduced to one stem–loop consisting of 11 nucleotides. In all other SRP-RNAs, the 5′ domain consists of at least two stem–loops, and sometimes involves pairing of the 5′ and the 3′ ends of the RNA. Base-pair interactions between the unpaired nucleotides in the two loops might be involved in the formation of an additional short helix to form a distinct tightly structured 5′ domain. Bases in the loops might interact either by parallel (as in tRNA) or by standard antiparallel pairing. Both pairing modes are possible in all SRP-RNAs (Fig. 13). This tRNA-like structure is missing the anticodon arm and it can be speculated that one of the functions of the SRP—the influence on the speed of translation—might be accomplished by a direct interaction of the 5′ domain of SRP-RNA with tRNA binding sites on the ribosome.

C. A Switch in the SRP-RNA

The meeting point of the three major double-helices in SRP-RNA is part of the conserved central domain. The mutagenesis of this region indicates that different conformers—or a switch—can be formed in the SRP-RNA (*3, 46, 49*). Different base-pairing schemes of

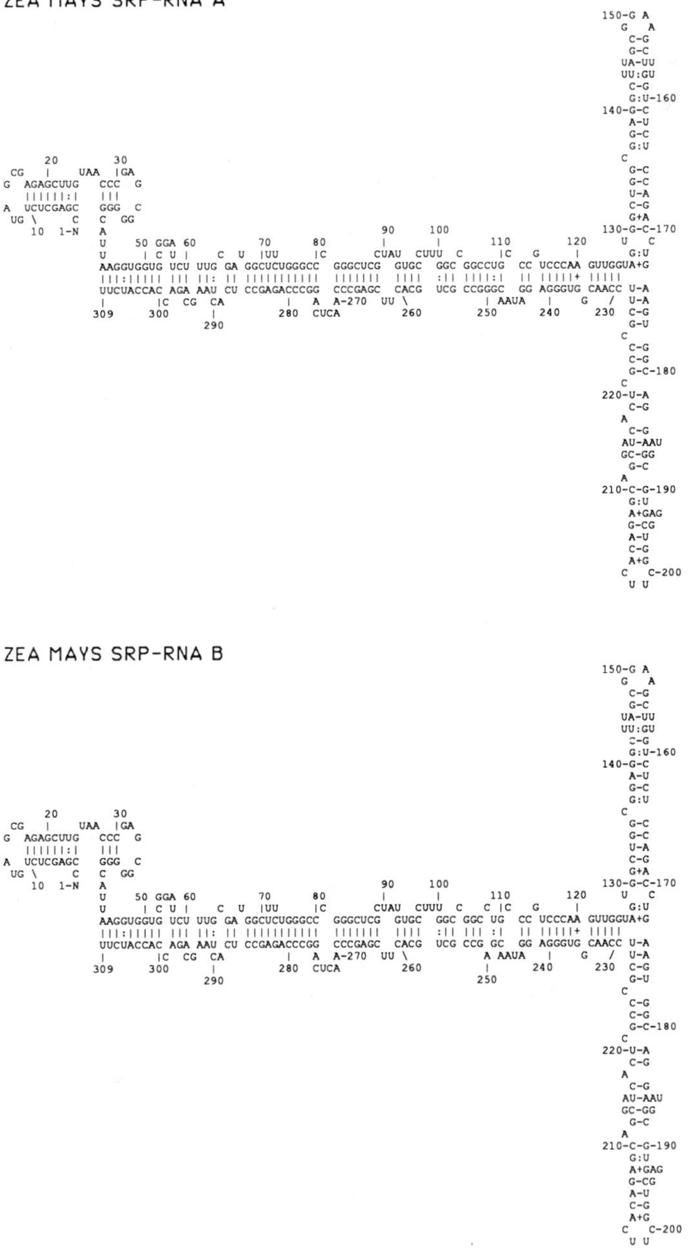

FIG. 2–11. Secondary-structure models for SRP-RNAs from plants (Figs. 2–6, top and bottom; Fig. 7, top), insects (Fig. 7, bottom), mammals (Fig. 8, top and bottom; Fig. 9, top), amphibia (Fig. 9, bottom), fungi (Fig. 10), and archaebacteria (Fig. 11). Secondary structures were constructed using the alignment of the primary sequence data shown in Fig. 1 and the compensatory base change approach. Maximal base-pairing was achieved by allowing for G–U (marked by colons) and A–G (marked by

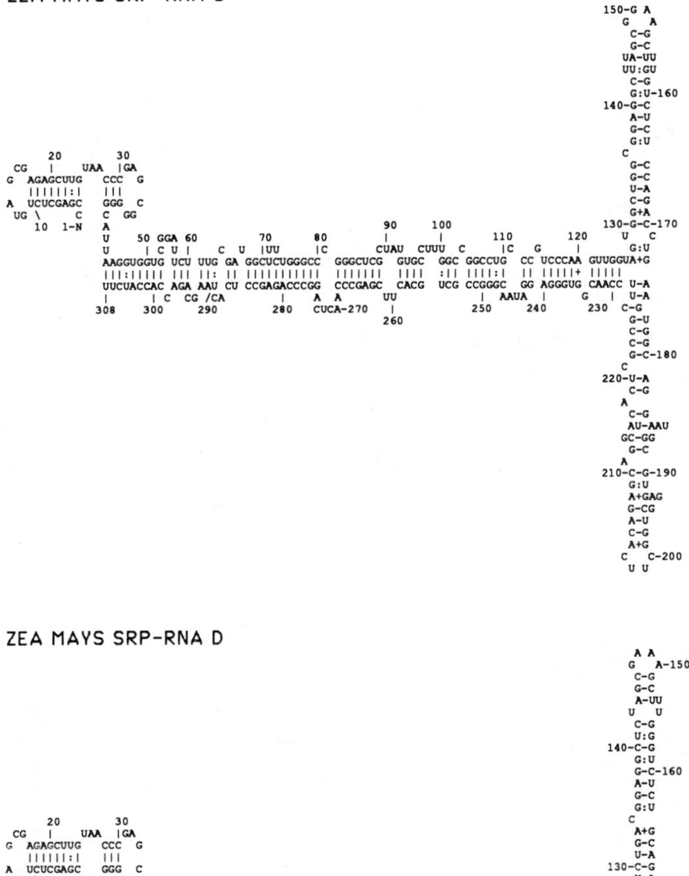

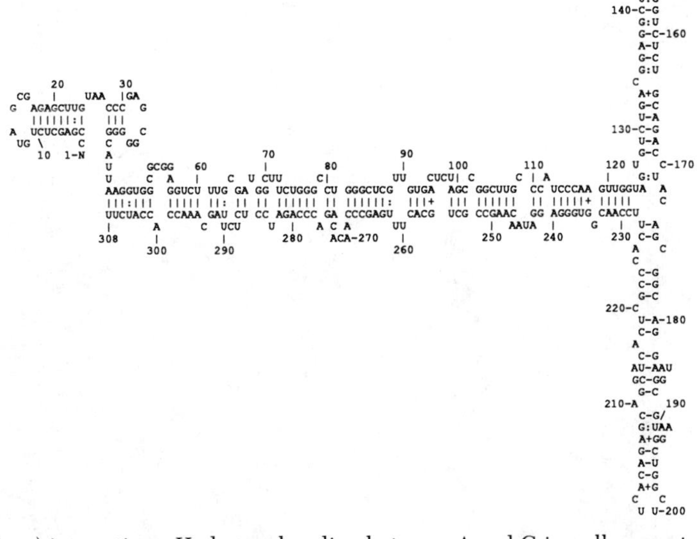

plus signs) interactions. Hydrogen bonding between A and G is well supported in the central domain, but not necessarily in other regions of the molecule (see text). Fig. 2: Structure of zmA and zmB; Fig. 3: zmC and zmD; Fig. 4: zmE and zmF; Fig. 5: zmG and zmH; Fig. 6: wheat and *C. hybrida*; Fig. 7: tomato and *Drosophila*; Fig. 8: human A and B; Fig. 9: rat and *Xenopus*; Fig. 10: *S. pombe* and *Y. lipolytica*; Fig. 11: *H. halobium*.

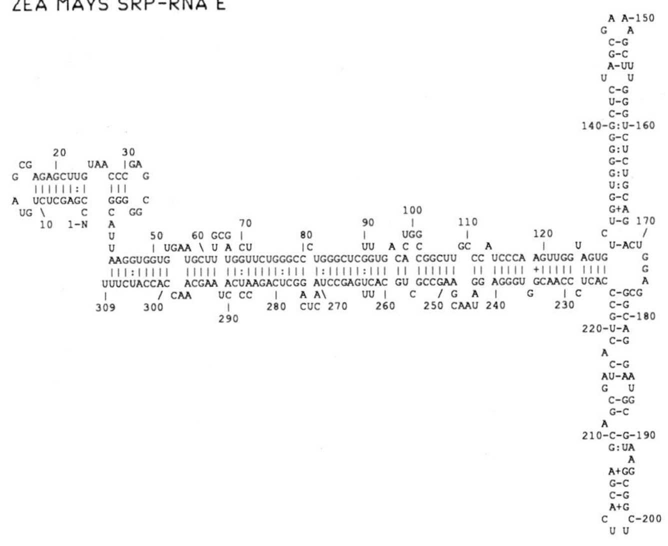

ZEA MAYS SRP-RNA F

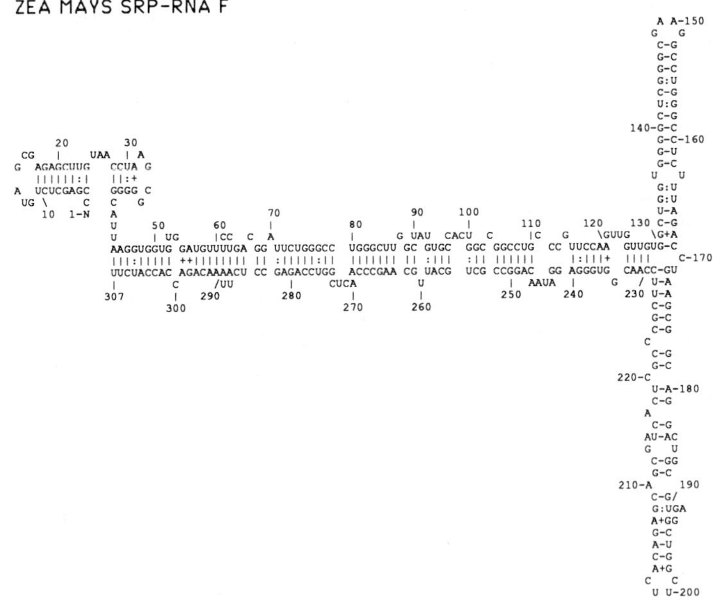

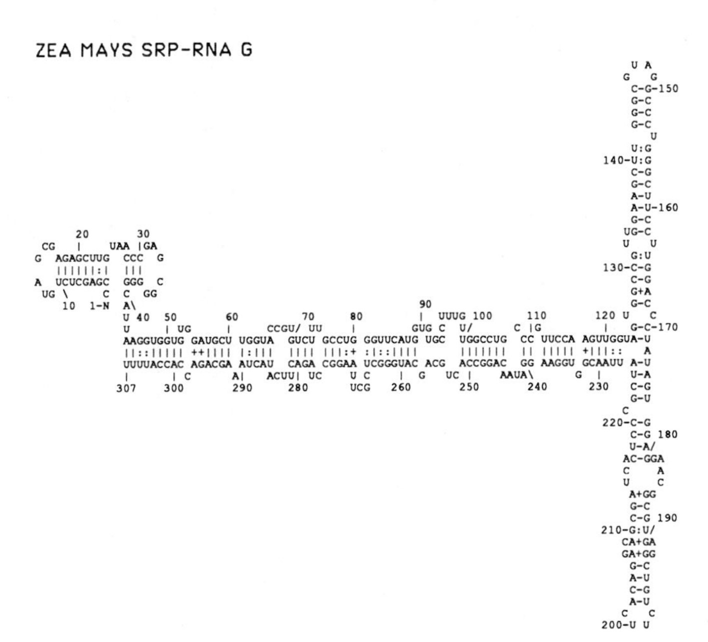
ZEA MAYS SRP-RNA G

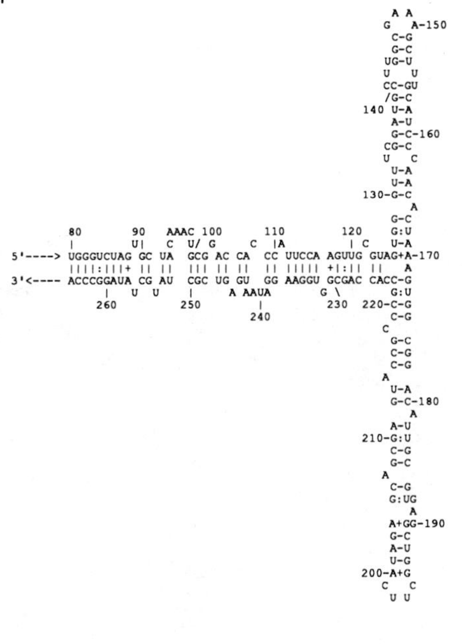
ZEA MAYS SRP-RNA H

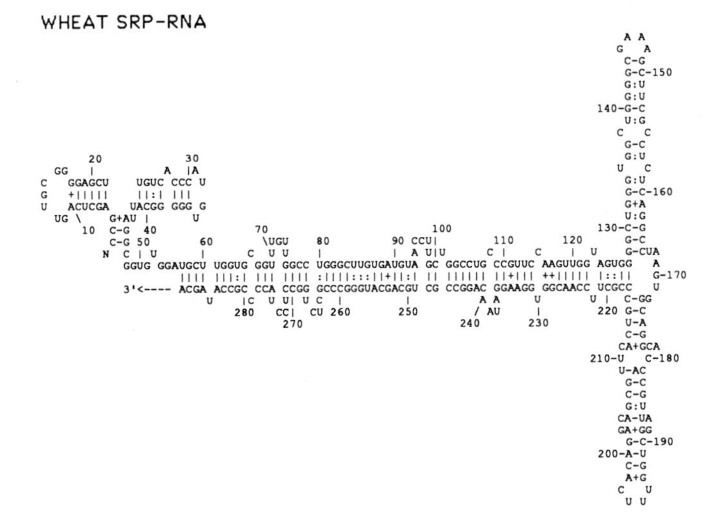
WHEAT SRP-RNA

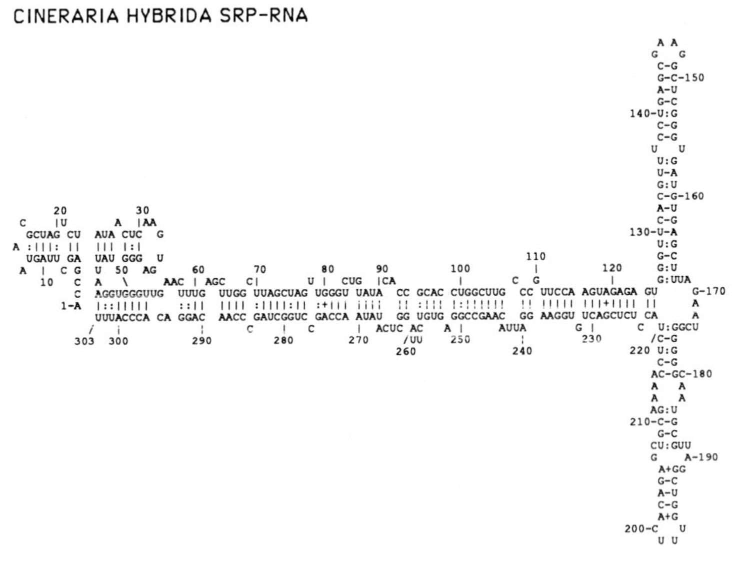
CINERARIA HYBRIDA SRP-RNA

TOMATO SRP-RNA

DROSOPHILA SRP-RNA

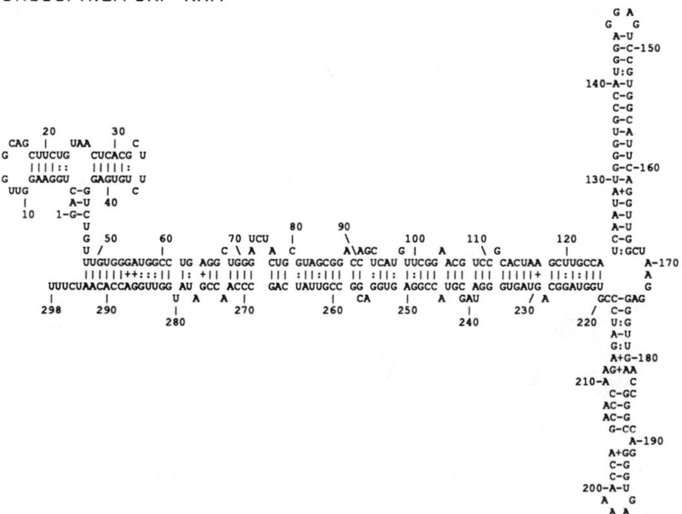

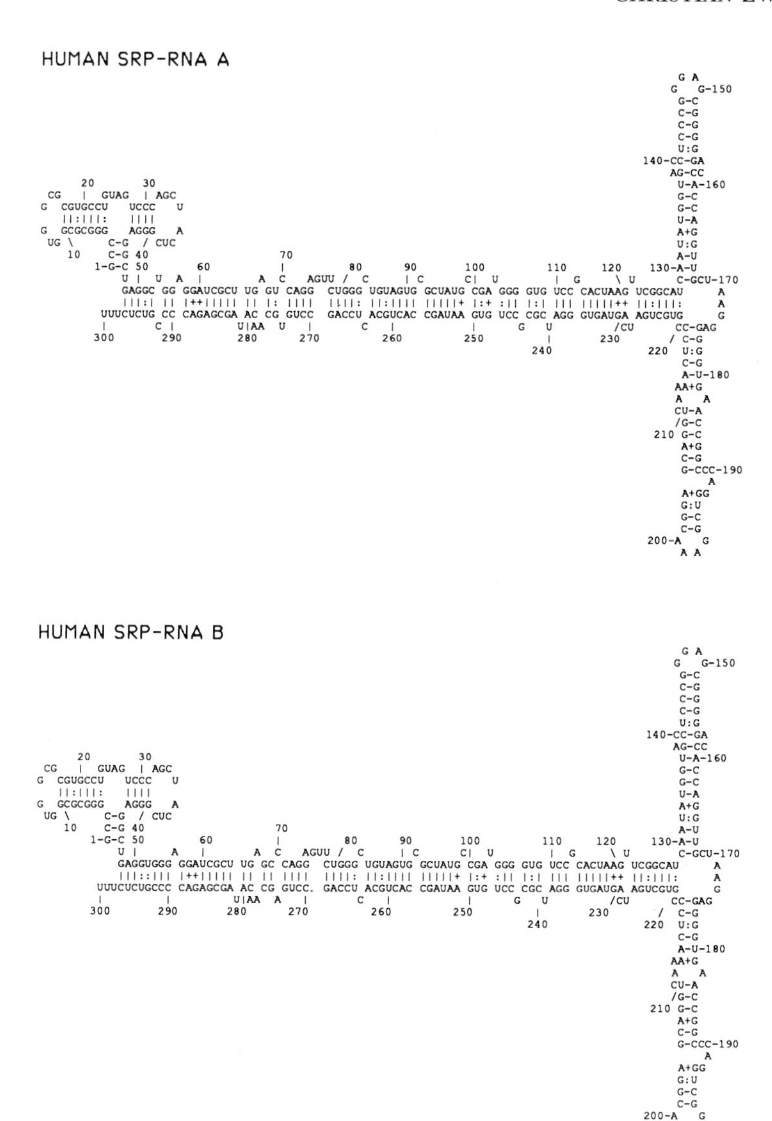

the Watson–Crick type can be suggested theoretically in most SRP-RNAs analyzed so far, but still need to be confirmed experimentally. If unusual base–base interactions take place in the switch structure, it will be difficult to prove this interaction by the classical phylogenetic approach using compensatory base changes. Coordinated base

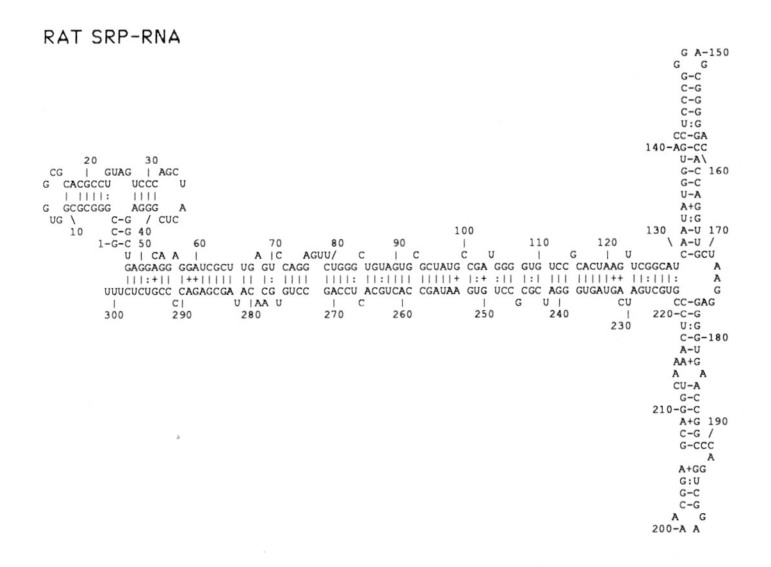

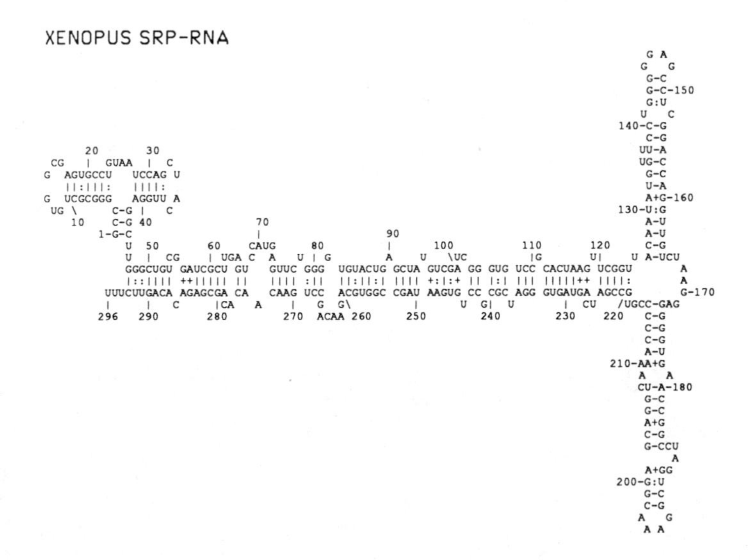

changes, which may be accompanied by sliding inside the switch structure, might cause a smooth structural change in the conserved central domain without the need for a significant input of energy. At present, it is not possible to understand in detail the role of the nucleotides involved in the switch.

SCHIZOSACCHAROMYCES POMBE SRP-RNA

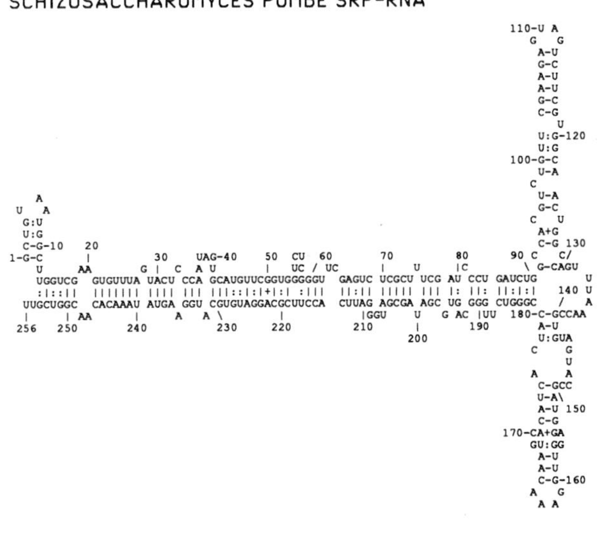

YARROWIA LIPOLYTICA SRP-RNA

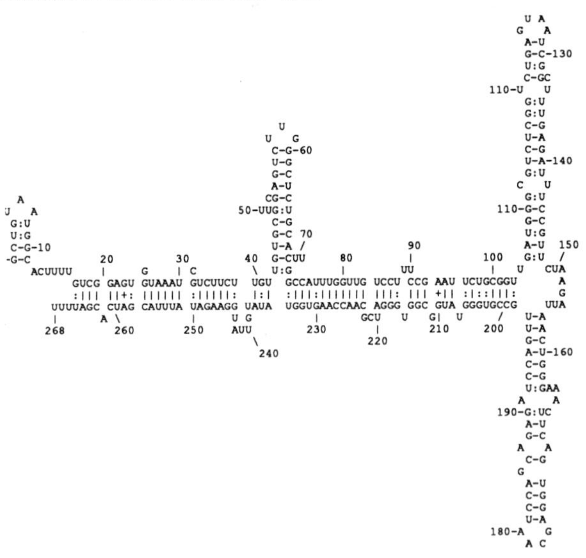

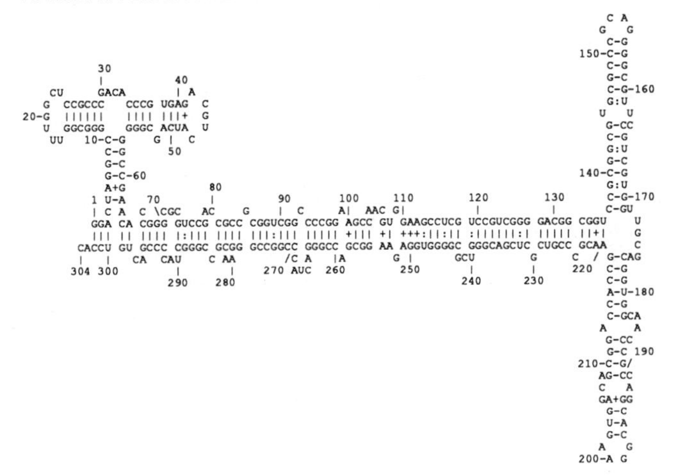

D. Proposed Three-Dimensional Structural Features

The folding of SRP-RNA is determined by the way the secondary-structure features of the RNA [5′ domain, adaptor, central domain, and finger (see Fig. 12, top)] participate in a three-dimensional fit to the known shape and size of the SRP. The number of possibilities by which SRP-RNA can be accommodated inside a cylinder of 50-Å diameter and 240-Å length are very much limited by the morphological limb and joint structure.

It is predicted that the three short double-helical segments located in the central domain are folded back into the main body of the SRP-RNA (see Fig. 12, bottom). If this is the case, the central domain will form a tightly defined structure composed solely of the conserved nucleotides of the SRP-RNA. As discussed above, the details of this tight structure are not understood yet, but probably involve several interactions not of the Watson–Crick type. The reason for the high degree of conservation might be related to the need for a large number of unusual interactions between the nucleotides in the central domain. In addition, the conserved central domain has dynamic properties that might limit the freedom for sequence variation considerably (*49*, *54*). An intramolecular cross-link in human SRP-RNA that involves nucleotides located around position 200 and nucleotides at the border between the adaptor and the central domain also suggests a structure that folds back upon itself (*40*). The finger region of the

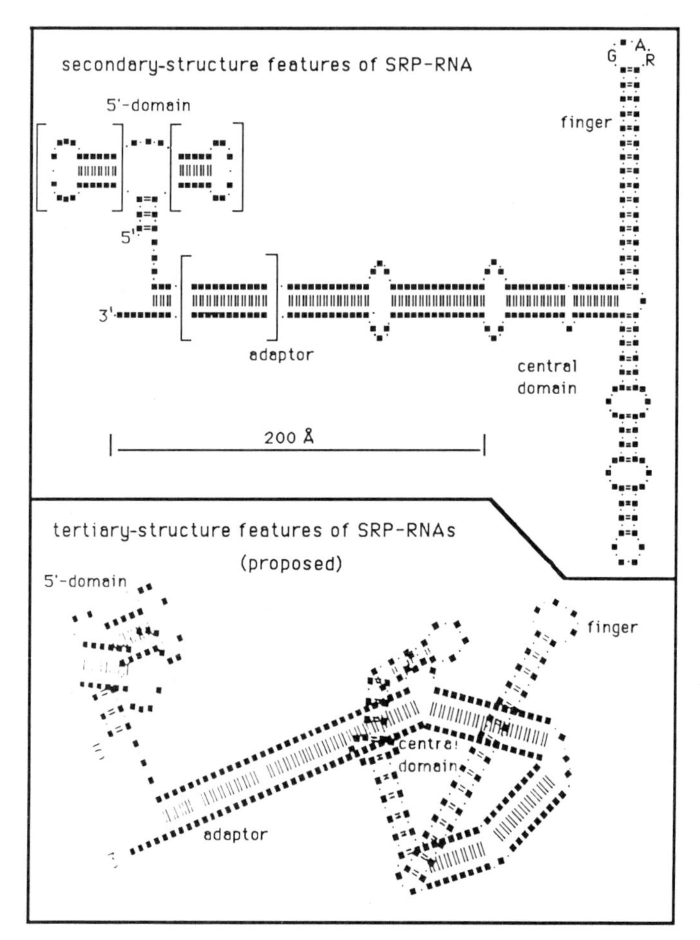

FIG. 12. Common structural features of SRP-RNAs. Four secondary-structure features are present in all SRP-RNAs: a 5′ domain, the adaptor, the central domain, and the finger. The number of double-helical turns in the adaptor varies as indicated by the brackets. The 5′ domain is usually composed of three double-helices but can be reduced to one stem–loop. The only strictly conserved stretch of nucleotides (G-N-A-R) is present in the loop of the finger domain. A more compact, but still elongated, molecule is obtained by tertiary interactions in the central domain of the molecule (bottom half of the figure) and by a possible interaction between the two loop tips of the 5′ domain in some SRP-RNAs.

SRP-RNA protrudes from the molecule, probably as a continuous double-helix consisting of at least two turns of stacked nucleotides. Some flexibility might exist at the meeting point of the finger with the central domain (55).

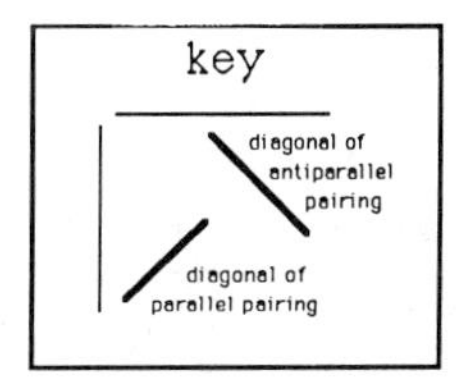

ZEA MAYS — G U A G C G / G G C G A G
WHEAT — G U U G C G G / U G U A
C. HYBRIDA — U A A C G / A G U G A A
TOMATO — G U A A C G / U G A A
DROSOPHILA — G U U G G C A G / U C U U C G
TRYPANOSOMA — G C A U U / G G G U U
HUMAN, RAT — G U G G C G / C U C A U C G A
XENOPUS — G U G G C G / U C A U C G
H. HALOBIUM — U U U G G C U / A C U G C A G

ZEA MAYS — G U A G C G / G G C G A G
WHEAT — G U U G C G G / U G U A
C. HYBRIDA — U A A C G / A G U G A A
TOMATO — G U A A C G / U G A A
DROSOPHILA — G U U G G C A G / U C U U C G
TRYPANOSOMA — G C A U U / G G G U U
HUMAN, RAT — G U G G C G / C U C A U C G A
XENOPUS — G U G G C G / U C A U C G
H. HALOBIUM — U U U G G C U / A C U G C A G

FIG. 13. Tertiary base-pair interactions between the two stem–loops of the 5′ domain are possible in all SRP-RNAs. Interactions are represented by diagonals in a matrix. Dots indicate G–C, A–U, or G–U pairing. A–G interactions are indicated by plus signs. The key diagram on top shows two possible diagonals, one for a parallel-pairing scheme (present in tRNA) and the other for antiparallel pairing. Possible interactions in the different species are shown in the middle panel. In the bottom panel, only the interactions involved in at least three consecutive base-pairs are marked. No bias toward one particular pairing scheme can be recognized.

SRP and SRP-RNA have been clearly identified in the electron microscope as possessing elongated structures with two domains. If the role of the adaptor is simply to keep these two domains apart by a certain distance, it seems peculiar that in none of the secondary structures is a perfect double-helix present. Possibly one or more joints must exist in the adaptor, which as a consequence enables the two domains of the RNA to approach each other. In the extreme case, SRP-RNA and SRP would appear as a globular RNP of about 160-Å diameter. The elongated appearance of SRP seen in the electron microscope might be the consequence of the rather harsh conditions of sample preparation. The adaptor seems relatively resistant to digestion by nucleases (*26*). The opposite would be expected if this region is exposed as suggested by an extended secondary- or tertiary-structure model.

In this context, it might be speculated that both appearances of the SRP might reflect its shape *in vivo*. SRP might change to an elongated state when it combines with a ribosome that is translating a secretory protein; after being released, it might adopt a more spherical appearance. Suggested RNA–RNA interactions might play a role in this alteration of the overall shape of the SRP (*31*, *40*). This speculation leads us to a possible scenario describing the function of SRP-RNA and SRP as discussed below.

V. Function of SRP-RNA

SRP cycles through a variety of interactions that involve, on one hand, components of cytosolic origin (the signal peptide and the translating ribosome), and on the other hand, membrane components (among others, the SRP-receptor). From cross-linking experiments, it has been discovered that the 54-kDa protein of ribosome-bound SRP is in close proximity to the signal peptide (*56–58*). The 54-kDa protein probably binds not to SRP-RNA, but rather associates with the 19-kDa RNA-binding protein (*45*). It follows that SRP-RNA is probably not directly involved in the signal-peptide recognition, but rather in the positioning of the 54-kDa protein. It can be speculated that SRP-RNA might function simply as a scaffold for the proper arrangement of the proteins of the SRP, and to keep the two functional domains of the SRP apart. One can imagine that RNA double helices might be suitable for this purpose, because a rather long distance can be bridged without the need to synthesize a special protein.

A. SRP-RNA and the Ribosome

The distinct possibility remains that SRP-RNA functions in the context of the translating ribosome, perhaps by direct interaction with ribosomal RNA or ribosomal proteins. Thus, SRP might not recognize the signal peptide directly, but rather a ribosome translating an mRNA for a signal-peptide-containing protein. It is conceivable that SRP-RNA acts in an even more indirect way, namely, by competing for or displacing components of the ribosomal machinery, e.g., tRNA or elongation factors. Although the evidence that SRP interacts with ribosomes is good (*2*, *4*, *12*), the suggestion of a direct interaction of SRP-RNA with the ribosome has not yet been confirmed. Two RNA–RNA interactions are proposed in the following scheme, one involving the 5′ domain, another the central domain of the SRP-RNA.

We have discussed above that the 5′ domain of SRP-RNA might resemble tRNA. Although in *S. pombe* and *Y. lipolytica* the 5′ domain of SRP-RNA is reduced to one stem–loop, this might still be sufficient to interact with a tRNA binding-site on the ribosome. The possibility exists that a translational effect is not present or not mediated through the 5′ domain of fungal SRPs. If the 5′ domain stacking pattern of double-helices is similar to a tRNA that has no anticodon arm, the influence on the speed of translation can be explained directly by a competition between tRNA and SRP-RNA for the same site(s) on the ribosome.

An isolated subparticle of the canine SRP consisting of the tRNA-like part of SRP-RNA can inhibit translation (*43*). This could be due to a direct interaction with the decoding site of the ribosome. By numerous criteria (e.g., *59*), this site is located in the "platform" between the "head" and the "body" of the 40-S subunit.

It is interesting to note that the distance between the site of tRNA interaction and the protein exit site is proper for an SRP molecule to bridge [about 150 Å, (*60–62*)]. The bridging distance might thus be expected to be reduced for a smaller ribosome from *H. halobium* and enlarged in ribosomes of higher eukaryotes. The analysis of the SRP-RNA secondary structures (Figs. 2–11) shows indeed that the length of the adaptor of the SRP-RNA varies accordingly. Although the number of nucleotides of the SRP-RNA of *H. halobium* is even larger than in most SRP-RNAs of higher eukaryotes, the distance between the 5′ domain and the meeting point of the three large double-helices is shortened by about one double-helical turn when compared to the *Z. mays* SRP-RNA (Figs. 2 and 11).

An oligodeoxynucleotide complementary to nucleotides at positions 96–110 in human SRP-RNA can serve as a primer in a reverse transcription reaction (*11*). Also, a product about 190 nucleotides long revealing a possible base complementarity between 18-S ribosomal RNA and SRP-RNA has been detected (*63–65*). It has been suspected that region 240–254 of SRP-RNA interacts with 18-S rRNA at positions 175–198 (the positions refer to human SRP-RNA and rat 18-S rRNA, respectively). This region of SRP-RNA might also be involved in dynamic structural changes (*49*), which might accompany the exposure of this critical sequence.

It must be stressed that the stage at which the switch might function has not yet been determined. The most attractive speculation concerns the function of the SRP itself. It is possible that at least two different base-pairing schemes exist in the conserved central domain of SRP-RNA, one of which is present when SRP is not interacting with the ribosome and/or the signal sequence of the secretory protein. Another might be induced by the binding of SRP to the ribosome or other components of the secretory machinery. This mechanism ensures that only the ribosome-associated SRP can occupy the binding sites of the membrane of the endoplasmic reticulum.

As a consequence of this hypothesis, the region of the central domain of SRP-RNA (labled "1" in Fig. 14) must be close to the ribosome. In fact, most of the central domain, the adaptor, and the tRNA-like 5′ domain (labeled "2" in Fig. 14) would be in contact with the ribosome in one or the other stage of the function of SRP. The finger would more or less protrude from the particle, making contact with the emerging signal peptide and protein components of the ER membrane (Fig. 14).

B. A Three-Legged Model

The proposed SRP–ribosome interactions (labeled "1" and "2" in Fig. 15) can take place at the same time in the context of a ribosome

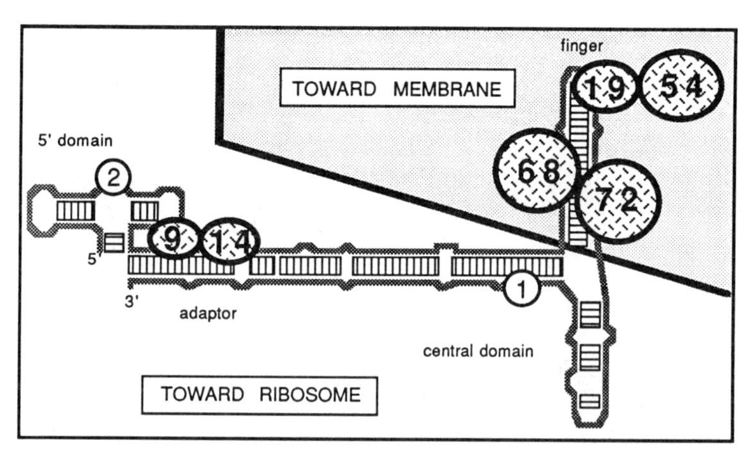

FIG. 14. Interaction of SRP-RNA with the SRP proteins. In this presentation, the 5′ domain, the adaptor, and a large part of the central domain are oriented toward the ribosome. Two regions (marked 1 and 2) might be in direct contact with ribosomal sites. A number of criteria suggest that the finger points toward the membrane: the binding site of the 19-kDa protein has been mapped to the tip of the finger; the 54-kDa protein, which is in close proximity to the emerging signal peptide, requires 19-kDa protein for binding to the SRP.

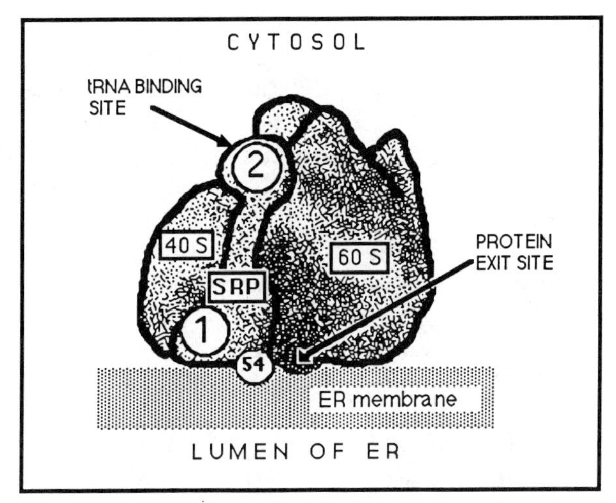

FIG. 15. A possible scheme of SRP-mediated protein translocation. Three parts are involved in the formation of a protein translocation junction: the protein-rich finger of the SRP; the bottom of the large ribosomal subunit; and the body of the small ribosomal subunit. The 54-kDa protein is shown close to the exit site of the secretory protein. The regions marked "1" and "2" (see Fig. 14) might be in direct contact with ribosomal sites. Interaction 2 might influence translation directly by competing for tRNA binding sites.

associated with the membrane. Figure 15 shows an arrangement of this situation, taking into account the approximate size and shape of SRP, and the 40-S and the 60-S ribosomal subunits. Interaction 1 is suggested by a possible base-pairing between SRP-RNA and 18-S rRNA. By analogy with its position in the prokaryotic ribosome, this part of the 18-S rRNA must be located in the body of the 40-S ribosomal subunit (*66*). Interaction 2 is assumed on the basis of a possible structural similarity of the 5′ end of SRP-RNA to tRNA as discussed above.

The exit site of the nascent polypeptide chain is located at the bottom of the large ribosomal subunit (*62*). The amino terminus of the nascent secretory protein is close to the 54-kDa protein of the SRP, since it can be cross-linked to the signal peptide (*56–58*). The 54-kDa protein requires the 19-kDa protein in order to become part of the SRP. If this is achieved by direct binding, the 54-kDa protein must be close to or even part of the finger to which the 19-kDa protein binds.

The three-legged model shown here proposes that three parts of the SRP-complexed ribosome are close to the membrane of the endoplasmic reticulum: a protein-rich finger of the SRP; the bottom of the large ribosomal subunit with the exit site for the nascent polypeptide; and the body of the small ribosomal subunit. Interestingly, the 5′ domain of 18-S rRNA is somewhat larger than its prokaryotic equivalent (67). The three legs shown might act together to create a suitable environment for the translocation of secretory proteins by exclusion of the nascent protein from the cytosolic compartment.

In summary, the role of the SRP-RNA in the SRP can therefore be defined as being 2-fold: first, the SRP would orient the eukaryotic ribosome properly toward the membrane of the endoplasmic reticulum, possibly using interactions with ribosomal sites, thus promoting a situation in which the secretory protein stays translocation-competent. Second, interactions between SRP and the ribosome, in conjunction with the dynamic property of the SRP-RNA, might be used to regulate the secretory activity of eukaryotic ribosomes by influencing basic functions of the ribosome itself.

The function of the SRP will only be well understood when the structural basis of the function of eukaryotic ribosomes has been defined. This situation might not be as hopeless as it seemed to be a year ago. Low-resolution, three-dimensional structure models of eubacterial ribosomes that are expected to be quite similar to their eukaryotic equivalent have been determined recently (68). Since many of the components of eukaryotic ribosomes have already been characterized, it is possible now to address some of the open questions related to the function of SRP. Cloning of SRP components and their manipulation by site-directed mutagenesis will continue to be an exciting area for exploring protein–RNA interactions on the molecular level and for understanding the functional potential of RNA.

Acknowledgments

I am indebted to many colleagues who have sent their results prior to publication, and to Ray Brown for a critical reading of the manuscript.

References

1. J. M. Bishop, W. E. Levinson, D. Sullivan, L. Fansier, N. Quintrell and J. Jackson, *Virology* **42,** 927 (1970).
2. T. A. Walker, N. R. Pace, R. L. Erikson, E. Erikson and F. Behr, *PNAS* **71,** 3390 (1974).
3. G. Zieve and S. Penman, *Cell* **8,** 19 (1976).
4. P. W. Gunning, P. Beguin, E. M. Shooter, L. Austin and L. Jeffrey, *JBC* **256,** 6670 (1981).
5. A. M. Weiner, *Cell* **22,** 209 (1980).
6. C. M. Houck, F. P. Reinhart and C. W. Schmid, *JMB* **132,** 289 (1979).
7. E. Ullu, S. Murphy and M. Melli, *Cell* **29,** 195 (1982).
8. E. Ullu and A. M. Weiner, *EMBO J.* **3,** 3303 (1984).

9. E. D. Gundelfinger, E. Krause, M. Melli and B. Dobberstein, *NARes* **11,** 7363 (1983).
10. E. Ullu and C. Tschudi, *Nature* **312,** 171 (1984).
11. E. Ullu and A. M. Weiner, *Nature* **318,** 371 (1985).
12. P. Walter, I. Ibrahimi and G. Blobel, *JCB* **91,** 545 (1981).
13. P. Walter and G. Blobel, *JCB* **91,** 551 (1981).
14. P. Walter and G. Blobel, *JCB* **91,** 557 (1981).
15. P. Walter and G. Blobel, *Nature* **299,** 691 (1982).
16. P. Walter and G. Blobel, *PNAS* **77,** 7112 (1980).
17. A. H. Erickson, P. Walter and G. Blobel, *BBRC* **115,** 275 (1981).
18. F. N. Katz, J. E. Rothman, V. R. Lingappa, G. Blobel and H. F. Lodish, *PNAS* **74,** 3278 (1977).
19. R. Gilmore, P. Walter and G. Blobel, *JCB* **95,** 470 (1982).
20. R. Gilmore and G. Blobel, *Cell* **35,** 677 (1983).
21. D. I. Meyer, E. Krause and B. Dobberstein, *Nature* **297,** 647 (1982).
22. M. Hortsch and D. I. Meyer, *Int. Rev. Cytol.* **102,** 215 (1986).
23. P. Walter and V. R. Lingappa, *Annu. Rev. Cell Biol.* **2,** 499 (1986).
24. W. Y. Li, R. Reddy, D. Henning, P. Epstein and H. Bush, *JBC* **257,** 5136 (1982).
25. A. Balmain, R. Krumlauf, J. K. Vass and G. D. Birnie, *NARes* **10,** 4259 (1982).
26. E. D. Gundelfinger, M. D. Carlo, D. Zopf and M. Melli, *EMBO J.* **3,** 2325 (1984).
27. S. Prehn, M. Wiedmann, T. A. Rapoport and C. Zwieb, *EMBO J.* **6,** 2093 (1987).
28. N. Campos, J. Palau, M. Torrent and M. D. Ludevid, *JBC* **263,** 9646 (1988).
29. N. Campos, J. Palau and C. Zwieb, *NARes* **17,** 1573 (1989).
29a. B. Haas, A. Klanner, K. Ramm and K. L. Sänger, *EMBO J.* **7,** 4063 (1988).
30. V. Ribes, P. Dehoux and D. Tollervey, *EMBO J.* **7,** 231 (1988).
31. P. Brennwald, X. Liao, K. Holm, G. Porter and J. A. Wise, *MCB* **8,** 1580 (1988).
32. M. A. Poritz, V. Siegel, W. Hansen and P. Walter, *PNAS* **85,** 4315 (1988).
33. C. Marshallsay, S. Prehn and C. Zwieb, *NARes* **17,** 1771 (1989).
34. A. Moritz, B. Lankat-Buttgereit, H. J. Gross and W. Goebel, *NARes* **13,** 31 (1985).
35. A. Moritz and W. Goebel, *NARes* **13,** 6969 (1985).
36. P. Walter and G. Blobel, *Cell* **34,** 525 (1983).
37. G. G. Brownlee, *Nature NB* **229,** 147 (1971).
38. C. Zwieb, *Endocytobiosis Cell Res.* **3,** 41 (1986).
39. C. A. Lee, M. J. Fournier and J. Beckwith, *JBact* **161,** 1156 (1985).
39a. J. C. R. Stuck, D. W. Vogel, N. Ulbrich and V. A. Erdmann, *NARes* **16,** 2719 (1988).
39b. L. M. Hsu, J. Zagorski and M. J. Fournier, *JMB* **178,** 509 (1984).
39c. C. Zwieb, *Endocytobiosis Cell Res.* **5,** 327 (1988).
39d. M. A. Poritz, K. Strub and P. Walter, *Cell* **55,** 4 (1988).
40. C. Zwieb and D. Schüler, *Biochem. Cell Biol.,* in press.
41. D. W. Andrews, P. Walter and F. P. Ottensmeyer, *PNAS* **82,** 785 (1985).
42. D. W. Andrews, P. Walter and F. P. Ottensmeyer, *EMBO J.* **6,** 3471 (1987).
43. V. Siegel and P. Walter, *Nature* **320,** 81 (1986).
44. V. Siegel and P. Walter, *JCB* **100,** 1913 (1985).
45. V. Siegel and P. Walter, *PNAS* **85,** 1801 (1988).
46. C. Zwieb, *NARes* **13,** 6105 (1985).
47. S. Boehm, *FEBS Lett.* **212,** 15 (1987).
48. G. Fox and C. R. Woese, *Nature* **312,** 171 (1975).
49. C. Zwieb and E. Ullu, *NARes* **14,** 4639 (1986).
50. L. Joshua-Tor, D. Rabinovich, H. Hope, F. Frolow, E. Appella and J. L. Sussman, *Nature* **334,** 82 (1988).
51. S. Roy, V. Sklenar, E. Appella and J. S. Cohen, *Biopolymers* **26,** 2041 (1987).

52. M. Miller, R. W. Harrison, A. Wlodawer, E. Appella and J. L. Sussman, *Nature* **334,** 85 (1988).
53. C. Zwieb, *Endocytobiosis Cell Res.* **3,** 167 (1986).
54. C. Marshallsay and C. Zwieb, *Endocytobiosis Cell Res.* **4,** 1 (1987).
55. R.-I. Ma, N. R. Kallenbach, R. D. Sheardy, M. L. Petrillo and N. C. Seeman, *NARes* **14,** 9745 (1986).
56. T. V. Kurzchalia, M. Wiedman, A. S. Girshovich, E. S. Bochkareva, H. Bielka and T. A. Rapoport, *Nature* **320,** 634 (1986).
57. U. C. Krieg, P. Walter and A. E. Johnson, *PNAS* **83,** 8604 (1986).
58. M. Wiedmann, T. V. Kurzchalia, H. Bielka and T. A. Rapoport, *JCB* **104,** 201 (1987).
59. H. M. Olson, P. D. Grant, B. S. Cooperman and D. H. Glitz, *JBC* **257,** 2649 (1982).
60. H. G. Wittmann, *in* "Structure, Function and Genetics of Ribosomes" (B. Hardesty and G. Kramer, eds.), pp. 1–27. Springer-Verlag, New York, 1986.
61. M. Oakes, E. Henderson, A. Scheinman, M. Clark and J. A. Lake, *in* "Structure, Function and Genetics of Ribosomes" (B. Hardesty and G. Kramer, eds.), p. 47. Springer-Verlag, New York, 1986.
62. C. Bernabeu, E. Tobin, A. Fowler, I. Zabin and J. A. Lake, *JCB* **96,** 1471 (1983).
63. Y. L. Chan, R. Gutell, H. Noller and I. Wool, *JBC* **259,** 224 (1984).
64. Y. L. Chan, J. Olvera and I. Wool, *NARes* **11,** 7819 (1983).
65. R. R. Gutell, B. Weiser, C. R. Woese and H. Noller, *This Series* **32,** 155 (1985).
66. R. Brimacombe, J. Atmadja, A. Kyriatsoulis and W. Stiege, *in* "Structure, Function and Genetics of Ribosomes" (B. Hardesty and G. Kramer, eds.), p. 184. Springer-Verlag, New York, 1986.
67. R. Brimacombe, J. Atmadja and W. Stiege, *JMB* **199,** 155 (1988).
68. D. Schüler and R. Brimacombe, *EMBO J.* **7,** 1509 (1988).

Eukaryotic DNA Polymerases α and δ: Conserved Properties and Interactions, from Yeast to Mammalian Cells

PETER M. J. BURGERS

Department of Biochemistry and Molecular Biophysics
Washington University School of Medicine
St. Louis, Missouri 63110

This review is prompted by a series of new understandings in the structure of the replicative DNA polymerases in eukaryotic cells. In the last 2 years, the gene for human DNA polymerase α has been isolated and characterized, and new high-fidelity forms of the enzyme have been purified, using rapid isolation techniques. DNA polymerase δ has gained considerable attention because of its interaction with a cell-cycle-regulated protein. This protein, proliferating cell nuclear antigen (PCNA), or cyclin, is also required for *in vitro* DNA replication of simian-virus-40 (SV40) DNA replication. We have isolated a yeast DNA polymerase δ and the yeast analogue of PCNA (cyclin). The emerging similarity between yeast and mammalian cells in DNA replication and the power of yeast genetic and molecular analyses make this organism an excellent choice for studying this fundamental process.

This review focuses mainly on DNA polymerases α and δ from mammalian cells and their accessory factors, and on the analogous

Progress in Nucleic Acid Research
and Molecular Biology, Vol. 37

DNA polymerases from *Drosophila melanogaster* and *Saccharomyces cerevisiae*.

DNA polymerase β and γ are briefly reviewed in Section I. For excellent earlier reviews on eukaryotic DNA polymerases, see ref *1* and *2*. Phylogenetic relationships of the DNA polymerases from various eukaryotic organisms are discussed in ref *2* and *3*.

I. Overview

A DNA-polymerizing activity in eukaryotic tissues (*4, 5*) was discovered not long after Kornberg's discovery of DNA polymerase I from *Escherichia coli* (*6–8*). The initial results indicating that the calf thymus enzyme would add deoxynucleotides to oligonucleotide primers in a nontemplated fashion (*9, 10*) were at odds with the known requirements and activities of the *E. coli* DNA polymerase (*6, 7, 11*): the DNA template directed covalent addition of complementary dNMPs from dNTPs to a base-paired oligo- or polynucleotide primer. Fractionation studies of the crude polymerase, however, led to separation of a template-independent terminal deoxynucleotidyl transferase from the DNA polymerase, which turned out to have characteristics similar to those of the *E. coli* enzyme (*12–14*). This DNA polymerase activity was later designated DNA polymerase α (Polα; *15, 15a*).

The isolation of an *E. coli* mutant deficient for DNA polymerase I (*16*) prompted the search for additional DNA polymerases in *E. coli* extracts, which resulted in the isolation of DNA polymerase II (*17–19*) and the replicative DNA polymerase III (*20–25*). With the example of multiple DNA polymerases in prokaryotes, a search for additional DNA polymerases in eukaryotes resulted in the isolation of the small DNA polymerase β from a variety of tissues and cell types (*26–29*). A third cellular DNA polymerase (DNA polymerase γ), which has considerable activity on RNA templates (*30, 31*), was later shown to be the mitochondrial DNA polymerase (*32–34*).

In 1975, the Greek-letter nomenclature system was proposed for the mammalian DNA polymerases to eliminate the confusion caused by the use of widely different laboratory nomenclatures (*15, 15a*).

In accordance with this nomenclature, a fourth mammalian DNA polymerase and the first nuclear polymerase to contain an associated 3′–5′-exonuclease activity was named DNA polymerase δ (*35*). The major differentiating characteristics of the four mammalian DNA polymerases are summarized in Table I.

The DNA polymerases from other animal cells are in general very similar to the mammalian enzymes, with a few exceptions. A β-like

TABLE I
DISTINGUISHING PROPERTIES OF MAMMALIAN DNA POLYMERASES α, β, γ, AND δ[a]

	DNA Polymerase			
Property	α	β	γ	δ
Catalytic polypeptide (kDa)	≥165	39	47, 135	125–220
Native size	>200	39	180–300	≥173
Associated 3′-5′ exonuclease	No	No	No	Yes
DNA primase	Yes	No	No	No
Catalytic properties				
Preferred template	Activated DNA	Activated DNA	Poly(rA)·oligo(dT)	Activated DNA
Divalent cation	Mg^{2+}	Mg^{2+}/Mn^{2+}	Mn^{2+}/Mg^{2+}	Mg^{2+}
Processivity	Intermediate	Low	High	Low/high[b]
Fidelity	High	Low	Intermediate	High
Inhibitors[c]				
NaCl (0.15 *M*)	++	−	−	++
Aphidicolin	++	−	−	++
N-ethylmaleimide	++	−	++	++
BuPhdGTP[d]	++	+[f]	−	+
ddTTP[e]	−	++	++	+

[a] Data for DNA polymerases β and γ taken from Refs. *1*, *2*, and *8*; for DNA polymerases α and δ, see the respective sections. (The nomenclature agrees with that in Ref. *15a*.)
[b] Depending on the form of Polδ isolated; see Section III.
[c] ++, Strong inhibition; +, weak inhibition; −, no inhibition.
[d] BuPhdGTP, N^2-(4-butylphenyl)-2′-deoxyguanosine 5′-triphosphate.
[e] ddTTP, 2′,3′-dideoxythymidine 5′-triphosphate.
[f] J. A. DiGiusseppe, S. H. Wilson, G. E. Wright and S. L. Dresler, personal communication.

DNA polymerase from *D. melanogaster* has a much higher M_r than the polymerases β from other animal cells, which are all ~40–50 kDa (Table II; *40*, *41*). The catalytic properties of this β-like polymerase, however, are very similar to those of the mammalian enzyme. These include sensitivity to inhibition by dideoxythymidine triphosphate and resistance to *N*-ethyl maleimide and aphidicolin (*41*).

The DNA polymerases γ from chick embryo and *D. melanogaster* are the most thoroughly studied mitochondrial polymerases from animal cells. The nearly homogeneous preparation from chick embryos showed two polypeptides of 47 and 135 kDa on SDS-PAGE[1]

[1] Abbreviations: Pol, DNA polymerase; MalNEt, *N*-ethylmaleimide; BuPhdGTP, N^2-(4-butylphenyl)-2′-deoxyguanosine 5′-triphosphate; PCNA, proliferating cell nuclear antigen; yPCNA, yeast PCNA; SDS-PAGE, sodium dodecyl sulfate/polyacrylamide gel electrophoresis; $A_{P_4}A$, P^1,P^4-bis(5′-adenosyl) tetraphosphate; SV40, simian virus 40.

TABLE II
ANIMAL CELL DNA POLYMERASES[a]

Organism	Type	Mass[b] (kDa)	3′-5′ Exo	Inhibition by			Reference
				MalNET	Aph	ddTTP	
Mammals	α	≥165	−	+	+	−	Section II
	β	39–43	−	−	−	+	*76–78a*
	γ	47, 140	−	+	−	+	*79–82*
	δ	125–220	+	+	+	+	Section III
Chick	α	133–155	−	+	+	−	*36, 83, 84*
	β	40	−	−	−	+	*36, 85*
	γ	47, 135	+	+			*36, 86, 87*
Drosophila	α	182	−	+	+	−	*37*
	α(δ)[c]	182	+	+	+	−	*38, 39*
	β	110	−	−	−	+	*40, 41*
	γ	125		+		+	*88, 89*
Xenopus	α	105, 178	−	+	+	−	*42–44*
	β	43	−	−	−	+	*42, 45*
	γ	(188)		+	−	+	*42*
Urchin	α	(6–8 S)		+	+	−	*46–48*
	β	50	−	−		+	*46*
	γ	(3.3 S)		+		+	*49*

[a] Abbreviations used: 3′–5′ Exo, 3′–5′ deoxyribonuclease activity; MalNEt, *N*-ethylmaleimide; Aph, aphidicolin; ddTTP, 2′,3′-dideoxythymidine 5′-triphosphate.

[b] Denatured mass of the catalytic polypeptide; in parentheses, mass or sedimentation coefficient of the native enzyme.

[c] A form of DNA polymerase α with 3′-5′-exonuclease activity.

(*36, 86*). Based on the separation of two multimeric forms of the enzyme by gel filtration and a correlation between specific activity and amount of the 47-kDa polypeptide in each form, it was concluded that the 47-kDa polypeptide contained the catalytic activity. Similarly, the catalytic activity was assigned to a 47-kDa polypeptide in a nearly homogeneous Polγ preparation from mouse cells, containing a 140-kDa polypeptide as a minor component in addition to the 47-kDa polypeptide (*82*). The homogeneous DNA polymerase γ from *D. melanogaster* contains two polypeptides of 125 and 35 kDa in a 1 : 1 molar ratio (*88, 89*). *In situ* gel assays for DNA polymerase activity after SDS-PAGE, however, showed that the 125-kDa polypeptide contained the catalytic site (*88*), a result opposite to that obtained with the mouse and chick enzymes (*82, 86*).

Both the chick and the *Drosophila* Polγ replicate natural DNA with a high fidelity [error rate 10^{-5} to 10^{-6} (*89, 97*)]. The chick enzyme was later shown to contain an associated 3′–5′-exonuclease activity with the properties of a proofreading enzyme (*98*). This determination has not been carried out for the *Drosophila* enzyme (*88, 89*). However, no exonuclease activity has been found in polymerases γ from mammalian cells (Table II; *99, 100*). In view of the low abundance of this enzyme and its high lability and microheterogeneity during purification, it has not been studied extensively in many different organisms. A more thorough investigation is necessary to determine whether the molecular weight of the catalytic polypeptide, as well as the presence of a proofreading 3′–5′ exonuclease, are conserved properties of Polγ in animal cells.

DNA polymerases have been isolated from a variety of unicellular eukaryotes (Table III). A comparison of the polymerases from these

TABLE III
DNA POLYMERASES FROM LOWER EUKARYOTES[a]

					Inhibition by		
rganism	Type	Class	Mass[b]	3′-5′ exo	MalNEt	Aph	Reference
accharomyces (yeast)	I,A	α	H	−	+	+	*50–54* (Section II)
	II,B	δ	H	+	+	+	*50–53*
	III	δ	H	+	+	+	*56, 57*
	mt	γ	I	−	+	−	*57, 58, 90*
eurospora	A		(H)	−	+		*59, 60*
	B		(H)	−	+		*59, 60*
	C	β	L		−		*60*
	mt		(I)		−		*60*
etrahymena	I,A	α	(H)	−	+	+	*61–63, 94*
	II,B	γ	I		+	−	*63, 64*
stilago	Major	δ	H	+	+		*65–67, 91, 92*
hlorella	I	α	(H)		+	+	*68*
	II(chl)	γ	(H)		+	−	*68*
hlamydomonas	A		(I)	−	+		*93*
	B		(H)	+	+		*93*
	C		(H)	+	+		*93*
uglena	A	α	(H)	−	+		*69, 70*
	B		(H)	+	+		*69, 70*
ictyostelium	A	α	(H)		+	+	*71, 72*
	B	β	(I)		−	−	*72*
rypanosoma	α	α	(H)		+		*73, 95*
	β	β	L		−		*73*
hysarum	α	α	H		+	+	*74, 75, 96*
	HSI, DEIII	β	H	−	−	−	*74, 75*

[a] Abbreviations (see Table II); mt, mitochondrial; chl, chloroplast.
[b] Denatured mass of catalytic polypeptide; mass of the native enzyme in parentheses; H, >100 kDa; I, 50–100 kDa; L, <50 kDa.

organisms with the prototypical mammalian DNA polymerases is problematic for several reasons: (1) From some organisms, only one (e.g., *Ustilago*) or two DNA polymerases have been isolated (Table III). (2) In many cases, a positive identification and thorough characterization of the mitochondrial enzyme have not been made. (3) DNA polymerases classified as distinct could be different forms of the same enzyme, related through differential proteolytic degradation or subunit composition. For instance, Pol-B and Pol-C from *Chlamydomonas* have very similar catalytic properties and sensitivities to inhibitors (*93*), and the same is true for Pol-A and Pol-B from *Neurospora crassa* (Table III; *59*), although the latter two appear to be distinct enzymes based upon lack of immunological cross-reactivity (*60*). (4) Not all DNA polymerases have been thoroughly characterized using inhibitors such as high salt concentrations, aphidicolin, 2′,3′-dideoxthymidine 5′-triphosphate, and N^2-(4-butylphenyl)-2′-deoxyguanosine 5′-triphosphate (BuPhdGTP) (*103*), which are diagnostic for distinguishing the mammalian polymerases. (5) Proteolytic degradation during purification of the enzymes not only makes it difficult to measure the size of the catalytic polypeptide, but can also alter the properties of the enzyme. Thus, extensive proteolysis of the yeast DNA polymerases makes these enzymes resistant to aphidicolin (*54, 101*). (6) Except for the enzymes from yeast and *Ustilago* (*50–53, 56, 67*), polymerase-associated exonuclease activities have not been characterized thoroughly enough to permit conclusions about their possible role in proofreading, i.e., whether they preferentially excise mismatched mononucleotides from the 3′ terminus of double-stranded DNA (*102*). The exonuclease activity copurifying with Pol-B from *Euglena* is more active on single-stranded than double stranded DNA, but more diagnostic data are lacking (*70*). The exonuclease activity copurifying with Pol-B from *Chlamydomonas* is as sensitive to heparin as the polymerase activity, suggesting a close interaction between the two activities, but the activity has not been further characterized (*93*).

In view of all these uncertainties, it is difficult to assess whether the DNA polymerases from these unicellular eukaryotes are very different from the mammalian enzymes and, secondly, to classify them according to the mammalian classification system (Table I). Thus, the assignments given in Table III must be considered very tentative. All organisms presumably have an α-type polymerase, i.e., an aphidicolin- and MalNEt-sensitive enzyme of high M_r lacking a 3′–5′-exonuclease activity. *Ustilago* is the exception, with a δ-like enzyme being the only DNA polymerase isolated from this organism

(*66*). A β-like DNA polymerase has been isolated from *Neurospora* (*60*), a trypanosome (*73*), *Dictyostelium* (*72*), and *Physarum* (*74, 75*), although the large sizes of the enzymes from the latter two organisms make them more similar to the *Drosophila* Polβ than to the mammalian enzymes (*40, 41*). No β-like polymerase has been found in yeast. The mitochondrial DNA polymerases from yeast and *Tetrahymena* are typical γ-type polymerases (*57, 58, 63, 64*), but the enzyme from *Neurospora* mitochondria is the only mitochondrial enzyme insensitive to MalNEt (quoted in *60*). The characterization by inhibitors and the classification of the *Tetrahymena* enzymes have been matters of dispute among various groups. Pol-II,B from *T. pyriformis*, characterized as MalNEt-resistant by one group, (*61*) was, under other reaction conditions, shown by others to be MalNEt-sensitive (*63*). This enzyme shows characteristics of Polβ and Polγ and may be similar to a mitochondrial enzyme isolated from the related *T. thermophylus*, which is induced up to 35-fold by exposure to agents that cause DNA damage (*64*). The DNA polymerase isolated from *Chlorella* chloroplasts is most similar to the mammalian Polγ (*68*). Finally, δ-like DNA polymerases have been isolated from yeast and *Ustilago* (*50–53, 56*).

At first sight, all of this indicates that no meaningful comparison can be made between the polymerases of unicellular eukaryotes and mammalian cells. However, an accumulating wealth of inforamtion from the best-studied lower eukaryote, *S. cerevisiae*, actually suggests that the opposite may be true and that, at least for Polα and Polδ, a remarkable conservation through evolution has been maintained. Since the other unicellular eukaryotes are evolutionarily not further removed from mammals than yeast is, it is likely that a large degree of conservation exists for the enzymes from these organisms as well, but that this is masked by the various experimental problems encountered in the study of these enzymes. A combined approach of genetic and biochemical methods is necessary in each of these organisms to reveal the true status of their replicative machinery. So far, this approach has been applied only to yeast.

II. DNA Polymerase α

In this section, the enzymes from mammalian cells, *D. melanogaster*, and *S. cerevisiae* (yeast) are the focus of discussion. More comprehensive reviews of DNA polymerases α from animal cells in general are in refs. *1, 2, 104*, and *105*, in addition to the overview in Section I. A detailed treatment of the catalytic mechanism of Polα has appeared before in this series (*104*) and elsewhere (*1*).

A. Purification and Subunit Structure

The emergence of a remarkable similar subunit structure of Polα from a variety of enzyme sources has vindicated the earlier work by Holmes and co-workers (Table IV; *115, 116, 154–156*), who purified several forms of Polα from calf thymus that were separable by DEAE-cellulose chromatography. Their form αA [actually, two very similar forms (*1, 2, 115, 154*)] resembles the prototypic Polα we know today. This form was separated by SDS-PAGE into a large polypeptide of 155 kDa and one or more smaller polypeptides of 50–70 kDa (*154*). Other forms of Polα could either be generated *in vitro* by removal of smaller subunits (*116, 154, 155*) or could be ascribed to proteolytic degradation of the large subunit (*155, 156*). Multiple forms of calf thymus Polα and their generation by proteolysis have also been observed by others (*117, 118, 157, 158*). Only recently has the introduction of immunoaffinity columns made feasible the purification of unproteolyzed Polα from calf thymus (*125, 126, 128, 129, 131*).

Similarly, proteolysis has been a problem in the purification of Polα from *D. melanogaster* (*159–161*). However, the enzyme could be isolated intact (*37, 113, 114*) by the application of three biochemical strategies: (1) the use of freshly harvested embryos, as the enzyme isolated from previously frozen embryos is primarily the degraded form (*161*); (2) a rapid purification procedure, to limit the exposure time of the enzyme to the endogenous proteases; (3) addition of several protease inhibitors.

Limiting proteolysis during the purification steps has allowed the isolation of nearly homogeneous, largely unproteolyzed preparations of Polα by conventional column chromatography from such diverse sources as yeast (*54, 162*), rat liver (*132, 133*), rabbit bone marrow (*134*), mouse Ehrlich ascites tumor cells (*136*), monkey CV-1 cells (*139*), and human HeLa cells (*148*; Table IV). The sizes of the large catalytic polypeptides in these preparations are in general agreement with the predicted sizes determined from *in situ* gel assays after SDS-PAGE of crude or partially purified cell extracts (*90–92*).

The introduction of immunopurification methods has occasioned a tremendous advance over the time-consuming conventional purification schemes of eukaryotic DNA polymerases. Polα has been immunopurified from yeast (*106–110*), calf thymus (*125, 126, 128, 129, 131*), and human KB cells (*150*). The subunit compositions of these immunopurified DNA polymerases are remarkably similar to each other as well as to those of the DNA polymerases extensively purified by conventional chromatographic methods (Table IV). The consensus eukaryotic DNA polymerase *α* consists of four subunits: the catalytic polymerase polypeptide of 180 kDa, a subunit of 70–75 kDa, and a pair of "primase" subunits of 60–55 and 50–48 kDa.

The gene sequence of the catalytic polymerase subunit from yeast predicts a molecular weight of 166,794 for this subunit (*111*). The gene sequence of the catalytic polymerase subunit from human KB cells predicts a molecular weight of about 165,000 for this subunit (*153*; T. S. F. Wang, personal communication). If the size of the catalytic polypeptide, which is almost equal for yeast and human cells, is preserved among other eukaryotes, then the occurrence of extremely large polypeptides (>200 kDa) in purified DNA polymerase preparations (*123, 127, 136, 158*) is either the result of posttranslational modification [e.g., glycosylation (*163–165*)] or artifactual. Glycosylation of the polymerase subunit from human KB cells has indeed been detected. It increases the apparent mass of this subunit by about 15 kDa (K. Hsi, A. F. Wahl and T. S. F. Wang, personal communication).

The function of the ~75-kDa polypeptide in most of these Pol*α* preparations, purified either by conventional or immunological methods, is unclear. For yeast, immunological evidence indicates that the 74-kDa polypeptide found in most immunopurified preparations of Pol-I (*106–109*) is derived, presumably by proteolysis, from a larger 86-kDa polypeptide (*110*). It has been postulated that this polypeptide might serve as an anchor connecting the polymerase with the primase subunits (*106*). A similar connecting and possibly regulatory role has been proposed for the 68-kDa polypeptide from calf thymus (*128*). In agreement with the proposed regulatory role of this subunit are experiments showing that a monoclonal antibody to the analogous 77-kDa subunit from human or mouse cells inhibits primase but not polymerase activity (*187*). The monoclonal antibody does not recognize the primase, and no primase activity is present in the 77-kDa subunit (*187*). The use of a panel of monoclonal antibodies and peptide mapping has clearly shown the consensus four-subunit structure of Pol*α* from yeast and human KB cells (*110, 152*). In Pol*α* from calf thymus a 68-kDa polypeptide (*128*) may be identical to the 73-kDa polypeptide identified by other investigators (*125, 129, 131*). Differences in the observed molecular weight could well be due to the use of different conditions for SDS-PAGE. In addition, a form of Pol*α* lacking this 74-kDa polypeptide and catalytically indistinguishable from the form containing this polypeptide has been purified (*129, 131*).

TABLE IV
DNA POLYMERASE α FROM DIFFERENT SOURCES

Source	SUBUNITS[a] (kDa)	Function	Purification method	Remarks	Reference
Yeast	180	Polymerase	Immun.[b]	Polymerase (M_r = 166,794) and 48-kDa primase (M_r = 47,623) genes sequenced	*106–112*
	86 (or 74)	Anchoring protein?			
	58, 48	Primase			
Drosophila embryo	182	Polymerase	Conv.[c]		*37, 113, 114*
	73	Exonuclease inhibitor			
	60, 50	Primase			
Calf thymus	140–170, 45–70		Conv.	Form A, 8–10 S	*115–118*
	118–200 (3)	Polymerase	Conv.	Form RC-α (11.5 S), contains 3′–5′ exonuclease	*120–124*
	116	Calmodulin binding			
	98, 87, 63				
	54, 49	Primase			
	47	Nuclear factor I?			
	150–180	Polymerase	Immun.	Native size: 335–404 kDa	*125–131*
	73	Anchoring protein?			
	59	Primase accessory factor			
	48	Primase			
Rat liver	156, 54–64 4 more subunits		Conv.	Contains primase	*132, 133*
Rabbit marrow	135, 54–66 (4)		Conv.		*134*

Mouse	115–183 48–70		Conv.	Contains primase	*135–137*
	185, 68		Conv.	No primase	*138*
Monkey	118–176 53–62, 30		Conv.	Contains primase	*139*
	190, 40–73 (5)		Immun.	Precipitation of polymerase subunits	*140*
Human	180 70 69 52 47 25	Polymerase Primase Exonuclease Protein C2 $A_{P_4}A$ binding Protein C1	Conv.	From synchronized HeLa cells α_2 form, native size = 640 kDa	*141–147*
	130–155 64, 52, 45		Conv.	From HeLa cells, contains primase	*148*
	180 77 55, 49	Polymerase Primase	Immun.	From KB cells, (partial) polymerase gene sequenced (M_r > 165,000)	*149–153*

[a] In parentheses: number of polypeptides in the indicated range.
[b] Immuno., purified using immune-affinity chromatography.
[c] Conv., purified using conventional column chromatography.

However, a clear role for the 73-kDa polypeptide has been found in the enzyme from *Drosphila*. Sedimentation of the four-subunit Polα on a glycerol gradient in the presence of 50% ethylene glycol, a mildly denaturing agent (*166*), separated the polymerase polypeptide from the 73-kDa polypeptide and the primase polypeptides (*38*, *167*). Upon removal of this 73-kDa polypeptide, a 3′–5′-exonuclease activity capable of "proofreading" was detected in the 182-kDa monomeric enzyme (*38*, *39*). In accordance with this observation, the 182-kDa monomeric enzyme displayed an increased fidelity of DNA replication (*168*).

The two subunits of 60–55 kDa and 50–48 kDa form the primase. They are discussed in Section II,B.

In addition to these preparations of Polα purified by immunoaffinity columns or conventional column chromatography (i.e., a large number of purification steps using cationic and anionic exchange resins, hydrophobic chromatography, sizing columns, and affinity columns such as DNA cellulose), attempts have been made to purify larger complexes containing the polymerase as well as accessory factors and enzymes likely present at the replication fork. Replisome-like Polα complexes have been isolated from calf thymus and HeLa cells (Table IV; 120–124, 141–147). In these approaches, purification steps such as hydrophobic chromatography and hydroxyapatite chromatography that tend to dissociate the loosely bound accessory factors from the Polα core were avoided (*120*). The isolation of a replication-competent holoenzyme form of Polα from calf thymus required the addition of ATP to all column steps, and purification was accomplished with only two ion-exchange resins and a sizing column (*123*). The large form of Polα from calf thymus contains about 10 polypeptides (*120*, *123*, *124*). In addition to a few polymerase polypeptides, primase activity, 3′–5′-exonuclease activity, and a calmodulin-binding protein were present in the purified complex (Table IV; *123*, *124*). In a slightly different fractionation scheme, other activities that could participate in the DNA replication process (i.e., DNA topoisomerase II, DNA helicase, DNA methylase, and RNase H) copurified with various complex forms of Polα separable by DEAE-cellulose chromatography (*121*, *122*).

For the isolation of the multiprotein complex from synchronized S-phase HeLa cells, preparative PAGE[1] was the cardinal step in obtaining a pure preparation (*143*). This complex from HeLa cells, called Polα_2, also contained primase activity, 3′–5′-exonuclease activity, a P^1,P^4-bis(5′-adenosyl) tetraphosphate ($A_{P_4}A$) binding protein (*119*), and the primer recognition factors C1 and C2 (Table IV; *143*,

144, 146, 147). Some of these accessory factors are discussed in Section II,B.

B. Accessory Factors

1. 3′-5′ Exonuclease

None of the conventionally or immunopurified four-subunit preparations of Pol*α* has an associated 3′–5′-exonuclease activity. However, upon removal of the 73-kDa subunit from *D. melanogaster* Pol*α*, a proofreading 3′–5′-exonuclease activity appeared in the 182-kDa polymerase polypeptide (*38, 39*). Because the altered properties of the monomeric enzyme agree more with a δ-like polymerase than an *α*-like polymerase, this form of *Drosophila* Pol*α* is discussed in Section III.

A form of calf thymus Pol*α* lacking the 73-kDa subunit has been obtained by immunopurification. No 3′–5′ exonuclease actually was detectable in this form and its biochemical properties were not different from those of the four-subunit enzyme (*129, 131*; F. Grosse, personal communication).

The multiprotein complexes from both calf thymus and HeLa cells do contain a 3′–5′-exonuclease activity (*121, 123, 143, 144*). Only the enzyme from HeLa cells has been further characterized (*144*). The exonuclease, contained in a 69-kDa polypeptide, is primarily a 3′–5′ single-stranded DNA-dependent exonuclease that can excise single nucleotide mismatches at the 3′ terminus of a double-stranded DNA substrate, as required for a proofreading enzyme (*102*). In addition, a small amount of a single-stranded DNA-dependent 5′–3′-exonuclease activity was also detected in this preparation. The exonuclease is similar to a polymerase-associated exonuclease activity detected in one of two forms of purified Pol*α* from mouse myeloma cells (*169*). Whether the exonuclease from HeLa cells increases the fidelity of DNA replication by Pol*α* has not yet been determined.

2. DNA Primase

DNA polymerases *α* from eukaryotic cells contain a tightly associated DNA-primase activity that cannot easily be dissociated from the polymerase by chromatographic methods. This association has been observed for such diverse organisms as sea urchin (*46, 47*), *Drosophila* (*170–173*), *Xenopus* (*174, 175*), mouse cells (*136, 137, 176, 177*), calf thymus (*126–129, 179, 180*), human cells (*148, 150, 151*), and yeast (*106, 109, 110, 181, 182*). The primase activity of the *Drosophila* Pol*α* has been resolved from the polymerase activity by sedimentation through a glycerol gradient in the presence of mildly denaturing agents such as 2.4 *M* urea or 50% ethylene glycol and shown to reside in the 60- and 50-kDa subunits (*38, 172, 173*). Free primase could also be eluted from a monoclonal antibody column containing bound yeast or calf thymus polymerase–primase complex by washing the column with 1–2 *M* KCl at neutral or alkaline pH (*106, 128, 130, 183*).

The yeast primase eluted in this manner showed two subunits of 58 and 48 kDa (*106*). An antiserum raised against the 48-kDa subunit inhibited primase activity (*107*). The 48-kDa subunit also contains an ATP binding site, indicating that the catalytic site is probably contained in this subunit (P. Plevani, personal communication). The *PRI1* gene coding for this subunit has been cloned and sequenced (*107, 112*). The predicted molecular weight of 47,623 agrees well with the mass of this subunit determined by SDS-PAGE (*112*). The *PRI1* gene is essential in yeast, as expected for a protein that has a vital function in the DNA replication machinery (*109*). The gene for the 58-kDa subunit has been cloned and sequenced as well (M_r = 61,200). This gene is also essential (P. Plevani, personal communication).

The calf thymus primase showed two subunits of 55 and 48 kDa in one report (*128*) and 59 and 48 kDa in another study (*130*). The 48-kDa monomer has been isolated by DEAE-cellulose chromatography of the heterodimer (*130*). This 48-kDA subunit has primase activity and contains a binding site for GTP, but is extremely labile. These and additional immunochemical results indicate that the 48-kDa subunit contains the catalytic site. The 59-kDa subunit presumably stabilizes the primase activity. In contrast to these results, a form of Polα from mouse cells showed two subunits of 115 and 58 kDa (*137*). The 58-kDa subunit was separated from the 115-kDa subunit, and contained primase activity (*137*).

These studies again show the remarkable evolutionary conservation of the Polα–primase subunit structure from yeast to mammals. More extensive documentation is required to determine the interaction between the two primase subunits and to establish the catalytic polypeptide in the various Polα preparations.

In addition to primase tightly complexed with the DNA polymerase α, free primase can also be purified to apparent homogeneity from yeast (*184–186*). This free primase, a monomer of 59–65 kDa, may either be a separate enzyme with, possibly, a function different from the Polα-bound primase, or be related to the 58-kDa subunit of the Polα-bound primase (*106–110*). In this case, the catalytic site for primase would reside in the 58- and not in the 48-kDa polypeptide. In light of the experiments discussed above, this seems unlikely.

However, since antibodies to the free primase (*186*) and the 58-kDa subunit (*107, 110*) have been characterized, an exchange of enzymes and reagents between the laboratories involved should help solve this interesting problem, at least for yeast.

Free monomeric (58 kDa; *137*) and heterodimeric (56 and 46 kDa; *178*) primase has also been purified from mouse cells, and free primase has been released through hydrophobic chromatography from the multiprotein Pol*α* complex of HeLa cells (70 kDa; *146*).

Regardless of the size or form of the DNA primases obtained from the various eukaryotic cells, no large differences have been observed in the catalytic mechanism of these enzymes (reviewed in *1, 188, 188a*). In the absence of concomitant DNA synthesis, the polymerase–primase complex synthesizes RNA products of discrete length (9–14 nucleotides), and often multimers thereof. Addition of dNTPs to the reaction suppresses the formation of these multimers, but does not change the length of the monomer RNA. Where the catalytic mechanism of the polymerase-free primase could be compared with that of the polymerase–primase complex, no major differences were detected (*130, 189*). One interesting difference between the free primase and the polymerase–primase complex of *Drosophila* is that the free primase shows enzymatic turnover whereas the complex does not (*189*).

3. Other Accessory Factors and Effector Molecules

Several accessory factors copurify with one or more forms of multiprotein complexes of Pol*α* (Table IV).

Proteins C1 and C2 were first identified as cofactors that would allow DNA synthesis by Pol*α* on DNA substrates with a low primer : template ratio, such as denatured DNA (*141, 190–192*). Factor C1 is a tetramer of identical 24-kDa subunits and factor C2 is a monomer of 52 kDa (*141, 143*). Their presence in the highly purified multiprotein complex from HeLa cells classifies them as subunits of DNA polymerase *α* (*143*). The C1C2 complex lowers the K_m values of the primer stem 20 to 30-fold, effectively eliminating nonproductive binding by Pol*α* to single-stranded DNA (*192*). Interestingly, the C1C2 complex interacts only with the Pol*α* from the same source. C1C2 from HeLa cells does not stimulate Pol*α* from monkey CV-1 cells, and vice versa (*192*).

Ribonucleases H (RNase H), which stimulate the homologous DNA polymerases, have been isolated from yeast and *D. melanogaster* (*333, 334*). The 70-kDa RNase H from yeast, but not any of the other known RNases H from the same organism, stimulate DNA synthesis on poly(dA)·oligo(dT) by Pol-I (*333*; U. Wintersberger, personal communication). Yeast Pol-II and *E. coli* Pol-I are not stimulated by the 70-kDa RNase H. RNase H probably exerts its stimulatory effect by increasing the affinity of Pol-I for the primer terminus (*333*). In contrast, the RNase H from *D. melanogaster*, a tetramer consisting of two 49- and two 39-kDa subunits, did not stimulate DNA synthesis by the homologous Polα on a primed DNA template, but stimulated DNA synthesis only in coupled priming-chain elongation reactions (*334*). The stimulation of DNA synthesis by the RNase H was primarily at the level of primase action by increasing the recycling rate of the polymerase–primase complex and, in this way, the rate of chain initiations (*334*).

Multiprotein complexes from calf thymus and HeLa cells contain an Ap_4A[1] binding protein (Table IV; *119, 142, 143, 147, 193, 194*). Ap_4A, a product of the reverse reactions in amino-acid activation by certain aminoacyl-tRNA synthetases (*195, 196*) is present at elevated levels in proliferating cells in general and in S-phase cells in particular (*197–199*). Equilibrium dialysis experiments show that the level of Ap_4A binding in lysates of differentiating neurons decreases as the level of Polα diminishes (*119*). Ap_4A could function as a primer for *in vitro* DNA synthesis on poly(dT) by Polα, although this dinucleotide was not much more efficient than the natural (3′–5′) oligoadenylates (*194*). On the other hand, no stimulatory effect of Ap_4A on DNA polymerase activity on natural DNA could be observed (*147, 200*) and Ap_4A does not prime DNA synthesis on viral single-stranded DNA (*201*). Thus, although this establishes a link between Polα and a proliferation-dependent small molecule via its binding protein, it does not provide further information as to how Polα is involved in this process.

A link has also been established between Polα and inositol 1,4-bisphosphate (*202–204*), a product of the phosphoinositidate turnover system in the generation of second messengers (reviewed in *205–208*). Inositol 1,4-bisphosphate can either be produced by phospholipase C-mediated breakdown of phosphatidylinositol 4-monophosphate or through hydrolysis of inositol 1,4,5-trisphosphate, formed similarly by phospholipase C-mediated breakdown of phosphatidylinositol 4,5-biphosphate (*207*). Breakdown of phosphatidylinositol 4,5-bisphosphate in response to specific agonists mediates a variety of cellular responses, including cell proliferation through the mobilization of intracellular calcium (*206, 208*). Inositol 1,4,5-trisphosphate, but not the 1,4-bisphosphate, is one of the second messengers in the mobilization of calcium.

Two forms of Polα from human fibroblasts, A_1 and A_2, have been separated by DEAE-cellulose chromatography, and further purified by immunoaffinity chromatography of each form (*204*). Form A_1 had a low specific activity and bound weakly to activated DNA-cellulose, whereas form A_2 had a high specific activity and bound tightly to DNA-cellulose. Incubation of form A_1 with phosphatidylinositol 4-phosphate or inositol 1,4-bisphosphate increased the specific activity of the enzyme about 10-fold and increased the binding affinity of the enzyme for DNA cellulose such that the activated form of A_1 became indistinguishable from form A_2. The phosphatidylinositol 4-phosphate added to form A_1 was hydrolyzed, and inositol 1,4-biphsophate but not diacylglycerol remained tightly bound to the enzyme (*204*). Free inositol 1,4-bisphosphate also bound strongly to form A_1, but not to form A_2. A number of other inositol phosphates including inisitol 1,4,5-trisphosphate did not activate form A_1 (*204*). These observations establish a link between Polα and cell proliferation, but clearly not through the main components involved in the messenger system. Activation of preexisting inactive Polα (*209, 210*) by small effector molecules at the appropriate time and location for DNA replication may be a common mechanism in the control or fine tuning of DNA replication in eukaryotic cells (*204, 211*).

C. Catalytic Mechanism and Fidelity

1. Catalytic Mechanism

The most complete enzymological description of Polα has been carried out with the homogeneous enzyme from human KB cells by Korn and co-workers (*211–217*; reviewed in *1, 104, 218*). Although these studies were performed with a proteolyzed form of Polα, a heterodimer of 76 and 66 kDa (*212*), a reassessment of some of the enzymatic properties with the immunopurified four subunit enzyme showed them to be virtually identical (*152*). The nucleic-acid-binding specificity of Polα was measured using a competition inhibition assay with nonsubstrate DNA (*211*). Single-stranded DNA, but not duplex DNA, was an effective competitor. The order of binding of homopolymers to Polα was poly(dT) $>$ poly(dC) $>>$ poly(dA) (*215*). In addition, competition inhibition experiments with single-stranded circular ϕX174 or M13 DNA showed positive cooperativity, indicating two interacting single-stranded DNA binding sites per polymerase molecule (*211*). Finally, an ordered sequential mechanism of substrate binding (i.e., template–primer followed by complementary dNTP) to Polα was determined from a series of kinetic studies (*152, 216, 217*). This ordered sequential mechanism, which is an important aspect of

DNA polymerase action because it relates to the fidelity of DNA synthesis, has also been determined by others (*219, 220*).

In contrast to the enzyme from KB cells, there are significant differences in enzymatic properties between the conventionally and immunopurified Polα from calf thymus (*117, 131*). Thus the K_m values for the dNTPs and primed single-stranded M13 DNA are all an order of magnitude lower with the immunopurified enzyme, indicating a much better interaction between the more native enzyme and the substrates. The immunopurified enzyme was also more aphidicolin-sensitive and less sensitive to salt inhibition (*131*). The decreased sensitivity to aphidicolin as a result of proteolytic enzyme degradation has also been observed by others for yeast DNA polymerase I (*54, 101*).

2. Processivity

Several methods have been developed for measuring the processivity of a DNA polymerase (reviewed in *8*). Of these, the direct analysis of the products of polymerase action on poly(dA)·oligo(dT) at very low primer : template ratios is the most convenient assay and most often used (*219, 221*). Although the processivity on this model template may be different from that on natural DNA, it provides a fast and easy assay for assessing conditions that determine processivity. Thus, the processivity of calf thymus Polα varies from very low to very high by adjusting parameters such as Mg^{2+} concentration, salt concentration, and pH (*222–224*). Because of the effect of temperature changes on the pH of the assays with the Tris buffer used in one of these studies, no conclusions can be drawn about temperature effects on the processivity of Polα (*223*). However, processivity measurements with *E. coli* DNA polymerase I in phosphate-buffered reactions, which have a negligible temperature coefficient, showed that a decrease in the temperature from 37°C to 5°C increases the processivity about 10-fold (*225*).

At a pH of 6.5, increasing the Mg^{2+} concentration from 1 to 10 m*M* decreases the processivity of calf thymus Polα from 500–2000 (nucleotides incorporated per binding event) to 100–200. At a pH of 8.0, a similar increase in the Mg^{2+} concentration decreases the processivity from 50–100 to 2–5 (*224*). The optimal rate of DNA synthesis was at pH 7.0 and 5 m*M* $MgCl_2$, indicating that, in addition to processivity, the kinetics of primer binding determines the overall rate of DNA synthesis (*224*). Due to the extreme sensitivity of the synthetic rate and the processivity of Polα to changes in pH and free Mg^{2+} concentration, the apparent stimulation of Polα by millimolar concentrations of ATP (*136, 100, 226–228*) can, at least in part, be attributed to the concomitant changes in free Mg^{2+} and, for some studies, to the pH with the type of buffer system used (*222*).

Because of the large variation in processivity measured for Pol*α* in response to the assay conditions (*222–224*), the primary value of measuring the processivity does not lie in its absolute number, but rather in determining the mechanism of action of effector molecules and accessory proteins. Thus, the C1C2 complex from human or monkey cells facilitates binding of Pol*α* to the primer terminus, but does not affect the processivity of the enzyme (*191, 192*). *E. coli* single-stranded DNA binding protein and mouse helix destabilizing protein-1 significantly increase the processivity of the monomeric Pol*α* from *D. melanogaster* and mouse Pol*α*, respectively (*167, 219*). Finally, the ability of Pol*α* from mouse cells to replicate long single-stranded DNA, such as viral M13 DNA, efficiently seems largely due to its interaction with an accessory factor (AF-1) (*229, 230*). Similarly, the efficient replication of single-stranded viral DNA by multiprotein complexes from calf thymus (*120, 123*), HeLa cells (*147*), and yeast (*231*) should proceed via a processive mechanism as well. In contrast, under appropriate conditions, the immunopurified four-subunit Pol*α* from calf thymus efficiently replicated single-stranded M13 DNA in a processive fashion without the need of added accessory factors (*129, 131*).

3. Fidelity of DNA Replication

The fidelity of DNA replication in yeast has recently been estimated at 2×10^{-9} errors per base-pair replicated (*232*). DNA replication errors in higher eukaryotes have been estimated to be even less frequent at 10^{-9} to 10^{-11} (*233;* reviewed in *1, 104, 234*). In *E. coli*, the overall fidelity of 10^{-9} to 10^{-10} is obtained in a three-step mechanism: (1) selection and insertion of the correct nucleotide by the polymerase; (2) exonucleolytic proofreading (recently reviewed in *235*); and (3) methylation-dependent mismatch repair (*236*). In the absence of a proofreading exonuclease in most preparations of Pol*α* (Table IV), the *in vitro* fidelity of DNA synthesis is still very high (10^{-5} to 10^{-6}), much higher than expected from simple base-pairing considerations. The free-energy difference of hydrogen-bond formation between correct and incorrect base-pairs in aqueous solution would predict a fidelity of about 10^{-2} (*237*), as is actually found for the non-enzymatic templated polymerization of suitably activated nucleotides (*238*).

Several mechanisms have been proposed by which Pol*α* might enhance this basic fidelity by 10^3 to 10^4. In the "K_m discrimination" model, an increase in the free energy of base-pairing on the enzyme surface due to its hydrophobic environment results in an increased free energy difference between correct and incorrect base-pairing (*239–242*). As a K_m value, or more precisely a K_D value, is a function of the free energy of base-pairing, error frequencies should be determined by differences in K_m values between correct and incorrect dNTPs. Actually measured K_m differences between correct and incorrect dNTP are 1100 to 2600-fold (*241*). However, V_{max} values for insertion of the incorrect dNTP are a quarter to an eighth lower as well, perhaps reflecting a suboptimal alignment of the incorrect dNTP (*241*). Thus, the kinetics of dNTP binding and insertion provide a fidelity factor of 10^{-4}.

Several models of non-exonucleolytic proofreading have been proposed. In the "energy-relay" model, the energy provided by phosphodiester-bond formation with release of pyrophosphate is used by the enzyme to proofread the insertion of the next nucleotide (*243*). This model is testable because it predicts a lower fidelity for the first nucleotide inserted by the DNA polymerase. No such "first-nucleotide-effect" was observed, however, indicating that, in its strictest sense, such a mechanism is not operative (*244, 245*).

Two models involving pyrophosphate-mediated proofreading have been proposed. In one model, the energy for proofreading would be provided by cleavage of the α–β phosphoanhydride bond in an enzyme·DNA·dNTP intermediate with release of the dNMP and PP_i (*246–248*). No such generation of dNMP has ever been observed with highly purified Polα (*213*). Also, this model predicts a decrease in replication fidelity upon addition of pyrophosphate to the assay, whereas the opposite was observed for Polα (*249*). A second model for pyrophosphate-mediated proofreading postulates a dNMP as an intermediate in phosphodiester-bond formation (*250, 251*):

$$\text{Pol}\cdot\text{DNA}_n\cdot\text{dNTP} \rightleftarrows [\text{Pol}\cdot\text{DNA}_n\cdot\text{dNMP}\cdot\text{PP}_i] \rightleftarrows \text{Pol}\cdot\text{DNA}_{n+1} + \text{PP}_i$$

In agreement with this model is the pyrophosphate-enhanced fidelity of Polα (*249*), the incorporation of added dNMPs into DNA under conditions of ongoing DNA synthesis, and the apparent exchange of both dNMP and PP_i into dNTP under conditions of ongoing DNA synthesis (*251*). To observe these effects, dNMPs must be present in a 10^3- to 10^4-fold excess over the analogous dNTPs (*251*). For an efficient, forward direction of DNA synthesis, however, the dNMP in the central complex must be present in an activated form, perhaps covalently linked to an enzyme amino acid side-chain.

The stereochemistry of phosphodiester-bond formation has been measured for several DNA polymerases (*252–254*). The inversion of configuration at the α-phosphorus atom favors an in-line mechanism of phosphodiester-bond formation, involving a nucleophilic attack of the primer terminus 3′-OH group on the α-P of the dNTP with the concomitant release of PP_i (*252, 255, 256*). However, the participation of a covalent enzyme-dNMP intermediate would lead to retention of configuration at the α-P, whereas the presence of a readily exchangeable non-covalently bound dNMP intermediate in the complex could well cause racemization of the α-P. Since most of these stereochemical studies have been done with prokaryotic DNA polymerases, a similar study with Polα should be very interesting.

Finally, dissociation of Polα after misincorporation should be considered as a fidelity mechanism. After the stable incorporation of an incorrect nucleotide, dissociation of the DNA polymerase is favored (*241, 257, 258*) and rebinding of the DNA polymerase to the mismatched primer is very unfavorable (*242*). In the sensitive *in vitro* fidelity assays, which depend upon transfection of the replicated DNA into *E. coli* for scoring replication errors (see below), the terminal mismatch would be efficiently repaired by the *E. coli* repair system and would go undetected (*244*). In *E coli, in vivo* repeated dissociations from the replication fork of the DNA polymerase III holoenzyme, which must replicate 5×10^6 nucleotides in a very short time, would probably be detrimental (*8*). In the eukaryotes such as yeast, replication origins are spaced every 36 kilobases apart (*262*); taking into consideration the rate of fork movement and the duration of the S phase (6 kilobase-pairs per minute and 50 minutes, respectively) (*263, 264*) an occasional dissociation of the polymerase from the replication fork in a small subset of the replication units could, in principle, allow repair by a mismatch repair system, followed by a reassociation of the replisome.

The most sensitive fidelity assays measure either: the reversion rate of a coliphage φX174 amber codon by transfection in *E. coli*, following *in vitro* replication of single-stranded φX174 DNA past the particular amber codon (*234*); or the forward mutation rate in a region of the lacZ gene by a plaque color assay in *E. coli*, following *in vitro* DNA replication of that portion of the *lac*Z gene in a single-stranded M13 vector (*259*).

Previous work established a link between proteolysis and fidelity. Proteolyzed forms of calf thymus Polα have a much lower fidelity than more intact forms (*261*). Similarly, the more native unproteolyzed form of Polα from *D. melanogaster* has a higher fidelity than previously reported for the proteolyzed form (*114*). Following up on these observations, it has recently been shown (*261*) that freshly immunopurified Polα from calf thymus, hamster V-79, and human TK-6 cells all have very high fidelities of 1–2 $\times$ 10^{-6} when measured in the ϕX174 amber reversion assay. Upon prolonged storage of the enzyme at −70°C, the fidelity decreased to about a tenth (*261*). Because the testing of the above-listed fidelity mechanisms were all carried out with the low-fidelity enzyme (~2 $\times$ 10^{-5}), it would be interesting to repeat those experiments with the high-fidelity enzyme to find out if one of these postulated mechanisms might have become inoperative upon enzyme storage or proteolysis.

D. The Polα Gene: Its Expression and Mutants

The gene for yeast DNA polymerase I was isolated by immunoscreening a genomic λgt11 library with polyclonal antibodies against purified Pol-I (*162, 265*). The M_r of the polymerase deduced from the gene sequence is 166,794 (*111*). The human gene for Polα was isolated from a cDNA library using oligonucleotide probes deduced from the amino-acid sequences of several peptides derived from the purified polymerase polypeptide (*153*). A cDNA clone of almost full length that contained a long open reading frame with the coding capacity for a 165-kDa protein was obtained (*153*). A comparison of the primary amino-acid sequence of the human Polα deduced from the nucleotide sequence showed large regions of homology with the sequence derived from the yeast gene (Fig. 1) (*111*), and with several other polymerase sequences. This family of DNA polymerases, which have conserved sequence domains, includes phage T_4 DNA polymerase (*266*), phage ϕ29 DNA polymerase (*267*), phage PRD1 DNA polymerase (*268*), a DNA polymerase encoded by the yeast "killer" plasmid pGKL1 (*269*), a DNA polymerase from the S1 plasmid in maize mitochondria (*270*), and DNA polymerases from such mammalian viruses as adenovirus Ad2 (*271*), vaccinia virus (*272*), herpes simplex virus (*273*), and distantly related viruses (*274–276*).

The complete amino-acid sequence of human Polα and yeast Pol-I are compared in Fig. 1. Extensive regions of homology between the two sequences occur predominantly in the carboxy-terminal half of the protein. In addition, these two cellular DNA polymerases are homologous to the above listed viral and phage polymerases in the specific regions indicated in Fig. 1. The numbering of these regions in order of degree of homology was established previously (*153, 268, 273*). Thus, region I contains the highly homologous YGDTDS domain, region II contains the SLYPSII domain, region III contains the NS-YG motif, and region IV, which is missing in ϕ29 DNA

```
  1' Yeast                           MSSKSEKLEKLRKLQAARNGTSIDDYEGDESDGDRIYDEI

  1" Human     MAPYHGDDSLSDSGSFVSSRARREKKSKKGRQEALERLKKAKAGEKYKYEVEDFTGVYEE

 41' DEKEYRARKRQELLHDDFVVDDDGVGYVDRGVEEDWREVDNSSSDEDTGNLASKDSKRKKNIKREKDHQI
                   .::..:::::.::    ::.. ::. ... ..:. . ..:... : . : ... .
 61" VDEEQYSKLVQARQDDDWIVDDDGIGY----VEDG-REIFDDDLEDDALDADEKGKDGKARNKDKRNVKK

111' TDMLRTQHSKSTLLA-HAKKSQKKSIPIDNFDDILGEFESGEVEKPNILLPSKLRENLNSSPTSEFKSSI
      .. .... :: ..:  .::...:.. ... :..::..      ... :.  :      .....:.  .:
126" LAVTKPNNIKSMFIACAGKKTADKAVDLSK-DGLLGDI----LQDLNTETP-----QITPPPVMILK--K

180' KRVNGNDESSHDAGISKKVKIDPDSSTDKYLEIESSPLKLQSRKLRYANDVQDLLDDVENSPVVATKRQN
     ::  :.. .. .. ... :  .  .:. .  :    .:. :.. ..  ..:::    .. :.... .  ..
191" KRSIGASPNPFSVHTATAVPSGKIASPVSRKEPPLTPVPLKRAEFA-GDDVQVESTEEEQESGAMEFEDG

250' VLQDTLLANPPSAQSLADEEDDEDSDEDIILKRRTMRSVTTTRRVNIDSRSNPSTSPFVTAPGT-PIGIK
     ...... ..  . ...:... :..:. .  .:. . .. .:.. .             ..:... .  :. .
254" DFDEPMEVEEVDLEPMAAKAWDKESEPAEEVKQEADSGKGTVSYL----------GSFLPDVSCWDIDQE

319' GLTPSKSLQSNTDVATLAVNVKKEDVVDPETDTFQMFWLDYCEVNNTLILFGKVKL-KDDNCVSAMVQIN
     : .. .  . ..: . :..     ..     .     .. .  .: .  ....::::: . .... :: : ..
313" GDSSFSVQEVQVDSSHLPLVKGADEEQVFHFYWLDAYEDQY-NQPGVVFLFGKVWIESAETHVSCCVMVK

388' GLCRELFFLPRE-------GK---TP---TDIHEEIIPLLMDKYGLDNIRAKPQKMKYSFELPDIPSESD
     .. :.:.:::::         ::    ::    .:..::.   . .:: . ....:: ...:.::.::.:..:.
382" NIERTLYFLPREMKIDLNTGKETGTPISMKDVYEEFDEKIATKYKIMKFKSKPVEKNYAFEIPDVPEKSE

445' YLKDLLPYQTPKSSRDTIPSDLSSDTFYHVFGGNSNIFESFVIQNRIM(G)PCWLDIKGADFNSIRNASHCA
     ::    . :    ....  .: ::...:: ::::.:.. .: :.....:.::::::..: .. .   . .: :
452" YL--EVKY---SAEMPQLPQDLKGETFSHVFGTNTSSLELFLMNRKIKGPCWLEVKKST-ALNQPVSWCK

515' VEVSVDKPQNITPTTTKTNPNLRCLSLSIQTLMNPKENKQEIVSITLSAYRNISLDSPIPENIKPDDLCT
     ::. . ::. .. ... . :  : ...:..:. :.:....::.... ......::.. :.    ....:.
516" VEAMALKPDLVNVIKDVSPPPLVVMAFSMKTMQNAKNHQNEIIAMAALVHHSFALDKAAPKPPFQSHFCV

585' LVRPPQSTSFPLGLAALAKQKLPGRVRLFNNEKAMLSCFCAMLKVEDPDVIIGHRLQNVYLDVLAHRMID
     . .: ... :: .. .. ..:    .: . ..:...:. : :...  :::.:.::.. .  :.:: .:.
586" VSKPKDCI-FPYAFKEVIEKK-NVKVEVAATERTLLGFFLAKVHKIDPDIIVGHNIYGFELEVLLQRINV
                                                    ------------IV----------
655' LNIPTFSSIGRRLRRTWPEKFGRGNSNMNHFFISDICSGRLICDIANEMGQSLTPKCQSWDLSEMYQVTC
      . : .:.:: ::.:.    :.: :.:....   ...: ::.:::.  :....    .:.:..:::. : .
654" CKAPHWSKIG-RLKRSNMPKLG-GRSGFGE--RNATC-GRMICDV--EISAKELIRCKSYHLSELVQQIL

725' EKEHKPLDIDYQNPQYQNDVNSMTMALQENITNCIVSAEVSYRIQLLTLTKQLTNLAGNAWAQTLGGTRA
     ..:.  . ..  .   : ..   ..         .. .  ...:.:. :.::.:::   ..:: :.:..:::.:::.
718" KTERVVIPMENIQNMYSESSQLLYL-LEHTWKDAKFILQIMCELNVLPLALQITNIAGNIMSRTLMGGRS

795' GRNEYILLHEFSRNGFIVPDKEGNRSRAQKQRQNEENADAPVN-----SKKAKYQGGLVFEPEKGLHKNY
     . . . :...:  :..:::::.  :...::  ...:. :. .:         .::. :.::::..:. :.....
786" ERNEFLLLHAFYENNYIVPDKQIFRKPQQKLGDEDEEIDGDTNKYKKGRKKGAYAGGLVLDPKVGFYDKF
                                                            ___---_---------
860' VLVMDFNSLYPSIIQEFNICFTTVDR---------NKEDIDELPSVPPSEVDQGVLPRLLANLVDRRREV
     .:..:::::::::::::::::::.:             .... ...:..: ....  :.:::  . .::.::.:
856" ILLLDFNSLYPSIIQEFNICFTTVQRVASEAQKVTEDGEQEQIPELPDPSLEMGILPREIRKLVERRKQV
     ---II----________-------
```

FIG. 1. Amino-acid comparison between yeast Pol-I and human Polα. Identical amino acids are indicated by a colon; amino-acid substitutions that retain the hydrophilic or hydrophobic character at a particular position are indicated by a single point (*359*). Regions of homology with phage and viral DNA polymerases are shown by a broken line; amino-acid sequences conserved in most of the DNA polymerases are shown by an uninterrupted line (see text for details). The circled glycine residue at position 493 of the yeast sequence was changed to an arginine residue in a Pol-I mutant defective in polymerase–primase interaction (*111, 288*).

```
921'  KKVMKTET-DPHKRVQCDIRQQALKLTANSMYGCLGYVNSRFYAKPLAMLVTNKGREILMNTRQLAESMN
      :..:: .. .:.  .:.::::.:::::::::::::::.  ::::::::: ::: :::::::.:......::
926"  KQLMKQQDLNPDLILQYDIRQKALKLTANSMYGCLGFSYSRFYAKPLAALVTYKGREILMHTKEMVQKMN
                     ---------—···—·—-----III---------------------

990'  LLVVYGDTDSVMIDTGCDNYADAIKIGLGFKRLVNERYRLLEIDIDNVFKKLLLHAKKKYAALTVNLDKN
      : :.:::::::.::.:...: ....:.:    :. ::. :.::::::::.:::.:::  :::::::.:. ...
996"  LEVIYGDTDSIMINTNSTNLEEVFKLGNKVKSEVNKLYKLLEIDIDGVFKSLLLLKKKKYAALVVEPTSD
       ---———————-I-

1060' GNGTTVLEVKGLDMKRREFCPLSRDVSIHVLNTILSDKDPEEALQEVYDYLEDIRIKVETNNIRIDKYKI
      :: .:  :.::::. ::..: :..:..  :.. ::::.... .... . : .:  .: ..........:
1066" GNYVTKQELKGLDIVRRDWCDLAKDTGNFVIGQILSDQSRDTIVENIQKRLIEIGENVLNGSVPVSQFEI

1130' NMKLSKDPKAYPGGKNMPAVQVALRM-RKAGRVVKAGSVITFVITKQDEIDNAADTPALSVAERAHALNE
      :. :.:::..::. :..: :.:::.. ...:: ::::.....::            .:...:....::.:  :
1136" NKALTKDPQDYPDKKSLPHVHVALWINSQGGRKVKAGDTVSYVI--------CQDGSNLTASQRAYA-PE

1199' VMIKSNNLIPDPQYYLEKQIFAPVERLLERIDSFNVVRLSEALGLDSKKYFRREGGNNNGEDINNLQPLE
       . : .::. :.::::..:: . :.:. :.::....: ... ::::... :: .    ...:. ..:
1197" QLQKQDNLTIDTQYYLAQQIHPVVARICEPIDGIDAVLIATWLGLDPTQ-FRVH-HYHKDEENDALLGGP

1269' TTITDVERFKDTVTLELSCPSCDKRFPFGGI---VSSNYYRVSY--NGLQCKHCEQLFTPLQLTSQIEHS
      . .:: :...:   .. .::.:..   ....    ...  .  :  ....:: .   :: .::....  .
1265" AQLTDEEKYRDCERFKCPCPTCGTENIYDNVFDGSGTDMEPSLYRCSNIDCKASPLTFT-VQLSNKLIMD

1334' IRAHISLYYAGWLQCDDSTCGIVTRQVSV-FGKR-CLNDGC-TGVMRYKYSDKQLYNQLLYFDSLFDCEK
      ::  :. ::.::: :...::    ::.... :..    :  .: ..... .:::: ::.:: ..  .:: :
1334" IRRFIKKYYDGWLICEEPTCRNRTRHLPLQFSRTGPLCPACMKATLQPEYSDKSLYTQLCFYRYIFDAEC

1401' NKKQELKPIYLPDDLDYPKEQLTESSIKALTEQNRELMETGRSVVQKYLNDCGRRYVDMTSIFDFMLN

1404" ALEKLTTDHEKDKLKKQFFTPKVLQDYRKLKNTAEQFLSRSGYSEVNLSKLFAGCAVK
```

FIG. 1. *(continued)*

polymerase (*267*), contains a number of conserved dipeptides separated by conserved spacings. Additional regions of less, but significant, homology, have been identified (*153, 266*). In all DNA polymerases in this family, these conserved sequence domains occur in the same order on the polypeptide chain. Their presence distinguishes this family of DNA polymerases from two other families, one including *E. coli* DNA polymerase I and phage T7 DNA polymerase (*277*), the other including mammalian DNA polymerase β and terminal deoxynucleotidyl transferase (*278*), as well as from *E. coli* DNA polymerase III, which so far is unique (*279*).

Functional domains in the DNA polymerase α polypeptide probably map in the conserved sequences. A number of mutations in herpes simplex virus conferring altered sensitivity to antiviral drugs have been isolated and mapped at the sequence level (*273, 280–286*). Mutants with altered sensitivities to the nucleoside analogue acyclovir and aphidicolin, and to the pyrophosphate analogue phosphonoacetic acid, map predominantly in regions II and III. Most mutants resistant to acyclovir also show altered sensitivities to phosphonoacetic acid, indicating interacting and/or overlapping binding sites for the nucleotide and pyrophosphate moieties (*273, 285, 286*).

All of these studies with the DNA polymerase of herpes simplex virus strongly suggest that regions II and III contain the dNTP binding site for the herpes DNA polymerase and probably also for all other DNA polymerases in the Polα family (*287*). No mutations have been found in the most conserved domain I. Possibly, this region is essential for catalysis. Histidine- and cysteine-rich motifs in the middle (residues 650–715) and in the carboxy-terminal portion (residues 1245–1376) of the polypeptide chain of human Polα are potential metal-binding domains ("zinc fingers") proposed to be involved in DNA binding (*153*). The carboxy-terminal but not the middle putative DNA-binding domain is conserved in yeast Pol-I (*111*; Fig. 1).

The presence of the putative dNTP and DNA binding sites, together with conserved region I in a small portion of the polymerase polypeptide, explains the previous observations that grossly proteolyzed preparations of Polα still retain catalytic activity (Section II,A). It then would be logical to assume that the amino-terminal half of the polymerase, which is apparently not involved in catalysis and is lacking in DNA polymerases from phages T4, ϕ29, and PRD1 (*266–268*), is required for interaction with accessory proteins. This view is supported by the observation that a glycine-to-arginine mutation at position 493 in yeast Pol-I has an *in vivo* temperature-sensitive phenotype and is defective in polymerase–primase interaction *in vitro* (*111, 288*). This domain, probably necessary for polymerase–primase interaction, is also present in human Polα (Fig. 1). Additional mutants in yeast DNA polymerase I that make the cell temperature-sensitive for growth have been isolated, but not yet mapped at the sequence level (*289*). One of them, *pol1–14*, is apparently an assembly mutant and might yield interesting information on the interaction of Pol-I with accessory factors (*289*).

With the genes for human Polα and yeast Pol-I in hand, expression studies have been greatly facilitated. Transcription of the human Polα gene is proliferation-dependent (*153*). This is in agreement with numerous observations showing that Polα could only be isolated from proliferating cells (reviewed in *1*). The yeast Pol-I gene is periodically expressed in the cell cycle, with the level of the transcript increasing about 100-fold in late G_1 and peaking around the G_1/S boundary (*290*). In addition, the Pol-I transcript is also regulated in meiosis, increasing about 20-fold in mid S-phase and thus indicating a function for Pol-I in meiotic DNA replication. Finally, transcription of the Pol-I gene, but not the PRI1 gene (coding for DNA primase) is also induced about

20-fold after irradiation of the cells with ultraviolet light, implying a role for this enzyme in DNA repair as well (*290*). However, X-ray-induced DNA damage was equally well repaired in a *pol1–17* mutant at the nonpermissive temperature, as in the wild-type strain, indicating either that other DNA polymerases can substitute or that Pol-I is not required for the repair of this type of damage (*291*).

III. DNA Polymerase δ

In 1976, Byrnes *et al.* isolated a novel DNA polymerase, Polδ, from rabbit bone marrow that, unlike the previously known DNA polymerases, possessed a potent 3′–5′-exonuclease activity (*35*). Similar proofreading DNA polymerases had previously been identified in *S. cerevisiae* and in *U. maydis* (*51, 67*). Initially, Polδ could only be purified in small amounts from a few sources such as rabbit bone marrow and calf thymus, and therefore was not considered to be a ubiquitous enzyme with an important role in DNA metabolism. Recently, however, it has been isolated from several sources, including yeast. For the purpose of this review, a DNA polymerase δ is defined as a high-molecular-weight DNA polymerase with a proofreading 3′-5′-exonuclease activity and an increased resistance to the nucleotide analogue BuPhdGTP[1] in comparison with Polα from the same source (*35, 297, 335*). Reviews of DNA polymerase δ discuss the enzyme from calf thymus in great detail (*292, 293*).

A. Purification and Subunit Structure

For the sake of convenience and to aid the discussion, DNA polymerase δ has been subdivided into two classes. Polδ_1 is relatively nonprocessive on the model template–primer poly(dA)·oligo(dT) with a low primer : template ratio, but its processivity is much enhanced by addition of proliferating cell nuclear antigen (PCNA), or cyclin (discussed in Section III,B). Polδ_2 is processive in the absence of PCNA and is not stimulated by the addition of this cofactor. Examples of each class of Polδ have been found in several organisms (Table V).

Yeast DNA polymerase II is the first eukaryotic DNA polymerase shown to have a tightly associated 3′–5′-exonuclease activity (*51–53*). Because of its low abundance in extracts, it was difficult to obtain homogeneous enzyme in significant quantities before the recent advent of high performance liquid chromatography (HPLC). The homogeneous enzyme is a single polypeptide with a mass of 145–170 kDa (R. Hamatake and A. Sugino, and J. L. Campbell, personal communications), indicating that both exonuclease and polymerase

TABLE V

DNA POLYMERASE δ FROM DIFFERENT SOURCES

Source	Form	Subunits (kDa)	Processivity	Stimulation by PCNA	Remarks	Reference
S. cerevisiae	δ_1	125, 55	Low	+	Pol-III	*56, 57, 294, 295*
	δ_2	145–170[a]	High	–	Pol-II,B	*54, 57, 294, 295*
U. maydis		140[b]			BuPhdGTP sensitivity not known	*65–67, 91, 92*
D. melanogaster	δ_2	182	High	–	Derived from Polα	*30, 39, 168*
Rabbit bone marrow	(δ_2)	122				*35, 296, 297*
Calf thymus	δ_1	125, 48	Low	+		*298, 299*
	δ_2	245, 135, 110, 60, 45	High	–	Form δII, contains primase	*224, 300, 301*
	δ_2	140, 40[c]	High	–		*302*
Monkey CV-1		(6 S)[d]			Partially purified	*303*
Human placenta	δ_1	170	Low	+		*304, 305*
HeLa	δ_2	215	High	–		*306, 308*

[a] R. Hamatake and A. Sugino, and J. L. Campbell, personal communications.

[b] DNA polymerase mass determined *in situ* after SDS-PAGE.

[c] U. Hübscher, personal communication.

[d] Sedimentation coefficient of partially purified enzyme.

domains are present in the same polypeptide. Crude Pol-II, when initially separated from Pol-I and Pol-III on DEAE-silica gel, is sensitive to BuPhdGTP (*57*). However, further purification steps render the enzyme resistant to BuPhdGTP, thus classifying it as a δ polymerase (P. M. J. Burgers, unpublished results). Pol-II is highly processive and is not stimulated by PCNA (*294, 295*). It is classified as a δ_2 enzyme (Table V).

That the problem in obtaining Polδ from a variety of organisms may have been the result of proteolytic degradation and inactivation is best illustrated in the case of yeast. DNA polymerase III was first identified as a minor activity (about 5% of the total polymerase activity) when extracts from log-phase wild-type yeast cells were fractionated on DEAE-silica gel columns (*56*). Addition of a large number of protease inhibitors increased the yield of this polymerase peak significantly, indicating that proteolysis had led to inactivation of the enzyme. Fortunately, genetic studies of proteolysis in yeast had yielded mutants defective in various proteases (*309*). One composite mutant strain carrying mutations in the structural genes for proteinases B and C and, in addition, a mutation (*pep4-3*) that failed to mature vacuolar zymogens, including inactive preproteases, was used as a source for the purification of Pol-III. Additional protease inhibitors were still required to obtain the pure enzyme in satisfactory yield.

Pol-III consists of two subunits of 125 and 55 dKa and is classified as a δ_1 enzyme (Table V; *294, 295*). Inhibition studies with aphidicolin, BuPhdGTP, and salt showed minor but definite differences between Pol-II and Pol-III (*57*). More significantly, however, polyclonal antibodies raised against pure Pol-III failed to cross-react with either native or denatured Pol-II (*57*). Moreover, changes in the isolation procedure, such as inclusion of protease inhibitors, that had a large effect on the detectable level of Pol-III, did not affect the level of Pol-II.

Finally, extracts from temperature-sensitive yeast strains with a mutation in the *cdc2* gene (*295a*) are deficient in Pol-III but not Pol-II (*366, 367*). From these studies it was concluded that Pol-II and Pol-III are structurally different (*57*).

The situation is more complex for the mammalian Polδ, as exemplified for the enzyme from calf thymus. Whereas the purification protocol of Lee *et al.* produced Polδ_1 as the sole polymerase δ (*299*), those of Focher *et al.* and Crute *et al.* only resulted in the isolation of Polδ_2 (*224, 300–302*). Those opposing results are probably ascribable to the use of different buffers as well as different column matrices. Calf thymus Polδ_1 consists of two subunits of 125 and 48 kDa and is nonprocessive in the absence of PCNA (*299*). One form of Polδ_2 consists of two subunits of 140 and 40 kDa and is processive (U. Hübscher, personal communication; *302*). The second form of Polδ_2 of calf thymus consists of five subunits and is the only δ polymerase with DNA primase activity (Table V; *300*). Whether the catalytic polymerase activity resides in the 245- or 135-kDa polypeptide is not known. Also not known is whether the associated primase activity is identical to the primase activity associated with Pol*α* (*300, 301*). This Polδ_2 is very processive over a wide range of pH values and Mg^{2+} concentrations (*224*).

A Polδ_1, consisting of a 170-kDa catalytic subunit, and possibly two or three small accessory proteins, was the only polymerase δ isolated from human placenta (*304, 305*). In contrast, the enzyme isolated from HeLa cells is a single polypeptide of 215 kDa with the characteristics of a δ_2 enzyme (Table V; *306–308*). Furthermore, the enzyme isolated from rabbit bone marrow, a single polypeptide of 122 kDa, should probably be classified as a δ_2 enzyme, because its high synthetic activity on poly(dA)·oligo(dT) is indicative of a processive polymerase (*296, 297*). Finally, the 182-kDa subunit of Pol*α* from *D. melanogaster*, when separated from the 73-kDa subunit, shows a potent 3′–5′-exonuclease activity, is resistant to BuPhdGTP[1] and very processive, and is not stimulated by PCNA, qualifying it as Polδ_2 (*38, 39, 168*).

The effective conversion of the *D. melanogaster* Pol*α* into a Polδ activity invites the suggestion that this may be a general phenomenon and that Pol*α* and Polδ of other organisms may also be structurally related. However, all evidence so far supports the contention that Pol*α* and Polδ in yeast and in mammalian cells are separate enzymes. Immune studies with yeast Pol-I and Pol-III antibodies failed to detect an immunological cross-reactivity (*57*). In addition, extracts from a *cdc17* mutant (a temperature-sensitive allele of *pol1*) were deficient for Pol-I but not Pol-III and extracts from *cdc2* mutants were deficient for Pol-III but not for Pol-I (*367*). There is a lack of immunological cross-reactivity between Pol*α* and Polδ_1 from human cells and calf thymus (*304, 358*). Furthermore, tryptic peptide maps of the polymerase subunit of calf thymus Pol*α* and the polymerase subunit of calf thymus Polδ_2 showed little or no common peptides (U. Hübscher, personal communication). All of these studies show that the *D. melanogaster* enzyme has unique properties not shared by the DNA polymerases of other organisms.

In conclusion, with one exception, Polδ is a one-subunit or two-subunit enzyme with the catalytic subunit ≥ 122 kDa (Table V).

Two different forms, Polδ_1 and Polδ_2, have been isolated from several organisms. Although it is well-established that yeast Polδ_1 and Polδ_2 are different enzymes with different genes (*56, 57, 366, 367*), this cannot be said for the mammalian enzymes. Since the form of the enzyme isolated is dependent on the specific isolation protocol, i.e., no protocol allows the simultaneous isolation of both calf thymus Polδ_1 and Polδ_2, as can be done for yeast, it is possible that these isolation and/or purification protocols can generate either form from a more complex Polδ present in the cell. In support of this view are recent experiments showing inhibition of Polδ_2 from HeLa cells by polyclonal antibodies directed against calf thymus Polδ_1 (S. W. Wong, J. Syvaoja, C. K. Tan, K. M. Downey, A. G. So, S. Linn and T. S. F. Wang, personal communication).

B. Proliferating Cell Nuclear Antigen (Cyclin) as an Accessory Factor

A subset of patients with the autoimmune disease, systemic lupus erythematosis, produce antibodies to a nuclear protein detectable only in proliferating, but not in terminally differentiated cells (*310*). This protein, called proliferating cell nuclear antigen (PCNA), is identical to cyclin (*311*). Cyclin was initially identified by two-dimensional gel electrophoresis as a protein abundant in the S phase after serum stimulation of quiescent cells (*312, 313*). [Because the name cyclin may lead to confusion, especially since a family of cell cycle control proteins related to the *S. pombe cdc13* gene has also been named cyclin (*314, 315*), we prefer to use the abbreviation PCNA.]

PCNA (recently reviewed in *316*) from mammalian cells is a small acidic protein with an SDS-PAGE mass of 36 kDa (*317*). Stimulation of quiescent cells, which contain no detectable PCNA, by serum or growth factors induces synthesis of PCNA slightly before the start of DNA synthesis (*318–320*). In contrast, immunoblotting of total cellular protein revealed that the level of PCNA remains unchanged through the cell cycle of proliferating cells (*321*). More significantly however, cytoimmunochemistry showed that, in the S phase, PCNA is localized in the nucleus at sites of ongoing DNA synthesis (*320, 322–324*). The nuclear distribution of PCNA, but not its synthesis, is dependent upon ongoing DNA synthesis (*322*). The cDNAs for the rat and the human PCNA are highly homologous (*325, 326*). PCNA mRNA levels increase many fold upon serum stimulation of quiescent cells, peak in the S phase and then decrease again in G_2/M (*325, 326*). The constant level of PCNA in all phases of exponentially growing cells may be explained, however, by the high stability of the protein (*322*). The importance of PCNA in DNA replication was further shown *in vivo* by inhibition of DNA synthesis and mitosis with anti-sense oligonucleotides to PCNA (*327*). The requirements for PCNA for *in vitro* DNA replication of SV40 DNA are discussed in Section IV,B.

One form of calf thymus Polδ (form δ_1) is virtually inactive on ply(dA)·oligo(dT) with a low primer : template ratio, which provides a convenient assay for the identification and purification of an auxiliary factor (*328*), subsequently shown to be identical to PCNA (*329–331*). The stimulatory effect of PCNA is primarily due to a large increase in the processivity of Polδ_1 (*30*). Similarly, the Polδ from human placenta has a low activity on poly(dA)·oligo(dT), which was greatly stimulated by a protein with the chromatographic properties of PCNA (*304*).

The low activity of yeast DNA polymerase III on poly(dA)·oligo(dT) classifies this enzyme as Polδ_1. An accessory factor that stimulates Pol-III activity on this template–primer has been purified to homogeneity (*294*). Because of its similarity to the mammalian PCNAs, it was called yPCNA (*294*). yPCNA is a small acidic protein with a molecular weight in SDS-PAGE of 26,000. It stimulates Pol-III by increasing the processivity of the enzyme (*294*). In the absence of yPCNA, Pol-III has a processivity of about 35. Addition of yPCNA increases the processivity to the full length of the available template (>400) (*294, 295*). Addition of subsaturating amounts of yPCNA to the processivity assays gave two distinct product size classes, one identical to that obtained in the absence of yPCNA, and a second identical to that obtained in the presence of saturating levels of yPCNA (*294*). Together with the knowledge that yPCNA does not bind to single-stranded or double-stranded DNA, it was concluded that yPCNA stimulates Pol-III via a protein–protein interaction. In agreement with this conclusion are gel filtration experiments showing that yPCNA co-elutes only with the template–primer DNA if, in addition, Pol-III was included in an incubation of yPCNA with the DNA prior to gel filtration (*332*).

Apparently, the peptide domains on PCNA and on Polδ necessary for their interaction were largely conserved through evolution. Thus, yeast Pol-III is also stimulated by calf thymus PCNA (*295*). Even more astonishing is the observation that calf thymus PCNA and yPCNA interact equally well with Pol-III and increase the processivity of Pol-III in an identical fashion (*294*). However, calf thymus Polδ requires about 10-fold more yPCNA than calf thymus PCNA to make the enzyme fully processive (*294*). These results indicate that calf thymus PCNA and yPCNA share similar contact points for interaction with yeast Pol-III, but that additional contact points necessary for interaction with calf thymus Polδ_1 are lacking in yPCNA.

C. Catalytic Mechanism and Proofreading

The study of Polδ has primarily focused on the structural aspects of the enzyme and on the catalytic mechanism of the exonuclease activity. No detailed mechanistic study of the polymerase activity of Polδ has been carried out, as has been done for Polα (Section II,C,1). Template preference studies with most preparations of Polδ, however, allow us to subdivide Polδ into two distinct classes (Section III,A; Table V).

The 3′–5′-exonuclease activity associated with Polδ has been studied by many investigators. With all preparations of Polδ, this exonuclease can excise single nucleotide mismatches, providing a base-paired primer–template that can be elongated by the DNA polymerase (*56, 292, 304, 306, 308, 336*). Determination of the ratio of polymerase : exonuclease activity and turnover studies (incorporation followed excision, resulting in a net conversion of dNTP to dNMP) indicate that Polδ has a more potent proofreading activity than *E. coli* Pol-I (*56, 296*).

A more detailed examination of the exonuclease activity, however, shows significant differences among the different preparations of Polδ. Thus, yeast Pol-III (δ_1) completely degrades single-stranded DNA to mononucleotides, whereas calf thymus Polδ_2 fails to degrade oligonucleotides smaller than 6 or 7 residues (*56, 337*). Also, the exonuclease activity of yeast Pol-III on single-stranded DNA is sensitive to aphidicolin (*57*). However, the nuclease activity of calf thymus Polδ_1 or rabbit bone marrow Polδ_2 is not inhibited by aphidicolin when single-stranded DNA is the substrate, but is very sensitive to aphidicolin with double-stranded DNA as the substrate (*299, 338*). Finally, the exonuclease activity of calf thymus Polδ_2 is resistant to inhibition by aphidicolin on both single- and double-stranded DNA substrates (*337*). Some caution is required in the interpretation of these results, because sensitivity to aphidicolin can vary with the nature as well as the concentration of the DNA substrate (*57*).

Proofreading exonucleases associated with DNA polymerases are expected to be nonprocessive enzymes. Because their function is to remove single nucleotide mismatches, release of the exonuclease domain from the primer terminus after hydrolysis of a single nucleotide would allow subsequent continuation of DNA synthesis by the polymerase domain. The nonprocessive mode of action of calf thymus

Polδ_2 is in agreement with this rationale (*337*). The exonuclease activity of yeast Pol-III is distributive as well (*332*). Moreover, yPCNA, which greatly increases the processivity of polymerization on poly(dA)·oligo(dT), has no effect on the processivity of the exonuclease activity (*332*). This indicates that yPCNA fails to stabilize the Pol-III·DNA complex when Pol-III is in the exonuclease mode.

The *in vitro* fidelity of calf thymus Polδ_2 was determined in a M13*lacZα* nonsense codon reversion assay (*339*). Error rates were about 10^{-6}, indicating that Polδ is highly accurate. Addition of nucleoside monophosphates, which inhibit the exonuclease activity but not the polymerization activity (*340*), increased the error frequency. Also, addition of high concentrations of the dNTP to be incorporated after the position in the amber codon used for detection favors polymerase over exonuclease action and causes an increased error frequency for a proofreading but not for a non-proofreading enzyme ("next nucleotide effect") (*341*). A strong next nucleotide effect was indeed observed for calf thymus Polδ_2, indicating that proofreading contributed significantly to the overall fidelity of DNA replication by this enzyme (*339*).

D. A Role for DNA Polymerase δ in DNA Replication and Repair

Several lines of evidence suggest that DNA polymerase δ is a major enzyme active at the eukaryotic replication fork. First, in yeast, *cdc2* [encoding the large subunit of Pol-III (*367*)] is essential; temperature-sensitive *cdc2* mutants arrest in S phase at the restrictive temperature, indicating a required function for this enzyme during DNA replication (*295a*). Second, the fact that PCNA, an accessory factor to Polδ_1 is found, by immunocytochemical technique, at sites of ongoing DNA replication in the nucleus suggests that Polδ is present at these sites as well (*320, 322–324*). Additional evidence derives from permeable cell studies and *in vitro* DNA replication studies of SV40 DNA.

1. Permeable Cell Assays

Mammalian cells can be made permeable by incubation in slightly hypotonic medium with or without low concentrations of non-ionic detergents (*306, 342, 343*). Small molecules and even proteins, e.g., antibodies, can freely diffuse into such permeable cells. Because these cells become depleted of their nucleotide pools, the effect of nucleotide analogues on semiconservative DNA replication (exponentially growing cells were permeabilized) or DNA repair (cells grown to confluence were damaged, e.g., by ultraviolet light, and permeabilized) can easily be assessed (*342*). Semiconservative DNA replication is completely inhibited by aphidicolin in permeable cells as well as in intact cells, clearly implicating Polα and/or Polδ in this process (*344*). Monoclonal antibodies to Polα inhibited DNA replication in permeable cells 40–70%, indicating that the remaining 30–60% was either due to a form of Polα inaccessible to the monoclonal antibodies or due to another polymerase, e.g., Polδ (*303*; M. G. Frattini and S. L. Dresler, personal communication).

BuPhdGTP, when added at levels that completely inhibit purified Polα but do not significantly inhibit purified Polδ, did not inhibit DNA replication (*345*). The lack of inhibition by BuPhdGTP is a little curious. It suggests that Polα is not involved in semiconservative DNA replication (*345*). More likely, BuPhdGTP fails to inhibit Polα in the replication complex, at least at the low concentrations that are effective for the purified Polα. In agreement with this explanation are *in vitro* DNA replication studies of SV40 chromatin (*347*). Replication of SV40, which depends upon Polα (*348*), was about 10-fold more resistant to BuPhdGTP than is purified Polα with "naked" SV40 DNA as substrate, but 100- to 1000-fold more resistant to this inhibitor when SV40 chromatin was the substrate (*347*). Inhibition experiments with 2′,3′-dideoxythymidine 5′-triphosphate, a more potent inhibitor of Polδ than of Polα, suggest the presence of Polδ at the replication fork as well. This nucleotide analogue inhibits DNA replication at inhibitor concentrations similar to those required to inhibit purified Polδ (*348*).

To assess accurately the role of Polδ in DNA replication in permeable cell systems, a potent, specific inhibitor of the enzyme is essential. For instance, monoclonal antibodies to Polδ would complement the studies with Polα monoclonal antibodies. These two reagents could then be used to map the site of action and the function of the two polymerases, e.g., in the synthesis and maturation of Okazaki fragment.

Permeabilization of UV-irradiated diploid human fibroblasts in the presence of the non-ionic detergent Brij-58 allows soluble protein to leak out of the cell and makes these cells repair-deficient (*306*, *307*). An enzyme that restores conservative DNA repair synthesis in these damaged cells has been separated from the culture supernatant and shown to be Polδ_2 (*306–308*). These experiments are direct proof for the involvement of Polδ in DNA repair. The use of nucleotide analogues in permeable cell assays has also shown a requirement for Polδ in the repair of certain types of DNA damage (*345*, *348*).

2. SV40 DNA Replication

The only protein encoded by SV40 that is required for viral DNA replication is the large tumor antigen. Because the replication of SV40 is very similar to the replication of chromosomal replicons and is largely dependent on the cellular replication machinery, this is an attractive system for studying enzymes and factors required for chromosomal DNA replication (reviewed in *105, 349, 350*). Through the efforts of several laboratories, host protein factors required for *in vitro* SV40 DNA replication have largely been defined and purified (*321, 347, 351–357*). PCNA is an essential component of the SV40 replication machinery (*321, 352–354*). In the absence of PCNA, total DNA synthesis was greatly reduced. The small amount of replicated DNA contained only short nascent strands about 100–300 nucleotides long, synthesized on the lagging strands of the two replication forks (*353, 354*). No DNA resulting from leading-strand DNA synthesis was detected in the absence of PCNA. These results indicate the PCNA is absolutely required for leading-strand DNA replication and that, in its absence, lagging-strand DNA synthesis is more or less impaired as well (*321, 353, 354*). By implication, because PCNA is a cofactor of Polδ, these studies suggest that Polδ is required for leading-strand DNA synthesis.

3. A Model for Polα and Polδ at the Replication Fork

The problem of coordinating leading-strand DNA replication proceeding in the direction of fork movement with lagging-strand DNA synthesis in a direction opposite to fork movement has invited proposals for its solution in prokaryotes and in eukaryotes. A topological rearrangement in the structure of the lagging strand, such as looping, would result in a reversal of the polarity of this strand at the point of DNA synthesis and allow replication of this strand to proceed in the direction of fork movement (*8, 360–362*). Concurrent DNA replication of both strands could then be accomplished in *E. coli* by an asymmetric dimeric form of DNA polymerase III holoenzyme, with each asymmetric unit containing different subunits associated with Pol-III core suitable for either leading- or lagging-strand DNA synthesis (*362, 363*). In contrast, concurrent DNA replication in eukaryotes would require both Polα and Polδ (*358, 364, 365*). Polα in a complex with DNA primase would be ideally suited for priming and elongation of Okazaki fragments on the lagging strand of the replication fork. Furthermore, the semiprocessive nature of this enzyme would make periodic dissociations of this enzyme more favorable. Polδ_1, on the other hand, in a complex with PCNA would be more suited for processive continuous DNA synthesis on the leading strand. As discussed above, indirect evidence for this model follows from the SV40 studies (*353, 354*). Additional evidence indicating a similar configuration of the cellular replication fork is still lacking.

IV. Conclusions and Perspectives

One of the most remarkable conclusions that can be drawn from our increased knowledge of the eukaryotic DNA polymerases is the strong conservation of structural and functional properties from yeast to human cells. Immediately apparent is the high degree of sequence homology between the catalytic polymerase subunits of yeast and human Polα. But also the structure of the four-subunit Polα-primase and even the sizes of the individual subunits have been conserved through evolution.

Structural conservation of Polδ is less immediately obvious but, as proteolytic artifacts and variations in subunit composition are eliminated, this enzyme may turn out to be as conserved as Polα. One of the most interesting problems that need to be addressed for mammalian cells is whether the two forms of δ isolated, δ_1 and δ_2, are structurally related. In yeast, there is good evidence that Polδ_1 and Polδ_2 are not structurally related. Based upon SV40 DNA replication studies, Polδ_1 in a complex with PCNA has been proposed as the leading-strand DNA polymerase and the Polα–primase complex as the lagging-strand enzyme. Additional factors may be necessary for the coordination of leading- and lagging-strand DNA replication.

In addition to the conservation of enzyme structure, we have also observed the conservation of functional domains from yeast to mammalian cells. Thus, yeast PCNA interacts in a protein–protein fashion with calf thymus Polδ_1 and makes this enzyme very processive, and a similar interaction has been observed between calf thymus PCNA and yeast Pol-III (δ_1). In this light, it is curious that the primer stem recognition proteins C1C2 from monkey cells do not interact with Polα from such a closely related organism as humans, and vice versa.

New understandings and developments will arise out of the continued biochemical studies of Polα and Polδ. However, complete understanding of the roles of these enzymes in eukaryotic DNA replication will require genetic analysis in addition to *in vitro* models for chromsomal DNA replication. I expect that in the next few years the contributions from yeast genetics and biochemistry, together with the excellent mammalian *in vitro* systems based on viruses such as SV40, will considerably clarify the picture of the eukaryotic replication fork and provide tantalizing insights into the control of proliferation of eukaryotic cells.

Acknowledgments

The work in the author's laboratory is supported by a grant from the National Institutes of Health and a salary award from the American Heart Association. I would like to thank John Majors, Glenn Bauer, and Steven Dresler for stimulating discussions.

References

1. M. Fry and L. A. Loeb, "Animal Cell DNA Polymerases." CRC Press, Boca Raton, Florida, 1986.
2. U. Hübscher, *Experientia* **39,** 1 (1983).
3. A. I. Scovassi, P. Plevani and U. Bertazzoni, *TIBS* **5,** 335 (1980).
4. F. J. Bollum and V. R. Potter, *JACS* **79,** 3603 (1957).
5. F. J. Bollum and V. R. Potter, *JBC* **233,** 478 (1958).
6. A. Kornberg, I. R. Lehman, M. J. Bessman and E. S. Simms, *BBA* **21,** 197 (1956).
7. I. R. Lehman, M. J. Bessman, E. S. Simms, and A. Kornberg, *JBC* **233,** 163 (1958).
8. A. Kornberg, "DNA Replication." Freeman, San Francisco, California, 1980, 1982 (suppl.).
9. F. J. Bollum, *JBC* **235,** PC18 (1960).
10. F. J. Bollum, *JBC* **235,** 1945 (1960).
11. I. R. Lehman, S. B. Zimmerman, J. Adler, M. J. Bessman, E. S. Simms and A. Kornberg, *PNAS* **44,** 1191 (1958).
12. J. S. Krakow, C. Coutsogeorgopoulos and E. S. Canellakis, *BBA* **55,** 639 (1962).
13. H. M. Keir and M. J. Smith, *BBA* **68,** 589 (1963).
14. M. Yoneda and F. J. Bollum, *JBC* **240,** 3385 (1965).
15. A. Weissbach, D. Baltimore, F. J. Bollum, R. C. Gallo and D. Korn, *Science* **190,** 401 (1975).
15a. D. Gillespie, W. C. Saxinger and R. C. Gallo, *This Series* **15,** 1 (1975).
16. P. DeLucia and J. Cairns, *Nature* **224,** 1164 (1969).
17. T. Kornberg and M. L. Gefter, *BBRC* **40,** 1348 (1970).
18. R. Knippers, *Nature* **228,** 1050 (1970).
19. R. E. Moses and C. C. Richardson, *BBRC* **41,** 1557 (1970).
20. T. Kornberg and M. L. Gefter, *PNAS* **68,** 761 (1971).
21. V. Nusslein, B. Otto, F. Bonhoeffer and H. Schaller, *Nature NB* **234,** 285 (1971).
22. D. M. Livingston, D. C. Hinkle and C. C. Richardson, *JBC* **250,** 461 (1975).
23. C. S. McHenry and W. J. Crow, *JBC* **254,** 1748 (1979).
24. S. Wickner, *PNAS* **73,** 3511 (1976).
25. C. McHenry and A. Kornberg, *JBC* **252,** 6478 (1977).
26. L. M. S. Chang and F. J. Bollum, *JBC* **246,** 5835 (1971).
27. A. Weissbach, A. Schlabach, B. Fridlender and A. Bolden, *Nature NB* **231,** 167 (1971).
28. E. F. Baril, O. E. Brown, M. D. Jenkins and J. Laszlo, *Bchem* **10,** 1981 (1971).
29. L. M. S. Chang, *Science* **191,** 1183 (1976).

30. B. Fridlender, M. Fry, A. Bolden and A. Weissbach, *PNAS* **69**, 452 (1972).
31. B. J. Lewis, J. W. Abrell, R. G. Smith and R. C. Gallo, *Science* **183**, 867 (1974).
32. R. R. Meyer and M. V. Simpson, *PNAS* **61**, 130 (1968).
33. G. F. Kalf and J. J. Ch'Ih, *JBC* **243**, 4904 (1968).
34. A. Bolden, G. Pedrali-Noy and A. Weissbach, *JBC* **252**, 3351 (1977).
35. J. J. Byrnes, K. M. Downey, V. L. Black and A. G. So, *Bchem* **15**, 2817 (1976).
36. A. Matsukage, M. Yamaguchi, K. Tanabe, Y. Taguchi, M. Hishizawa and T. Takahashi, *in* "New Approaches in Eukaryotic DNA Replication" (A. M. deRecondo, ed.). Plenum, New York, 1983.
37. L. S. Kaguni, J.-M. Rossignol, R. C. Conaway and I. R. Lehman, *PNAS* **80**, 2221 (1983).
38. S. M. Cotterill, M. E. Reyland, L. A. Loeb and I. R. Lehman, *PNAS* **84**, 5635 (1987).
39. M. E. Reyland, I. R. Lehman and L. A. Loeb, *JBC* **263**, 6518 (1988).
40. M. Furia, L. C. Polito, G. Locorotondo and P. Grippo, *NARes* **6**, 3399 (1979).
41. K. Sakaguchi and J. B. Boyd, *JBC* **260**, 10406 (1985).
42. E. M. Nelson, D. J. Stowers, M. L. Bayne and R. M. Benbow, *Dev. Biol.* **96**, 11 (1983).
43. H. Konig, H. D. Riedel and R. Knippers, *EJB* **135**, 435 (1983).
44. H. B. Kaiserman and R. M. Benbow, *NARes* **15**, 10249 (1987).
45. H. Joenje and R. M. Benbow, *JBC* **253**, 2640 (1978).
46. P. M. Racine and P. W. Morris, *NARes* **5**, 3945 (1978).
47. P. W. Morris and P. M. Racine, *NARes* **5**, 3959 (1978).
48. M. Shioda, *EJB* **160**, 571 (1986).
49. A. Habara, H. Nagano and Y. Mano, *BBA* **561**, 17 (1979).
50. U. Wintersberger and E. Wintersberger, *EJB* **13**, 11 (1970).
51. W. B. Helfman, *EJB* **32**, 42 (1973).
52. L. M. S. Chang, *JBC* **252**, 1873 (1977).
53. E. Wintersberger, *EJB* **84**, 167 (1978).
54. G. Badaracco, L. Capucci, P. Plevani and L. M. S. Chang, *JBC* **258**, 10720 (1983).
56. G. A. Bauer, H. M. Heller and P. M. J. Burgers, *JBC* **263**, 917 (1988).
57. P. M. J. Burgers and G. A. Bauer, *JBC* **263**, 925 (1988).
58. U. Wintersberger and H. Blutsch, *EJB* **68**, 199 (1976).
59. W. Joester, K. E. Joester, B. van Dorp and P. H. Hofschneider, *NARes* **5**, 3043 (1978).
60. G. Stauder, H. Riesemann, W. M. Joester and K. E. Joester, *BBA* **741**, 308 (1983).
61. Y. Furukawa, R. Yamada and M. Kohno, *NARes* **7**, 2387 (1979).
62. P. R. Ganz and R. E. Pearlman, *EJB* **113**, 159 (1980).
63. A. Sakai and Y. Watanabe, *J. Biochem.* (*Tokyo*) **91**, 845, 855 (1982).
64. E. Ostergaard, P. Brams, O. Westergaard and O. F. Nielsen, *BBA* **908**, 150 (1987).
65. P. A. Jeggo, P. Unrau, G. R. Banks and R. Holliday, *Nature NB* **242**, 14 (1973).
66. G. R. Banks, W. K. Holloman, M. V. Kairis, A. Spanos and G. T. Yarranton, *EJB* **62**, 131 (1976).
67. G. R. Banks and G. T. Yarranton, *EJB* **62**, 143 (1976).
68. J. Aoshima, M. Kubota, T. Nishimura and I. Tatsuichi, *J. Biochem.* (*Tokyo*) **96**, 461 (1984).
69. A. G. McLennan and H. M. Keir, *BJ* **151**, 227 (1975).
70. A. G. McLennan and H. M. Keir, *BJ* **151**, 239 (1975).
71. L. W. Loomis, E. F. Rossomando and L. M. S. Chang, *BBA* **425**, 469 (1976).
72. E. F. Baril, C. Scheiner and T. Pederson, *PNAS* **77**, 3317 (1980).
73. L. M. S. Chang, E. Cheriathundram, E. M. Mahoney and A. Cerami, *Science* **208**, 510 (1980).

74. W. Schiebel and A. Raffael, *FEBS Lett.* **121,** 81 (1980).
75. E. Holler, H. Fischer, C. Weber, H. Stopper, H. Steger and H. Simek, *EJB* **163,** 397 (1987).
76. L. M. S. Chang, *JBC* **248,** 3789 (1973).
77. T. S. F. Wang, W. D. Sedwick and D. Korn, *JBC* **249,** 841 (1974).
78. S. Wilson, J. Abbotts and S. Wider, *BBA* **949,** 149 (1988).
78a. L. M. S. Chang, F. J. Bollum, X. X. Xiu, L. C. Cheung, K. Huebner, C. M. Croce, B. K. Hecht, F. Hecht and L. A. Cannizzaro, *in* "Cancer Cells. 6: Eukaryotic DNA Replication" (T. Kelly and B. Stillman, eds.), p. 417. CSH Laboratory, Cold Spring Harbor, New York, 1988.
79. K. W. Knopf, M. Yamada and A. Weissbach, *Bchem* **15,** 4540 (1976).
80. A. Matsukage, E. W. Bohn and S. H. Wilson, *Bchem* **14,** 1006 (1975).
81. U. Hübscher, C. C. Kuenzle and S. Spadari, *EJB* **81,** 249 (1977).
82. A. Matsukage, K. Tanabe, M. Yamaguchi, Y. N. Taguchi, M. Nishizawa and T. Takahashi, *BBA* **655,** 269 (1981).
83. M. Yamaguchi, K. Tanabe, T. Takahashi and A. Matsukage, *JBC* **257,** 4484 (1982).
84. G. Brun, F. Rougeon, M. Lauber and F. Chapville, *EJB* **41,** 241 (1974).
85. M. Yamaguchi, K. Tanabe, Y. N. Taguchi, M. Nishizawa, T. Takahashi and A. Matsukage, *JBC* **255,** 9942 (1980).
86. M. Yamaguchi, A. Matsukage and T. Takahashi, *JBC* **255,** 7002 (1980).
87. T. A. Kunkel and A. Soni, *JBC* **263,** 4450 (1988).
88. C. Wernette and L. S. Kaguni, *JBC* **261,** 14764 (1986).
89. L. S. Kaguni, C. M. Wernette, M. C. Conway and P. Yang-Cashman, *in* "Cancer Cells. 6: Eukaryotic DNA Replication" (T. Kelly and B. Stillman, eds.), p. 425. CSH Laboratory, Cold Spring Harbor, New York, 1988.
90. A. I. Scovassi, S. Torsello, P. Plevani, G. F. Badaracco and U. Bertazzoni, *EMBO J.* **1,** 1161 (1982).
91. A. Spanos, S. G. Sedgwick, G. F. Yarranton, U. Hübscher and G. F. Banks, *NARes* **9,** 1825 (1981).
92. U. Hübscher, A. Spanos, W. Albert, F. Grummt and G. R. Banks, *PNAS* **78,** 6771 (1981).
93. C. A. Ross and W. J. Harris, *BJ* **171,** 231, 241 (1978).
94. M. Crerar and R. E. Pearlman, *JBC* **249,** 3123 (1974).
95. D. K. Dube, R. O. Williams, G. Seal and S. C. Williams, *BBA* **561,** 10 (1979).
96. A. M. McNicol, G. R. Banks and R. A. Cox, *FEBS Lett.* **221,** 48 (1987).
97. T. A. Kunkel, *JBC* **260,** 12866 (1985).
98. T. A. Kunkel and A. Soni, *JBC* **263,** 4450 (1988).
99. A. Matsukage, E. W. Bohn and S. H. Wilson, *Bchem* **14,** 1006 (1975).
100. S. Kraus and S. Linn, *Bchem* **19,** 220 (1980).
101. P. Plevani, G. Badaracco, F. Ginelli and S. Sora, *Antimicrob. Agents Chemother.* **18,** 50 (1980).
102. D. Brutlag and A. Kornberg, *JBC* **247,** 241 (1972).
103. N. N. Khan, G. E. Wright, L. W. Dudycz and N. C. Brown, *NARes* **12,** 3695 (1984).
104. L. A. Loeb, P. K. Liu and M. Fry, *This Series* **33,** 57 (1986).
105. J. L. Campbell, *ARB* **55,** 733 (1986).
106. P. Plevani, M. Foiani, P. Valsasnini, G. Badaracco, E. Cheriothundam and L. M. S. Chang, *JBC* **260,** 7102 (1985).
107. G. Lucchini, S. Francesconi, M. Foiani, G. Badaracco and P. Plevani, *EMBO J.* **6,** 737 (1987).
108. P. Plevani, G. Lucchini, M. Foiani, P. Valsasnini, A. Brandazza, M. Bianchi, G. Magni and G. Badaracco, *Life Sci. Adv.* **6,** 53 (1987).

109. P. Plevani, M. Foiani, S. Francesconi, C. Mazza, A. Pizzagelli, P. Valsasnini and G. Lucchini, *in* "Cancer Cells. 6: Eukaryotic DNA Replication" (T. Kelly and B. Stillman, eds.), p. 341. CSH Laboratory, Cold Spring Harbor, New York, 1988.
110. M. H. Pausch, B. C. Peterson and L. B. Dumas, *in* "Cancer Cells. 6: Eukaryotic DNA Replication" (T. Kelly and B. Stillman, eds.), p. 359. CSH Laboratory, Cold Spring Harbor, New York, 1988.
111. A. Pizzagalli, P. Valsasnini, P. Plevani and G. Lucchini, *PNAS* **85,** 3772 (1988).
112. P. Plevani, S. Francesconi and G. Lucchini, *NARes* **15,** 7975 (1987).
113. L. S. Kaguni, J. M. Rossignol, R. C. Conaway and I. R. Lehman, *JBC* **258,** 9037 (1983).
114. L. S. Kaguni, A. A. DiFrancesco and I. R. Lehman, *JBC* **259,** 9314 (1984).
115. A. M. Holmes, I. P. Hesslewood and I. R. Johnston, *EJB* **43,** 487 (1974).
116. K. McKune and A. M. Holmes, *NARes* **6,** 3341 (1979).
117. F. Grosse and G. Kraus, *Bchem* **20,** 5470 (1981).
118. S. Masaki, O. Koiwai and S. Yoshida, *JBC* **257,** 7172 (1982).
119. F. Grummt, G. Waltl, H. M. Jantzen, K. Hamprecht, U. Hübscher and C. C. Kuenzle, *PNAS* **76,** 6081 (1979).
120. U. Hübscher, P. Gerschwiler and G. K. McMaster, *EMBO J.* **1,** 1513 (1982).
121. H. P. Ottiger and U. Hübscher, *PNAS* **81,** 3993 (1984).
122. U. Hübscher and H. P. Stalder, *NARes* **13,** 5471 (1985).
123. H. Ottiger, P. Frei, M. Haessig and U. Hübscher, *NARes* **15,** 4789 (1987).
124. U. Hübscher, M. Gassman, S. Spadari, N. C. Brown, E. Ferrari and H. J. Buhk, *Philos. Trans. R. Soc. London, Ser. B* **317,** 421 (1987).
125. A. F. Wahl, S. P. Kowalski, L. W. Harwell, E. M. Lord and R. A. Bambara, *Bchem* **23,** 1895 (1984).
126. L. M. S. Chang, E. Rafter, C. Augl and F. J. Bollum, *JBC* **259,** 14679 (1984).
127. S. Masaki, K. Tanabe and S. Yoshida, *NARes* **12,** 4455 (1984).
128. A. M. Holmes, E. Cheriathundam, F. J. Bollum and L. M. S. Chang, *JBC* **261,** 11924 (1986).
129. H. P. Nasheuer and F. Grosse, *Bchem* **26,** 8458 (1987).
130. H. P. Nasheuer and F. Grosse, *JBC* **263,** 8981 (1988).
131. F. Grosse and H. P. Hasheuer, *in* "Cancer Cells. 6: Eukaryotic DNA Replication" (T. Kelly and B. Stillman, eds.), p. 397. CSH Laboratory, Cold Spring Harbor, New York, 1988.
132. M. Mechali, J. Abadidebat and A. M. deRocondo, *JBC* **255,** 2114 (1980).
133. M. Philippe, J. M. Rossignol and A. M. deRecondo, *Bchem* **25,** 1611 (1986).
134. L. P. Goscin and J. J. Byrnes, *NARes* **10,** 6023 (1982).
135. E. M. Karawya and S. H. Wilson, *JBC* **257,** 13129 (1982).
136. E. A. Faust, G. Gloor, M. F. Macintyre and R. Nagy, *BBA* **781,** 216 (1984).
137. T. Yagura, T. Kozu and T. Seno, *BBA* **870,** 1 (1986).
138. C. E. Prussak and B. Y. Tseng, *JBC* **262,** 6018 (1987).
139. M. Yamaguchi, E. A. Hendrickson and M. L. DePamphilis, *JBC* **260,** 6254 (1985).
140. E. Karawya, J. Swack, W. Albert, J. Fedorko, J. D. Minna and S. H. Wilson, *PNAS* **81,** 7777 (1984).
141. P. Lamothe, B. Baril, A. Chi, L. Lee and E. Baril, *PNAS* **78,** 4723 (1981).
142. E. Baril, P. Bonin, D. Burstein, K. Mara and P. Zamecnik, *PNAS* **80,** 4931 (1983).
143. J. K. Vishwanatha, S. A. Coughlin, M. Wesolowsk-Owen and E. F. Baril, *JBC* **261,** 6619 (1986).
144. W. Skarnes, P. Bonin and E. Baril, *JBC* **261,** 6629 (1986).
145. E. Holler, H. Fischer and S. Helmut, *EJB* **151,** 311 (1985).

146. J. K. Vishwanatha and E. F. Baril, *NARes* **14,** 8467 (1986).
147. E. F. Baril, L. K. Molkas, R. Hickey, C. J. Li, J. K. Vishwanatha and S. Coughlin, *in* "Cancer Cells. 6: Eukaryotic DNA Replication" (T. Kelly and B. Stillman, eds.), p. 373. CSH Laboratory, Cold Spring Harbor, New York, 1988.
148. R. M. Gronostaiski, J. Field and J. Hurwitz, *JBC* **259,** 9479 (1984).
149. S. Tanaka, S. Z. Hu, T. S. F. Wang and D. Korn, *JBC* **257,** 8386 (1982).
150. T. S. F. Wang, S. Z. Hu and D. Korn, *JBC* **259,** 1854 (1984).
151. S. Z. Hu, T. S. F. Wang and D. Korn, *JBC* **259,** 2602 (1984).
152. S. W. Wong, L. R. Paborsky, P. A. Fischer, T. S. F. Wang and D. Korn, *JBC* **261,** 7958 (1986).
153. S. W. Wong, A. F. Wahl, P. M. Yuan, N. Arai, B. E. Pearson, K. Arai, D. Korn, M. W. Hunkapiller and T. S. F. Wang, *EMBO J.* **7,** 37 (1988).
154. A. M. Holmes, I. P. Hesslewood and I. R. Johnston, *EJB* **62,** 229 (1976).
155. A. M. Holmes, I. P. Hesslewood and I. R. Johnston, *Nature* **255,** 420 (1975).
156. I. P. Hesslewood, A. M. Holmes, W. F. Wakeling and I. R. Johnston, *EJB* **84,** 123 (1978).
157. F. Grosse and G. Krauss, *NARes* **8,** 5703 (1980).
158. S. Masaki, K. Tanabe and S. Yoshida, *NARes* **12,** 4455 (1984).
159. G. R. Banks, J. A. Boezi and I. R. Lehman, *JBC* **254,** 9886 (1979).
160. G. Villani, B. Sauer and I. R. Lehman, *JBC* **255,** 9479 (1980).
161. B. Sauer and I. R. Lehman, *JBC* **257,** 12394 (1982).
162. L. M. Johnson, M. Snyder, L. M. S. Chang, R. W. Davis and J. L. Campbell, *Cell* **43,** 369 (1985).
163. P. Bhattacharya, I. Simet and S. Basu, *PNAS* **76,** 2218 (1979).
164. P. Bhattacharya, I. Simet and S. Basu, *PNAS* **78,** 2683 (1981).
165. P. Bhattacharya and S. Basu, *PNAS* **79,** 1488 (1982).
166. M. Suzuki, T. Enomato, F. Hanaoka and M. Yamada, *J. Biochem.* (*Tokyo*) **98,** 581 (1985).
167. S. M. Cotterill, G. Chui and I. R. Lehman, *JBC* **262,** 16100 (1987).
168. S. M. Cotterill, M. E. Reyland, L. A. Loeb and I. R. Lehman, *in* "Cancer Cells. 6: Eukaryotic RNA Replication" (T. Kelly and B. Stillman, eds.), p. 367. CSH Laboratory, Cold Spring Harbor, New York, 1988.
169. Y. C. Chen, E. W. Bohn, S. R. Planck and S. H. Wilson, *JBC* **254,** 11678 (1979).
170. R. C. Conaway and I. R. Lehman, *PNAS* **79,** 2523 (1982).
171. R. C. Conaway and I. R. Lehman, *PNAS* **79,** 4585 (1982).
172. L. S. Kaguni, J. M. Rossignol, R. C. Conaway, G. R. Banks and I. R. Lehman, *JBC* **258,** 9037 (1983).
173. S. Cotterill, G. Chui and I. R. Lehman, *JBC* **262,** 16105 (1987).
174. M. Shioda, E. M. Nelson, M. L. Bayne and R. M. Benbow, *PNAS* **79,** 7209 (1982).
175. H. D. Riedel, H. Konig, H. Stahl and R. Knippers, *NARes* **10,** 5621 (1982).
176. T. Yagura, T. Kozu and T. Seno, *J. Biochem.* (*Tokyo*) **91,** 607 (1982).
177. T. Yagura, T. Kozu and T. Seno, *JBC* **257,** 11121 (1982).
178. B. Y. Tseng and C. N. Ahlem, *JBC* **258,** 9845 (1983).
179. S. Yoshida, R. Suzuki, S. Masaki and O. Koiwai, *BBA* **741,** 348 (1983).
180. F. Grosse and G. Kraus, *JBC* **260,** 1881 (1985).
181. P. Plevani, G. Badaracco, C. Augl and L. M. S. Chang, *JBC* **259,** 7532 (1984).
182. H. Singh and L. B. Dumas, *JBC* **259,** 7936 (1984).
183. Y. Murakami, C. R. Wobbe, L. Weissbach, F. B. Dean and J. Hurwitz, *PNAS* **83,** 2869 (1986).
184. S. M. Jazwinsky and G. M. Edelman, *JBC* **260,** 4995 (1985).

185. F. E. Wilson and A. Sugino, *JBC* **260,** 8173 (1985).
186. E. E. Biswas, P. E. Joseph and S. B. Biswas, *Bchem* **26,** 5377 (1987).
187. T. Yagura, T. Kozu, T. Seno and S. Tanaka, *Bchem* **26,** 7749 (1987).
188. L. S. Kaguni and I. R. Lehman, *BBA* **950,** 87 (1988).
188a. Y. F. Roth, *EJB* **165,** 473 (1987).
189. S. M. Cotterill, G. Chui and I. R. Lehman, *JBC* **262,** 16105 (1987).
190. B. Novak and E. Baril, *NARes* **5,** 221 (1978).
191. C. G. Pritchard and M. L. DePamphilis, *JBC* **258,** 9801 (1983).
192. C. G. Pritchard, D. T. Weaver, E. F. Baril and M. L. DePamphilis, *JBC* **258,** 9810 (1983).
193. E. Rapaport, P. C. Zamecnik and E. F. Baril, *PNAS* **78,** 838 (1981).
194. E. Rapaport, P. C. Zamecnik and E. F. Baril, *JBC* **256,** 12148 (1981).
195. P. C. Zamecnik, M. L. Stephenson, C. M. Janeway and K. Randerath, *BBRC* **24,** 91 (1966).
196. K. Randerath, C. M. Janeway, M. L. Stephenson and P. C. Zamecnik, *BBRC* **24,** 98 (1966).
197. E. Rapaport and P. C. Zamecnik, *PNAS* **73,** 3984 (1976).
198. F. Grummt, *CSHSQB* **43,** 649 (1978).
199. C. Weinman-Dorsch, A. Hedl, I. Grummt, W. Albert, F. J. Ferdinand, R. R. Friis, G. Pierran, W. Moll and F. Grummt, *EJB* **138,** 179 (1984).
200. K. G. Lawton, J. V. Wierowski, S. Schechter, R. Hilf and R. A. Bambara, *Bchem* **23,** 4294 (1984).
201. F. Grosse and G. Kraus, *EJB* **141,** 109 (1984).
202. V. L. Sylvia, C. O. Joe, J. O. Norman, G. M. Curtin and D. L. Busbee, *BBRC* **135,** 880 (1986).
203. V. L. Sylvia, J. O. Norman, G. M. Curtin and D. L. Busbee, *BBRC* **141,** 60 (1986).
204. V. Sylvia, G. Curtin, J. Norman, J. Stec and D. Busbee, *Cell* **54,** 651 (1988).
205. M. J. Berridge, *BJ* **220,** 345 (1984).
206. L. E. Hokin, *ARB* **54,** 205 (1985).
207. P. W. Majerus, T. M. Connolly, H. Deckmyn, T. S. Ross, T. E. Bross, H. Ishii, V. S. Bansal and D. B. Wilson, *Science* **234,** 1519 (1986).
208. M. J. Berridge, *ARB* **56,** 159 (1987).
209. A. B. Pardee, D. L. Coppock and H. C. Yang, *J. Cell Sci. Suppl.* **4,** 171 (1986).
210. M. K. Zierler, N. J. Marini, D. J. Stowes and R. M. Benbow, *JBC* **260,** 974 (1986).
211. P. A. Fischer and D. Korn, *JBC* **254,** 11033 (1979).
212. P. A. Fisher and D. Korn, *JBC* **252,** 6528 (1977).
213. P. A. Fisher, T. S. F. Wang and D. Korn, *JBC* **254,** 6128 (1979).
214. P. A. Fisher and D. Korn, *JBC* **254,** 11040 (1979).
215. P. A. Fisher, J. T. Chen and D. Korn, *JBC* **256,** 133 (1981).
216. P. A. Fisher and D. Korn, *Bchem* **20,** 4560 (1981).
217. P. A. Fisher and D. Korn, *Bchem* **20,** 4570 (1981).
218. D. Korn, P. A. Fisher and T. S. F. Wang, *in* "New Approaches in Eukaryotic DNA Replication" (A. M. deRecondo, ed.), p. 17. Plenum, New York, 1983.
219. S. D. Detera, S. P. Becerra, J. A. Swack and S. H. Wilson, *JBC* **256,** 6933 (1981).
220. K. Tanabe, Y. N. Taguchi, A. Matsukage and T. Takahashi, *J. Biochem. (Tokyo)* **88,** 35 (1980).
221. S. K. Das and R. K. Fujimura, *JBC* **254,** 1227 (1979).
222. C. K. Tan, M. J. So, K. M. Downey and A. G. So, *NARes* **15,** 2269 (1987).
223. K. T. Hohn and F. Grosse, *Bchem* **26,** 2870 (1987).
224. R. D. Sabatino, T. W. Myers, R. A. Bambara, O. K. Shin, R. L. Morraccino and P. H. Frickey, *Bchem* **27,** 2998 (1988).

225. R. A. Bambara, D. Uyemura and T. Choi, *JBC* **253**, 413 (1978).
226. K. G. Lawton, J. V. Wierowski, S. Schechter, R. Hilf and R. A. Bambara, *Bchem* **23**, 4294 (1984).
227. E. A. Faust and C. D. Rankin, *NARes* **10**, 4181 (1982).
228. H. D. Riedel, H. König and R. Knippers, *BBA* **783**, 158 (1984).
229. M. Goulian, C. Carton, L. De Granpre, B. Olinger and S. Richards, *UCLA Symp. Mol. Cell. Biol., New Ser.* **47**, 101 (1987).
230. M. Goulian, C. Carton, L. de Granpre, C. Heard, B. Olinger and S. Richards, *in* "Cancer Cells. 6: Eukaryotic DNA Replication" (T. Kelly and B. Stillman, eds.), p. 393. Cold Spring Harbor Laboratory, Cold Spring Harbor, New York, 1988.
231. E. E. Biswas and S. B. Biswas, *NARes* **16**, 6411 (1988).
232. C. N. Giroux, J. R. Mis, M. K. Pierce, S. E. Kohalmi and B. A. Kunz, *MCB* **8**, 978 (1988).
233. J. W. Drake, *Nature* **221**, 1132 (1969).
234. L. A. Loeb and T. A. Kunkel, *ARB* **51**, 429 (1982).
235. T. A. Kunkel, *Cell* **53**, 837 (1988).
236. P. Modrich, *ARB* **56**, 435 (1987).
237. M. D. Topal and J. R. Fresco, *Nature* **263**, 285 (1976).
238. R. Lohrmann, P. K. Bridson and L. E. Orgel, *J. Mol. Evol.* **17**, 303 (1981).
239. S. M. Watanabe and M. F. Goodman, *PNAS* **79**, 6429 (1982).
240. J. Petruska, L. C. Sowers and M. F. Goodman, *PNAS* **83**, 1559 (1986).
241. M. S. Boosalis, J. Petruska and M. F. Goodman, *JBC* **262**, 14689 (1987).
242. M. F. Goodman, *Mutat. Res.* **200**, 11 (1988).
243. J. J. Hopfield, *PNAS* **77**, 5248 (1980).
244. F. Grosse, G. Krauss, J. W. Knill-Jones and A. R. Fersht, *EMBO J.* **2**, 1515 (1983).
245. J. Abbotts and L. A. Loeb, *JBC* **259**, 6712 (1984).
246. J. J. Hopfield, *PNAS* **71**, 4135 (1974).
247. J. Ninio, *Biochimie* **57**, 587 (1975).
248. E. Bernardi and J. Ninio, *Biochimie* **60**, 1083 (1978).
249. J. Abbots and C. A. Loeb, *NARes* **13**, 261 (1985).
250. O. P. Doubleday, P. Lecomte and M. Radman, *in* "Cellular Responses to DNA Damage" (E. C. Friedberg and B. A. Bridges, eds.), p. 489. Liss, New York, 1983.
251. P. Lecomte, O. P. Doubleday and M. Radman, *JMB* **189**, 643 (1986).
252. P. M. J. Burgers and F. Eckstein, *JBC* **254**, 6889 (1979).
253. P. J. Romaniuk and F. Eckstein, *JBC* **257**, 7684 (1982).
254. P. A. Bartlett and F. Eckstein, *JBC* **257**, 8879 (1982).
255. F. Eckstein, *ARB* **54**, 367 (1985).
256. V. Mizrahi and S. J. Benkovic, *Adv. Enzymol.* **61**, 437 (1988).
257. G. G. Hillebrand, A. H. McClusky, K. A. Abbot, G. G. Revich and K. L. Beattie, *NARes* **12**, 3155 (1984).
258. G. G. Hillebrand and K. L. Beattie, *JBC* **260**, 3116 (1985).
259. S. Brosius, F. Grosse and G. Kraus, *NARes* **11**, 193 (1983).
261. M. E. Reylandt and L. A. Loeb, *JBC* **262**, 10824 (1987).
262. C. S. Newlon and W. Burke, *ICN–UCLA Symp. Mol. Cell. Biol.* **19**, 399 (1980).
263. C. J. Rivin and W. L. Fangman, *J. Cell Biol.* **85**, 96 (1980).
264. C. J. Rivin and W. L. Fangman, *J. Cell Biol.* **85**, 108 (1980).
265. G. Lucchini, A. Brandazza, G. Badaracco, M. Bianchi and P. Plevani, *Curr. Genet* **10**, 245 (1985).
266. F. K. Spicer, J. Rush, C. Fung, L. J. Reha-Krentz, J. D. Karam and W. H. Konigsberg, *JBC* **263**, 7478 (1988).
267. H. Yoshikawa and J. Ito, *Gene* **17**, 323 (1982).

268. G. Jung, M. C. Laevitt, J. C. Hsieh and J. Ito, *PNAS* **84,** 8287 (1987).
269. G. Jung, M. C. Laevitt and J. Ito, *NARes* **15,** 9088 (1987).
270. E. V. Kuzmin and I. Levchenko, *NARes* **15,** 6758 (1987).
271. T. R. Gingeras, D. Sciaky, R. E. Gelinas, J. Bing-Dong, L. E. Yeu, M. M. Kelly, P. A. Bullock, B. L. Parsons, K. E. O'Neill and R. J. Roberts, *JBC* **257,** 13475 (1982).
272. P. L. Earl, E. V. Jones and B. Moss, *PNAS* **83,** 3659 (1986).
273. B. A. Larder, S. D. Kemp and G. Darby, *EMBO J.* **6,** 169 (1987).
274. R. Baer, A. T. Bankier, M. D. Biggin, P. L. Deininger, P. J. Farrell, T. J. Gibson, G. Hatfull, G. S. Hudson, S. C. Satchwell, C. Seguin, P. S. Tuffnell and B. G. Barrell, *Nature* **310,** 207 (1984).
275. A. J. Davison and J. E. Scott, *J. Gen. Virol.* **67,** 1759 (1986).
276. A. Kouzarides, A. T. Bankier, S. C. Satchwell, K. Weston, P. Tomlinson and B. G. Barrell, *J. Virol.* **61,** 125 (1987).
277. D. L. Ollis, C. Kline and T. A. Steitz, *Nature* **313,** 818 (1985).
278. A. Matsukage, K. Nishikawa, T. Ooi, Y. Seto and M. Yamaguchi, *JBC* **262,** 8960 (1987).
279. H. G. Tomasiewicz and C. S. McHenry, *JBact* **169,** 5735 (1987).
280. K. W. Knopf, E. R. Kaufman and C. Crumpacker, *J. Virol.* **39,** 746 (1981).
281. D. M. Coen, P. A. Fuman, D. Aschman and P. A. Schaffer, *NARes* **11,** 5287 (1983).
282. J. P. Quinn and D. J. McGeoch, *NARes* **13,** 8143 (1985).
283. K. W. Knopf, *NARes* **14,** 8225 (1986).
284. T. Tsurumi, K. Maeno and Y. Nishiyama, *J. Virol.* **61,** 388 (1987).
285. K. W. Knopf, *J. Gen. Virol* **68,** 1429 (1987).
286. J. S. Gibbs, H. C. Chiou, K. F. Bastow, Y. C. Cheng and D. M. Coen, *PNAS* **85,** 6672 (1988).
287. J. D. Hall, *Trends Genet.* **4,** 42 (1988).
288. G. Lucchini, C. Mazza, E. Scacheri and P. Plevani, *MGG* **212,** 459 (1988).
289. M. Budd and J. L. Campbell, *PNAS* **84,** 2838 (1987).
290. L. M. Johnston, J. H. M. Whie, A. L. Johnson, G. Lucchini and P. Plevani, *NARes* **15,** 5017 (1987).
291. M. Budd, C. Gordon, K. Sitney, K. Sweder and J. L. Campbell, *in* "Cancer Cells. 6: Eukaryotic DNA Replication" (T. Kelly and B. Stillman, eds.), p. 347. Cold Spring Harbor Laboratory, Cold Spring Harbor, New York, 1988.
292. M. Y. W. T. Lee, C. K. Tan, K. M. Downey and A. G. So, *This Series* **26,** 82 (1981).
293. A. G. So and K. M. Downey, *Bchem* **27,** 4591 (1988).
294. G. A. Bauer and P. M. J. Burgers, *PNAS* **85,** 7506 (1988).
295. P. M. J. Burgers, *NARes* **16,** 6297 (1988).
295a. J. R. Pringle and L. H. Hartwell, *in* "The Molecular Biology of the Yeast Saccharomyces" (J. N. Strathern, E. W. Jones and J. R. Broach, eds.), Vol. 2, p. 97. CSH Laboratory, Cold Spring Harbor, New York, 1981.
296. L. P. Goscin and J. J. Byrnes, *Bchem* **21,** 2513 (1982).
297. J. J. Byrnes, *BBRC* **132,** 628 (1985).
298. M. Y. W. T. Lee, C. K. Tan, A. G. So and K. M. Downey, *Bchem* **19,** 2096 (1980).
299. M. Y. W. T. Lee, C. K. Tan, K. M. Downey and A. G. So, *Bchem* **23,** 1906 (1984).
300. J. J. Crute, A. F. Wahl and R. A. Bambara, *Bchem* **25,** 26 (1986).
301. A. F. Wahl, J. J. Crute, R. D. Sabatino, J. D. Bodner, R. L. Marraccino, L. W. Harwell, E. M. Lord and R. A. Bambara, *Bchem* **25,** 7821 (1986).
302. F. Focher, S. Spadari, B. Ginelli, M. Hottiger, M. Gassman and U. Hübscher, *NARes* **16,** 6279 (1988).

303. R. A. Hammond, J. J. Byrnes and M. R. Miller, *Bchem* **26,** 6817 (1987).
304. M. Y. W. T. Lee and N. L. Toomey, *Bchem* **26,** 1076 (1987).
305. M. Y. W. T. Lee, *Bchem* **27,** 5193 (1988).
306. C. Nishida, P. Reinhard and S. Linn, *JBC* **263,** 501 (1988).
307. C. Nishida and S. Linn, *in* "Cancer Cells. 6: Eukaryotic DNA Replication" (T. Kelly and B. Stillman, eds.), p. 411. CSH Laboratory, Cold Spring Harbor, New York, 1988.
308. J. Syvaoja and S. Linn, *JBC* **264,** 2489 (1989).
309. E. W. Jones, *Genetics* **85,** 23 (1977).
310. K. Miachi, M. J. Frilzler and E. M. Tan, *J. Immunol.* **121,** 2228 (1978).
311. M. B. Mathews, R. M. Bernstein, B. R. Franza and J. I. Garrels, *Nature* **309,** 374 (1984).
312. R. Bravo and J. E. Celis, *J. Cell Biol.* **84,** 795 (1980).
313. R. Bravo, S. J. Fey, J. Bellatin, P. Mose Larsen, J. Arevalo and J. E. Celis, *Exp. Cell Res.* **136,** 311 (1981).
314. M. Solomon, R. Booher, M. Kirschner and D. Beach, *Cell* **54,** 738 (1988).
315. M. Goebl and B. Byers, *Cell* **54,** 739 (1988).
316. J. E. Celis, P. Madsen, A. Celis, H. V. Nielsen and B. Gesser, *FEBS Lett.* **220,** 1 (1987).
317. Y. Takasaki, D. Fischwild and E. M. Tan, *J. Exp. Med.* **159,** 981 (1984).
318. R. Bravo and H. MacDonald-Bravo, *EMBO J.* **3,** 3177 (1984).
319. P. Kurki, M. Vanderlaan, F. Dolbeare, J. Gray and E. M. Tan, *Exp. Cell Res.* **166,** 209 (1986).
320. R. Bravo, *Exp. Cell Res.* **163,** 287 (1986).
321. M. S. Wold, J. J. Li, D. H. Weinberg, D. M. Virshup, J. L. Sherley, E. Verheyen and T. Kelly, *in* "Cancer Cells. 6: Eukaryotic DNA Replication" (T. Kelly and B. Stillman, eds.), p. 133. CSH Laboratory, Cold Spring Harbor, New York, 1988.
322. R. Bravo and H. MacDonald-Bravo, *EMBO J.* **4,** 655 (1985).
323. J. E. Celis and A. Celis, *PNAS* **82,** 3262 (1985).
324. P. Madsen and J. E. Celis, *FEBS Lett.* **193,** 5 (1985).
325. J. M. Almendral, D. Huebsch, P. A. Blundell, H. MacDonald-Bravo and R. Bravo, *PNAS* **84,** 1575 (1987).
326. K. Matsumoto, T. Moriuchi, T. Koji and P. K. Nakane, *EMBO J.* **6,** 637 (1987).
327. D. Jaskalski, J. K. de Riel, W. E. Mercer, B. Calabretta and R. Baserga, *Science* **240,** 1544 (1988).
328. C. K. Tan, C. Castillo, A. G. So and K. M. Downey, *JBC* **261,** 12310 (1986).
329. R. Bravo, R. Frank, P. A. Blundell and H. MacDonald-Bravo, *Nature* **326,** 515 (1987).
330. G. Prelich, C. K. Tan, M. Kostura, M. B. Mathews, A. G. So, K. M. Downey and B. Stillman, *Nature* **326,** 517 (1987).
331. C. K. Tan, K. Sullivan, X. Li, E. M. Tan, K. M. Downey and A. G. So, *NARes* **15,** 9299 (1987).
332. G. A. Bauer and P. M. J. Burgers, *BBA* **951,** 274 (1988).
333. R. Karwan, H. Blutsch and U. Wintersberger, *Biochemistry* **22,** 5500 (1983).
334. R. A. DiFrancesco and I. R. Lehman, *JBC* **260,** 14764 (1985).
335. M. Y. W. T. Lee, N. L. Toomey and G. E. Wright, *NARes* **13,** 8623 (1985).
336. L. M. S. Chang, *JBC* **252,** 1873 (1977).
337. R. D. Sabatino and R. A. Bambara, *Bchem* **27,** 2266 (1988).
338. J. J. Byrnes, *Mol. Cell. Biochem.* **62,** 13 (1984).
339. T. A. Kunkel, R. D. Sabatino and R. A. Bambara, *PNAS* **84,** 4865 (1987).

340. J. J. Byrnes, K. M. Downey, B. G. Que, M. Y. W. T. Lee, V. L. Black and A. G. So, *Bchem* **16,** 3740 (1977).
341. T. A. Kunkel, R. M. Schaaper, R. A. Beckman and L. A. Loeb, *JBC* **256,** 9883 (1981).
342. S. L. Dresler and M. W. Lieberman, *JBC* **258,** 12269 (1983).
343. M. R. Miller, C. Seighman and R. G. Ulrich, *Bchem* **24,** 7440 (1985).
344. J. A. Huberman, *Cell* **23,** 647 (1981).
345. S. L. Dresler and M. G. Frattini, *NARes* **14,** 7093 (1986).
346. R. S. Decker, M. Yamaguchi, M. Rossenti, M. K. Bradley and M. L. DePamphilis, *JBC* **262,** 10863 (1987).
347. Y. Murakami, C. R. Wobbe, L. Weissbach, F. B. Dean and J. Hurwitz, *PNAS* **83,** 2369 (1986).
348. S. L. Dresler and K. S. Kimbro, *Bchem* **26,** 2664 (1987).
349. M. L. DePamphilis and P. M. Wasserman, *in* "Organization and Replication of Viral DNA" (A. S. Kaplan, ed.), p. 37. CRC Press, Boca Raton, Florida, 1982.
350. M. D. Challberg and T. J. Kelly, *ARB* **51,** 901 (1982).
351. C. R. Wobbe, C. Weissbach, J. A. Borowiec, F. B. Dean, Y. Murakami, P. Bullock and J. Hurwitz, *PNAS* **84,** 1834 (1987).
352. G. Prelich, M. Kostura, D. R. Marshak, M. B. Mathews and B. Stillman, *Nature* **326,** 471 (1987).
353. M. P. Fairman, G. Prelich, T. Tsurimoto and B. Stillman, *in* "Cancer Cells. 6: Eukaryotic DNA Replication" (T. Kelly and B. Stillman, eds.), p. 143, CSH Laboratory, Cold Spring Harbor, New York, 1988.
354. G. Prelich and B. Stillman, *Cell* **53,** 117 (1988).
355. J. J. Li and T. J. Kelly, *MCBiol* **51,** 1238 (1985).
356. L. Yang, M. S. Wold, J. J. Li, T. J. Kelly and L. F. Liu, *PNAS* **84,** 950 (1987).
357. M. S. Wold and T. Kelly, *PNAS* **85,** 2523 (1988).
358. K. M. Downey, C. K. Tan, D. M. Andrews, X. Li and A. G. So, *in* "Cancer Cells. 6: Eukaryotic DNA Replication" (T. Kelly and B. Stillman, eds.), p. 403. CSH Laboratory, Cold Spring Harbor, New York, 1988.
359. W. J. Wilbur and D. J. Lipman, *PNAS* **80,** 726 (1983).
360. N. K. Sinha, C. F. Morris and B. M. Alberts, *JBC* **255,** 4290 (1980).
361. C. S. McHenry, *Mol. Cell. Biochem.* **254,** 1748 (1985).
362. C. S. McHenry, *ARB* **57,** 519 (1988).
363. H. Maki, S. Maki and A. Kornberg, *JBC* **263,** 6570 (1988).
364. J. Blow, *Nature* **326,** 441 (1987).
365. F. Focher, E. Ferrari, s. Spadori and U. Hübscher, *FEBS Lett.* **229,** 6 (1988).
366. K. C. Sitney, M. E. Budd and J. L. Campbell, *Cell* **56,** 599 (1989).
367. A. Boulet, M. Simon, G. Faye, G. A. Bauer and P. M. J. Burgers, *EMBO J.* **8,** in press (1989).

Structure and Regulation of the Multigene Family Controlling Maltose Fermentation in Budding Yeast[1]

MARCO VANONI*,
PAUL SOLLITTI,†
MICHAEL GOLDENTHAL‡
and JULIUS MARMUR†

** Dipartmento di Fisiologia e Biochimica Generali*
Sezione Biochimica Comparata
Universitá degli Studi di Milano
Milano, Italy
† Department of Biochemistry
Albert Einstein College of Medicine
Bronx, New York 10461
‡ Department of Biology
Rutgers University
Camden, New Jersey 08103

[1] Glossary: *MAL1–MAL4*, *MAL6*: Five unlinked loci each controlling the utilization of maltose. *MALp*, *MALg*: Two tightly linked complementation groups at the *MAL* loci, first described by Naumov (*19–23*). *MALp* encodes a regulatory function (*MALR*), while the *MALg* function comprises information for both maltase and maltose permease (*MALS* and *MALT*). *MAL6S* (*MAL62*): The gene encoding maltase at the *MAL6* locus. *MAL6T* (*MAL61*): The gene encoding maltose permease at the *MAL6*

Progress in Nucleic Acid Research and Molecular Biology, Vol. 37

The budding yeast *Saccharomyces cerevisiae* and closely related species can utilize several sugars, besides glucose, as carbon and energy sources. These include galactose, starch, sucrose, melibiose, and maltose. Indeed, the ability to ferment particular sugars has often been used as a taxonomic tool; for example, the starch-degrading (diastatic) yeast has been assigned to a separate species, *S. diastaticus* (*1, 2*), despite the fact that it interbreeds readily with *S. cerevisiae* to produce fertile progeny. The ability to ferment various sugars depends on the genetic background of each particular species or strain and is governed by complex and partially interacting regulatory mechanisms, such as induction, carbon-catabolite repression (*3*), and inactivation (*4*). Maltose hydrolysis is catalyzed by the enzyme maltase 1,4-α-glucosidase, EC 3.2.1.20), which hydrolyzes the glycosidic bond of the disaccharide, releasing two glucose molecules (*5–7*). Maltose fermentation has been studied since the early 1950s, and four unlinked loci, *MAL1* through *MAL4*, were originally described (*8, 9*). Subsequently, three more *MAL* loci were identified: *MAL5* through *MAL7* (*10, 11*). Two of these, namely, *MAL5* and

---

locus. *MAL6R* (*MAL63*): The gene encoding the *MAL* regulatory protein at the *MAL6* locus (R for "regulatory"). *MAL6RR* (*MAL64*): A sequence homologous and adjacent to *MALR* of the *MAL6* locus. Standard yeast genetic nomenclature designates each genetic locus by three letters followed by a number. This system cannot fully describe the complexity of the *MAL* loci. We designate the three genes at each locus by a single capital letter after the locus abbreviation, indicating the gene function. This nomenclature system permits unambiguous functional identification of genes, irrespective of the locus to which they belong (e.g., *MALR* is a generic regulatory gene, and *MAL1R* is the regulatory gene of the *MAL1* locus). Another proposal (*28a*) is to name the genes by adding a digit to the locus name; however, this nomenclature may be misleading. For instance, *MAL63*, the regulatory gene of the *MAL6* locus, was so named because it encodes the smallest of the three transcripts identified at the *MAL6* locus. In this context, 63 is not a single number but actually two digits, the first referring to the locus and the second to the gene within the locus. A disadvantage of this nomenclature is that it is awkward to refer to a gene irrespective of its locus position, because of the ambiguities that would arise between designating *MAL* loci and *MAL* genes. Therefore, throughout this review we use the *MALS*, *MALT*, and *MALR* designation system. mal^0: Naturally occurring strain that does not ferment maltose. Mal^+ (Mal^-): Maltose fermenting (nonfermenting) phenotype of yeast strains. MALR (MALT): Protein encoded by the regulatory (transport) gene. *mal1S* (*mal1S*-Δ): Mutant (disrupted) allele of the *MAL1S* gene. *mal1T* (*mal1T*-Δ): Mutant (disrupted) allele of the *MAL1T* gene. *mal1R* (*mal1R*-Δ): Mutant (disrupted) allele of the *MAL1R* gene. bp (kbp): Base-pairs (kilobase-pairs), referring to double-stranded nucleic acids. kb: Kilobases referring to single-stranded nucleic acids. UAS: Upstream activation site (usually includes a subscript with an abbreviated locus name, e.g., UAS_{MAL}). RFLP: Restriction-fragment-length polymorphism. IG: Intergenic region between the divergently transcribed *MAL6S* and *MAL6T* genes.

MAL7, were later found to encode amylomaltase[2] (*12, 13*), *MAL5* being allelic to the *STA1* gene that encodes glucoamylase (EC 3.2.1.3) (*14*). The functional *MAL* loci have been mapped on the following chromosomes: *MAL1*, VII; *MAL2*, III; *MAL3*, II; *MAL4*, XI; *MAL6*, VIII (*15, 16*).

Until recently, the structural organization of the five *MAL* loci has been controversial, since experimental evidence suggested that the *MAL* loci contain either maltase structural genes or regulatory genes. The first hypothesis was supported by gene-dosage studies showing that the amount of induced maltase activity is proportional to the number of *MAL* loci present in a given yeast strain (*13, 17*). Support for the regulatory-gene model came from the observation that strains carrying different *MAL* loci produce biochemically indistinguishable maltase enzymes (*12*), and that, although several temperature-sensitive mutants that do not ferment maltose had been isolated, none of them displayed temperature-sensitive maltase activity (*18*). Subsequently, Naumov showed that two distinct complementation groups, *MALp* and *MALg*, were present (and linked) at the *MAL1*, *MAL3*, and *MAL6* loci in naturally occurring Mal$^-$ strains (*19–23*). The available evidence suggests that *MALp* is functionally equivalent to the complentation group identified by mutational analysis at the *MAL6* locus, and that this gene, designated *MAL6R*, encodes a regulatory protein involved in the coordinate induction of the synthesis of both maltase and maltose permease.

Since maltase is a cytosolic enzyme, yeast must transport maltose into the cell to utilize it. Maltose uptake and maltase activity increase coordinately following induction by maltose (*24*), and both maltase and maltose permease are also coordinately carbon-catabolite-repressed by glucose. The first mapping of a gene for maltose transport to a *MAL* locus was achieved in a constitutive *MAL1* strain that, by both physical and genetic criteria, had no other *MAL*-related sequences (*25*). The strain with the wild-type *MAL1T* gene that was isolated and designated *mal1T-1* displayed a 10-fold increase in the K_m for maltose uptake but no alteration in its V_{max}, thus suggesting that the mutation alters the structure of the maltose permease itself rather than its regulation.

The cloning of the *MAL6* locus in this laboratory made possible more detailed genetic, physical, and molecular analyses of the *MAL* loci. The *MAL6* locus was cloned by transformation of a maltose nonfermenting *malp* recipient strain with a plasmid-based genomic

[2] Now 4-α-glucanotransferase, EC 2.4.1.25 [Eds.].

library derived from the maltose-fermenting *MAL6pMAL6g* *S. carlsbergensis* strain CB11 (*26*). A cloned DNA insert of about 12.5 kbp carried not only *MAL6p* complementing activity, but also the structural gene for maltase as detected by *in vitro* translation of hybrid-selected mRNA; this finding supported, at the molecular level, Naumov's previous analysis of strain CB11 as carrying the closely linked *MAL6p* and *MAL6g* genes (*23, 26*). The cloned *MAL6* locus was later shown to transform the *mal1T-1* strain by complementation, thus identifying a gene encoding a transport function at this locus (*MAL6T*). Thus, three functional genes were detected at the *MAL6* locus: *MAL6R*, *MAL6S*, and *MAL6T*.

In this review we concentrate on the structure and regulation of the *MAL* loci, with particular emphasis on the work conducted in our own laboratory. References will be made to other sugar-metabolizing systems, such as the *SUC* multigene family that encodes the sucrose-hydrolyzing enzyme invertase (EC 3.2.1.26) (*27–29*) and the genes of the galactose/melibiose regulon (*30*), as well as to carbon-catabolite repression and inactivation. Both the regulation of *GAL* gene expression (*30*) and carbon-catabolite repression in yeast have been reviewed recently (*29, 31, 32*).

I. Structural Organization of the *MAL* Loci

A. Genetic and Physical Characterization of the *MAL1* and *MAL6* Loci

1. Identification and Functional Characterization of Three Genes at the *MAL1* Locus

A *MAL1* strain carrying no other *MAL*-related sequences detectable by Southern hybridization (see Section I,C,1) was constructed and used in mutagenesis studies. In order to isolate yeast mutants defective in maltose metabolism, a positive selection scheme was devised. It took advantage of the ability of strains defective in pyruvate decarboxylase (EC 4.1.1.1) (*pdc*) to grow in media supplemented with glycerol and ethanol as carbon sources (*33*). The addition of maltose to the glycerol- and ethanol-containing media prevented the utilization of these nonfermentable carbon sources by the mitochondria, except in strains that were Mal^{-}. About 100 ethyl methanesulfonate (EMS)-induced mutants were isolated by this method. Only one was found to map outside the *MAL1* locus.

All the above mutations, except one, were recessive to the *MAL1* locus and fell into two major complementation groups. These two classes of mutants identified both a pleiotropic gene involved in the regulation of maltose fermentation (*MAL1R*) and a gene involved in maltose transport (*MAL1T*) (*25*). These two classes of mutants could be readily distinguished by the ability of *mal1T*, but not *mal1R*, mutants to grow on a medium supplemented with 10% maltose. The *MAl1T* mutants failed to complement the previously characterized mutant *mal1T-1*. Complementation analysis with *malp* and *malg* tester strains confirmed the nature of Naumov's *MALp* function as structural genes, including a regulatory gene, and identified *MALg* as including a transport function. Surprisingly, no single mutation in the *MAL1S* gene was detected. Instead, one was able to isolate double mutants such as *mal1T mal1S* and *mal1R mal1S*. Strains carrying such multiple mutations were diploidized and reverted to Mal^+ at 24°C. From the revertants, after sporulation, there was isolated a temperature-sensitive mutant whose maltase was thermolabile when assayed in cell-free extracts. The mutant allele was designated *mal1-*S^{ts}. The failure of both the *mal1T* and *mal1*S^{ts} mutants to complement Naumov's *malg* tester strains implies that the *malg* deficiency involves at least two genes (*34*).

Extensive sequence homology of the *MALR*, *MALT*, and *MALS* genes at the *MAL1* and *MAL6* loci has been demonstrated by Southern blot hybridization of endonuclease-restricted genomic DNA isolated from tetrads (*35, 36*). Such a physical homology between the two loci, monitored with a *MAL6* probe, made it possible to use restructured *MAL6* genes to create null mutations in their counterpart at the *MAL1* locus. A series of isogenic deletion/disruption mutants at each *MAL1* gene has been constructed, by standard methods, in different laboratories (*37–40*). The deletion/disruption mutants at the regulatory, permease, and maltase genes are referred to as *mal1R*-Δ, *mal1T*-Δ, and *mal1S*-Δ, respectively. In genetic crosses, each disrupted mutant was capable of complementing mutants with lesions in the two other genes (as well as mutants obtained by standard chemical mutagenesis), but not itself or strains carrying a point mutation in the corresponding gene. Interestingly, *MAL6* strains disrupted in their regulatory or permease gene failed to complement each other (cited in *38*). *mal1R*-Δ, as well as *mal6R*-Δ (*41*) disruption mutants are uninducible for both maltase and maltose permease activities and for their mRNAs. These observations, together with gene-dosage experiments and molecular analyses of *mal1R* mutants, indicate that *MAL1R* acts as a positive activator of *MAL1S* and *MAL1T* transcription (*38–40*). Disruption of *MAL1T* (*MAL11*) results in a failure to induce both maltase and maltose permease, but has no effect on transcription of the regulatory protein (*38*). This pleiotropic effect of *mal1T*-Δ disruption that includes a reduction in *MAL1S* transcription, is most likely due to the inability of the maltose regulatory gene to activate maltase transcription in the absence of intracellular maltose. Finally, disruption of the *MAL1S* gene affects only the expression of maltase, while having no effect on transcription of the maltose permease and maltose regulatory genes (*37–40*).

Disruption of the *MAL6S* did not always lead to a Mal$^-$ phenotype, if the *MAL6* strains also carried *MAL*-related information at the *MAl1* locus, i.e., the *mal1*^{0} allele. However, the disrupted strain had a more thermolabile maltase activity than the nondisrupted strain. Synthesis of the thermolabile maltase was due to a structural gene located at the *mal1*^{0} locus present in the strain. Disruption of the maltase gene at both the *MAL6* and *mal1*^{0} loci produced a strain unable to ferment maltose (*42*).

2. Genetic and Physical Characterization of the Cloned *MAL6* DNA

Because it was difficult to generate mutations in all three genes at the *MAL6* locus using classical genetic manipulations, due to the presence of ubiquitous cryptic *mal1*^{0} sequences, the previously described *MAL1* mutants were used as recipients for the functional complementation analysis of the subcloned DNA fragments of the *MAL6* locus. Earlier genetic analysis had shown that the *MAL6* and the *MAL1* loci from mutants not fermenting maltose could complement each other, giving rise to Mal$^+$ phenotypes (*19–23*). This functional identity and the extensive sequence homology between these two loci (see Section I,C,2) suggested that *MAL6* and *MAL1* genes could be used interchangeably in most analyses. Using multicopy yeast plasmid vectors, various segments of the *MAL6* locus were subcloned (*37*). The functional characterization of the *MAL6* subclones was determined by (i) analyzing the levels of activity of the *MAL*-encoded structural proteins (maltase and the maltose permease) in cells transformed with various *MAL6* subclones, and (ii) monitoring the ability of these subclones to complement the maltose fermentation defects of well-characterized Mal$^-$ mutations in the highly homologous *MAL1* locus. The functional identification of each subcloned component of the *MAL6* locus allowed unequivocal assignment of functions to the three transcripts (*34, 37, 43*). They have been referred to as *MAL61*, *MAL62*, and *MAL63*. By correlating the functions and locations of the three transcripts at the *MAL6* locus, it became apparent that the *MAL6R*, *MAL6T*, and *MAL6S* genes encode, respectively, the *MAL63*, *MAL61*, and *MAL62* mRNAs.

B. Transcriptional Organization, Sequence, and Biochemical Properties of the *MAL* Genes and Their Products

1. Transcriptional Organization of the *MAL* Genes

The transcription of the *MAL* genes has been studied in both *MAL6* and *MAL1* strains (*38, 40, 41, 43*). The *MAL6S* (*MAL1S*) and *MAL6T* (*MAL1T*) genes are divergently transcribed from opposite strands following induction by maltose, the size of each mRNA being about 1.9 kb. The putative ATG translation initiation codons for the *MAL6*-encoded maltase and maltose permease are 884 bp apart, while the major transcription initiation sites under inducing conditions are separated by 786 bp (*44*). It is worth noting that *MALT* mRNAs of approximately 2.4 kb are constitutively expressed at very low levels in both *MAL1* and *MAL6* strains (*38, 40, 43, 45*), but are absent in *mal1T*-Δ strains, thus indicating that both the 1.9- and the 2.4-kb transcripts are made from the *MAL1T* gene. Putative TATA boxes for the *MAL6S* and *MAL6T* genes are found about 70 and 28 bp upstream from their respective transcription initiation sites. Both elements have the canonical TATAAA sequence proposed to define an "inducible" TATA element, as opposed to "constitutive" elements whose sequence deviates from the perfect consensus and are unable to respond to upstream activating sites (UASs) (*46, 47*). Imperfect palindromic sequences are located between the TATA boxes and the ATG translational start sites of *MAL6R*, *MAL6S*, and *MAL6T*. Similar palindromes can be found in homologous positions upstream from the *GAL1*, *GAL10*, and *SUC* genes (*48*) that, like the *MAL* structural genes, are catabolite-repressed by glucose. A 10-bp sequence (consensus TTCYTTTYRS) is found within each of the aforementioned palindromes and lies in a similar position in all of the predicted loops.

The putative TATA sequence of *MAL6R* (*16*) is located at position −119 relative to the AUG translation initiation codon. The TAATAA sequence observed differs from the TATAAA consensus observed in inducible promoters. This variant sequence may be more consistent with the constitutive expression of the *MALR* gene.

A large palindrome exists near the transcription initiation site of *MAL6R* that shows about 50% homology to the UAS_{MAL} palindrome.

Both palindromes contain the sequence TTTAAGTT . . . TACAGG, although the spacing of these two cores is different (6 bp in UAS and 12 bp at *MAL6R*). Interestingly, this TTTAAGTT sequence can also be detected near the transcription initiation site of *GAL10* at a position nearly identical to that found in *MAL6R*, although the orientation is reversed. It is not yet known if this sequence is functional in the *MAL6R* or *GAL10* contexts. There are many other large imperfect palindromic sequences (~60%) scattered throughout both the 5′ upstream and coding regions. These sequences might be involved in the modulation of the expression of *MAL6R* via the formation of stable DNA or RNA secondary structures (*16*). Translation of *S. cerevisiae* is highly sensitive to secondary structure (*48a*). A detailed restriction map, as well as a summary of the major transcriptional features of the *MAL6* locus is shown in Fig. 1.

2. Sequence Analysis of the *MAL* Genes and Biochemical Properties of Their Encoded Proteins

The complete nucleotide sequences of the maltase (*MAL6S*), maltose permease (*MAL6T*), and maltose regulatory (*MAL6R*) genes of the *MAL6* locus of *S. carlsbergensis* have been determined (*16, 44, 48b*). The major characteristics deduced from the analysis of the sequences are summarized and compared with available biochemical information on the respective proteins.

a. MAL6R. The nucleotide sequence of the *MAL6R* gene has a single major open reading frame (ORF) of 473 codons that predicts a protein with a calculated M_r of 54,892 (*16*). Its codon-bias index (*49*) is very low, 0.04, indicating almost random usage of degenerate codons, a feature that is often associated with a low level of expression in yeast. This is not unexpected, since regulatory proteins often have a low level of expression. It should be noted that direct determinations of the level of *MALR* mRNA suggest that it is 50-fold more abundant than that of other yeast regulatory gene transcripts (*43*) such as *PPR1* (*50*), *GAL4* (*51*), and *PHO4* (*52*). However, the exceptionally low codon-bias index of *MAL6R* may reduce the translation of its transcripts into protein.

The NH_2-terminal region of the deduced MAL6R protein contains many basic amino acids and cysteine residues. The sequence ^{8}Cys-Asp-Cys-Cys-Arg-Val-Arg-Arg-Val-Lys-Cys-Asp-Arg-Asn-Lys-Pro-Cys-Asn-Arg-Cys-Ile (*16*) demonstrates a striking similarity to the consensus sequence determined for the putative cysteine–zinc-associated

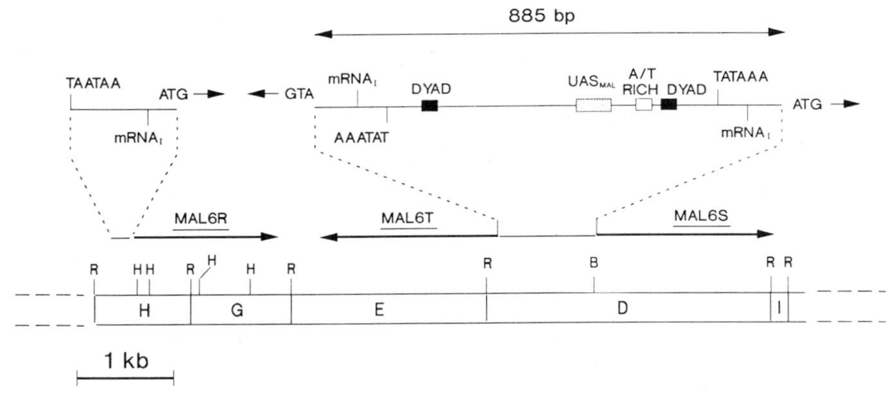

FIG. 1. Transcriptional and physical map of the *MAL6* locus. The *Eco*RI restriction map of the *MAL6* locus, with the locations of some of the *Hin*dIII (H) and *Bgl*II (B) sites is shown. Arrows indicate the direction of transcription and the approximate size of each gene. The expanded diagram at the top shows the upstream region of *MAL6R* and the intergenic region of *MAL6T–MAL6S*. The intergenic region has some interesting features, including two direct repeats showing dyad symmetry with unknown function, an A-T-rich region believed to be involved in catabolite repression and an upstream activation site (UAS_{MAL}) involved in the inducibility of the *MAL6S* structural genes. The *MAL6T* and *MAL6S* genes have the consensus TATAAA sequences observed in many inducible systems. *MAL6R* has a variant TAATAA sequence that is consistent with its constitutive expression.

DNA-binding "fingers" found in other fungal transcriptional regulatory proteins (*30*, *53*, *54*). The consensus sequence determined for fungal "zinc fingers" is ala-CYS-asp-x-CYS-ARG-HYD-(lys/arg)-(lys/-arg)-HYD-LYS-CYS-asp-(arg/lys)-x-x-PRO-x-CYS-x-(arg/lys)-CYS [capitalized amino acids are required; HYD = hydrophobic amino acid; x = any amino acid; parentheses indicate that either of the enclosed amino acids is frequently found; lower-case amino acids appear most of the time in these positions (*30*)].

Adjacent to the single "zinc-finger" domain is a helix-turn-helix motif composed of predominantly basic amino acids. The cysteine–zinc coordination domain may have a role in DNA binding as inferred by its isology to other regulatory protein domains known to bind to DNA regulatory sequences. The α-helices containing basic amino acids may have a role in specific *cis*-element recognition (*55*). This interpretation is supported by results obtained using the GAL4/MAL6R chimeric proteins (*56*). An acidic-residue-rich, α-helical "tail" region, analogous to other DNA-binding regulatory proteins, may interact with other transcription factors to control the expression of the regulated genes (*57–59*). It is at present unclear whether the activating regions must form amphipathic α-helices to regulate transcription (*60*), or whether acidic residue-rich sequences suffice (*61*). The role of MALR in sequence-specific DNA binding is supported by the finding that a UAS_{MAL}-binding protein of about 50 kDa has been prepared from a strain overexpressing the *MAL6R* gene (M. Goldenthal, unpublished).

b. MAL6T. The presence of a specific maltose permease, typical of carrier-mediated transport has been inferred from a kinetic analysis of maltose transport (*24, 62–64*). Maltose is actively transported by a proton symport (stoichiometric ratio 1 mol of protons per mol of maltose) mechanism that is sensitive to uncoupling agents, independent of intracellular ATP levels, and coupled to the protein electrochemical gradient (*64–67*). Electrochemical balance is maintained through K^+ ion extrusion, with a stoichiometric ratio of about 1 mol of K^+ released per mol of H^+ taken up (*68*). The nucleotide sequence of the *MAL6T* gene contains a single ORF of 641 codons that predicts a protein of 68,206 M_r (*48b*). Its calculated codon-bias index is 0.17, suggesting that *MAL6T* is a low to moderately expressed gene. The overall structure of the predicted protein indicates that the MAL6T protein is most likely an integral membrane protein as suggested by the presence of putative multiple transmembrane domains and putative glycation sites. This structural organization of the MAL6T protein strongly reinforces (although it does not prove) the genetic assignment of *MAL6T* as the structural gene for maltose permease.

c. MAL6S. The maltase purified from a *MAL6* strain has been shown to be a cytosolic 63,000-Da monomer with K_m and V_{max} for maltose of 1.66×10^{-2} M and 44.8×10^{-6} mol min^{-1}, respectively (*69*). The enzyme has an absolute anomeric specificity for α-glycosidic linkages and appears to recognize a glucosyl residue in α linkage on the nonreducing end of the substrate. The pH optimum is between 6.7 and 6.8. Monophasic heat inactivation profiles with a decay constant of 2.42×10^{-1} min^{-1} at 48°C were obtained for the purified enzyme (*69*), suggesting that the multiphasic decay kinetics reported in crude extracts (*70*) might reflect the occurrence of either proteolytic degradation of a single enzyme form, or the presence of multiple maltase isozymes with intrinsically different thermal stability (*42*). The nucleotide sequence of the *MAL6S* gene (*44*) contains one major ORF of 584 codons that predicts a protein of 68,107 M_r. Molecular weights in the range of 66,000–67,000 have previously been reported for yeast maltases by other groups (*71*). A good correlation in the fraction of each amino acid per mole of enzyme was observed between the value derived from the translated DNA sequence and the one obtained by analyzing the hydrolyzed purified maltase. The derived animo-acid sequence of maltase shows a limited regional homology with other α-glucosidases (*73*), including the *STA*-encoded glucoamylase (*74*). There is significant homology (pointed out by S. Henikoff) between maltase and each of the three genes of the *HDL* locus of *Drosophila melanogaster* whose function is unknown, but for which a role in digestion has been proposed (*75*). The calculated codon-bias index for the MAL6S protein is 0.24, a slightly higher value than that observed for permease, suggesting that *MAL6S* is a moderately expressed gene.

C. Cloning and Physical Organization of the Family of *MAL* Loci in *Saccharomyces*

1. Identification of Restriction Fragments Containing Specific *MAL* Loci

Southern blot analyses of *Bam*HI-restricted genomic DNAs prepared from standard laboratory strains did not allow the unambiguous identification of the components of the complete *MAL* locus present in each of the yeast strains (*35*). Thus, in order to identify maltase gene sequences, Mal^+ strains, each carrying a genetically defined *MAL* locus, were crossed with a mal^0 strain and the co-segregation of the functional locus (scored by the Mal^+ phenotype) and of sequences complementary to the maltase structural genes at that locus were analyzed among the meiotic products. The maltase structural gene sequences of each *MAL* locus were detected by Southern blot hybridization of *Bam*HI-restricted genomic DNA of the meiotic products. *Bam*HI was chosen because it cleaves outside the confines of each of the *MAL* loci. Each *MAL* locus encompassed at least one maltase gene and could be identified as a characteristic *Bam*HI fragment of genomic DNA; subsequently, each *MAL* locus was found to include the *MALT* and *MALR* genes as well (*35, 39, 76*).

The characteristic fragments generated by *Bam*HI digestion represent restriction-fragment-length polymorphisms (RFLPs) resulting from the variation in the sequences flanking the different *MAL* loci (*35, 39*). The *MAL1* locus is ubiquitous, i.e., it or one of its nonfunctional or partially active alleles can be detected in every laboratory strain examined. Thus, it is the only locus that can easily be reduced by genetic crosses to a single copy per cell. The finding that a functional *MAL1* locus (or one of its partially active alleles) is present in all of the yeast strains examined has a precedent in similar studies with the *SUC* loci (*27, 28*), in which it was found that a *SUC2*-associated restriction fragment was present in all of the *Saccharomyces* strains examined.

2. Molecular Cloning and Analysis of the *MAL* Loci

The *Bam*HI RFLPs of the *MAL* loci allowed their individual identification as well as their cloning. The *MAL2* (15 kbp), the constitutive *MAL4* (19 kbp) as well as the naturally occurring, partially active *mal1*^{0}-12 and *mal1*^{0}-15 loci (located on 12- and 15-kbp *Bam*HI restriction fragments, respectively) were cloned into a λ 1059 replacement vector. The *MAL1* (24 kbp) and *MAL3* (>30 kbp) loci were cloned into a *E. coli*/*S. cerevisiae* shuttle cosmid vector as a *Bam*HI fragment (*MAL1*) or by a quasi-random distribution approach using *Sau*3A partial digestion (*MAL3*). After the successful cloning of each *MAL* locus was verified by Southern blotting—taking advantage of their characteristic *Bam*HI RFLP except for *MAL3*—the loci cloned in phage λ were subsequently excised and inserted into an *Escherichia coli*/yeast shuttle plasmid vectors. The extent of homology of the *MAL* loci was then studied by Southern blot hybridization, restriction mapping, and Southern cross hybridization (*39, 76a*).

The restriction maps and homology data shown in Figs. 2 and 3 suggest that *MAL1* through *MAL4*, *MAL6*, *mal1*^{0}-12, and *mal1*^{0}-15 are all remarkably similar, especially in the sequences of their maltase genes. The conservation of DNA sequence diminishes as one approaches *MALR* and its 5′ end. In fact, using a *MAL6R* probe, it was found that only the functional *MAL* loci (but neither *mal1*^{0}-12 nor *mal1*^{0}-15) displayed homologous *MALR* sequences; however, the organization of this gene is not as highly conserved as the maltase gene sequences. The absence of homology to the regulatory sequence in *mal1*^{0}-12 appeared to be due to a deletion and/or other rearrangements of its regulatory gene, whereas the absence of homology in *mal1*^{0}-15 appeared to be due to sequence divergence. These deletions and/or rearrangements are not limited to the regulatory region, but may extend into the permease gene region, as demonstrated by the recent identification of two strains that show no homology to wild-type *MAL1T* sequences (*28a*). These may represent previously unidentified *mal1*^{0} alleles.

In general, the cluster of the three *MAL* genes comprising a functional *MAL* locus is contained within about 7 kbp of DNA. However, nearly all of the cloned *MAL* DNAs have sequences 2–3 kbp beyond the 3′ ends of their maltase genes, which exhibit homology to one another. Such homology in some cases ends abruptly

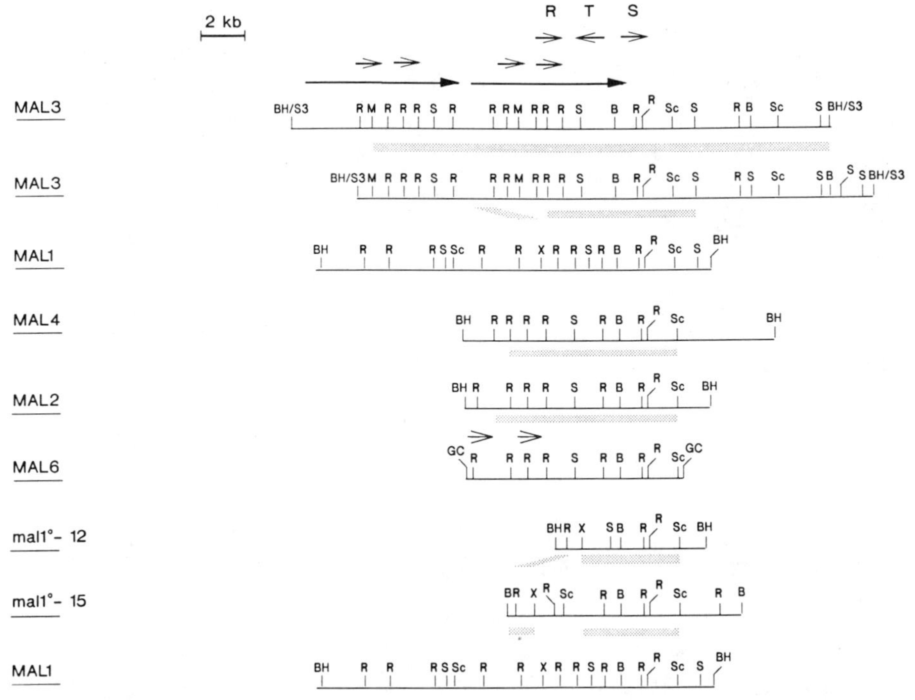

FIG. 2. Restriction maps and structural homologies of the *MAL* loci. Restriction maps of the cloned *MAL* loci. Arrows at the top of the figure indicate the locations and directions of transcription for each of the three essential genes at each *MAL* locus (R = *MALR*, T = *MALT*, and S = *MALS*). Arrows over the *MAL3* and *MAL6* restriction maps indicate the approximate locations and extents of the direct repeats at these loci, as discussed in the text. Stippled boxes indicate regions of homology among the various *MAL* loci. BH = *Bam*HI, B = *BglII*, GC = GC tailing, M = *Mlu*I, R = *Eco*RI, S = *Sal*I, Sc = *Sac*I, S3 = *Sau*3A, and X = *Xho*I. Vector sequences have been omitted and not all known restriction sites are shown.

and in other cases could not be compared because the *Bam*HI fragments carrying the *MAL* clones did not extend far enough.

The absence of homology at the 3′ ends of some maltase genes could be explained in some cases by the abrupt appearance of another gene in that region. For example, in the cases of *MAL3* and *mal1*^{0}-15 DNA, the 5′ proximal region of invertase gene of a *SUC* locus borders the 3′ end of the maltase genes of these *MAL* loci. Restriction mapping and hybridization data using a 5′ *SUC2* probe indicate that the 3′ end of *MAL3S* is 6 kbp from the 5′ end of *SUC3*. Restriction mapping of the *MAL1–SUC1* region indicates a physical distance of 4.5 kbp between *MAL1S* and *SUC1*. The proximity of these genes is

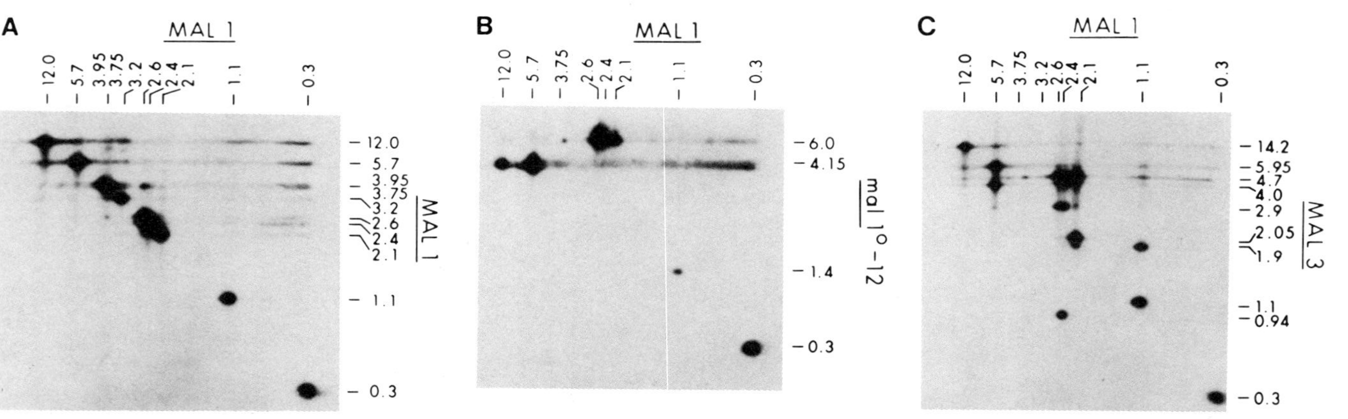

FIG. 3. Southern-cross analysis of the cloned *MAL* locus. *Eco*RI-digested DNA from the clones of the respective loci were used in all Southern-cross experiments. Panel A shows a Southern cross of *MAL1* × *MAL1*. All hybridization signals form a diagonal array, indicative of perfect homology without repeats. Panel B shows a Southern cross of *MAL1* × *mal1*^{0}-12. Some of the hybridization signals are shifted to fragment sizes other than those found in *MAL1* × *MAL1*, indicating some polymorphism between these loci. Some other spots are completely absent, which is the result of a deletion of the regulatory region in *mal1*^{0}-12 (vertical axis) corresponding to the *Eco*RI fragments 2.4 and 1.1 kbp on the *MAL1* (vertical) axis. The shift in some of the hybridization spots is the result of new *Eco*RI restriction-fragment sizes caused by the "fusion" of the ends of the deletion. Panel C shows a Southern cross of *MAL1* × *MAL3*. Hybridization signals appear along the diagonal in a pattern very similar to *MAL1* × *MAL1* but the restriction-fragment sizes are not exactly the same, indicating some polymorphism at this locus as well. Additionally, there are hybridization signals that are displaced from the diagonal plane. These additional hybridization signals are caused by multiple *MAL3 Eco*RI restriction fragments showing homology to single *MAL1* restriction fragments. This type of pattern is diagnostic of repeated sequences at the *MAL3* locus. These fragments correspond to *MAL1R* (2.4- and 1.1-kbp fragments) and *MAL1T* (2.1-kbp fragment). The *MAL3* fragments that show hybridization are, relative to the left-to-right orientation of the restriction map shown in Fig. 2, the *MAL3R* sequence repeats (in kbp): 1.9 (doublet) (fourth copy), 0.94 (doublet) and 1.1 (triplet) (third copy), 1.1 (triplet) and 1.9 (doublet) (second copy) and 0.94 (doublet) and 1.1 (triplet) (the expected *MAL3R*); *MAL3T* sequence repeats (in kbp): 2.05 plus a portion of 2.9 (second copy) and 4.7 (expected *MAL3T* plus a part of *MAL3S*). The extents of the repeated sequences can be estimated by adding together the sizes of the restriction fragments whose signals fall outside of the diagonal plane.

not surprising, since classical mapping studies have shown that *SUC1* and *MAL1* behave as though they are allelic to each other and that *MAL3* and *SUC3* are closely linked (*15*). In addition, the organization of the restriction sites typical of the *MAL6S* region was detected 4.5 kbp upstream from the cloned *SUC1*, *suc1*^{0}, and *suc3*^{0} loci (*28*). The spatial relationship between the *SUC* and *MAL* loci at the telomeres of chromsomes II and VII suggests that the events that brought these genes into close proximity are most likely independent from one another, i.e., that the *MAL* and *SUC* loci evolved and became transposed independently.

The *MAL1* and *SUC1* loci have recently been mapped as the most centromeredistal markers on the right arm of the *Saccharomyces* chromosome VII, very close to the telomere, with the following order: centromere–*MAL1*–*SUC1*–telomere. Similarly, *suc3*^{0}, and probably *SUC3* as well, and *MAL3* can be ordered as follows (*28*): centromere–*MAL3*–*suc3*^{0} (*SUC3*)–telomere (*76*). Such an assignment is based mainly on the detection within the cloned *SUC* genes of telomere-associated X and Y sequences (*28, 28a, 78, 79*).

The sequence divergence at the 5′ end of some *MAL* regulatory genes could be explained in part by the presence of a duplication of sequences composed of parts of *MALT* and *MALR*. For example, the *MAL3* and the *MAL6* loci exhibit direct repeats centered around the regulatory gene of each locus, resulting in a doubling or quadrupling of the *MALR* sequences of the respective *MAL* locus. These direct duplications are unlikely to be cloning artifacts because independent genomic clones confirm the restriction maps of DNA sequences at the duplicated regions. The independent isolation and characterization of a *MAL6R*-homologous repeated sequence (*MAL64*) (*80*) corroborate our results concerning the existence and position of the repeated sequence at the *MAL6* locus (*39*). The duplicated region at the *MAL3* locus consists of a pair of approximately 10 kbp direct repeats containing all of *MAL3R* and parts of *MAL3T*. Furthermore, within each 10 kbp of direct repeat of *MAL3*, a second pair of direct repeats exists that appears to be the result of a smaller, direct duplication of *MAL3R*. Such a structure appears to be the result of two independent duplication events, the first being a small duplication encompassing only *MAL3R* and part of *MAL3T* and a subsequent duplication encompassing the entire 10-kbp region of the first duplication. Similar studies with the *MAL6* locus showed that the duplication of the *MAL6R* gene is more modest, including a direct repeat of about 2.5 kbp of DNA encompassing part of *MAL6T* and most, if not all, of *MAL6R*.

It is not known whether the *MAL3R*-homologous repeats are functional, although, upon mutation, (MAL64) (*MAL6RR*) is transcribed, giving rise to the constitutive expression of maltase and maltose permease, even in the absence of a functional *MAL6R* (*MAL63*) gene (*80*). With this in mind, different hypotheses can be proposed to account for the generation and maintenance of sequences homologous to *MALR* genes.

If the repeats are functional, an increased copy number of the regulatory gene might confer a selective advantage by allowing increased synthesis of maltase and/or maltose permease as well as by allowing a yeast cell to respond more rapidly to the presence of maltose. A related hypothesis is that the tandem duplications allow the yeast cell to generate enough regulatory protein to regulate *MALS* and *MALT* genes at other homologous loci, especially those that lack a functional *MALR* gene due to its inactivation, deletion, or sequence divergence.

Results based on the introduction of the *MAL6R* gene on a multicopy plasmids into *MAL1* strains are only partially consistent with the notion that an increase in gene dosage has driven *MALR* duplication, since *MAL1* strains transformed with the *MAL6R* gene on a multicopy plasmid show no increase in maltose uptake and only a modest 1.5- to 2-fold increase in the levels of maltase and its mRNA (*45*). Moreover, when introduced into a *mal1*^{0} strain, the multicopy regulatory gene is unable to increase significantly the levels of the *mal1*-encoded maltase and maltose permease activities (*45*), thus suggesting that the *cis*-acting sequences responsible for the MALR-mediated induction (see II,B,1) may have diverged to a point where they are no longer recognizable by the MALR protein.

Alternatively, it is conceivable that the duplications could allow *MALR* to diverge in sequence and function, thereby allowing the regulation of other sugar-metabolizing gene systems. In several duplicated yeast genes, they are regulated differentially. For example, the enolase genes *ENO1* and *ENO2* are expressed differently when exposed to glucose (*81*); the genes of the acid phosphatase gene family (*82*) and alcohol dehydrogenase family (*83*) in yeast are also regulated differently. Since the *MAL6* regulatory determinant is also able to regulate the expression of α-methylglucosidase (oligo-1,6-glucosidase, EC 3.2.1.10) (*84, 85*) and since some of the genetic determinants controlling α-methylglucosidase expression (*MGL*) have been shown to be closely linked to the *MAL* loci (e.g., *MGL2* is 0.4 centiMorgans upstream from the 5′ end of *MAL3R*), it is not unreasonable to speculate that the duplication of the *MAL3* regulatory gene may represent a case of multiple copies that could lead to a differentially regulated gene family, and that *MGL2* might possibly be a duplicated but divergent copy of the *MAL3R* gene.

3. Evolution and Transposition of *MAL* Loci

There is insufficient evidence to propose specific mechanisms for the evolution and transposition of the *MAL* loci. The following findings, based on the study of likely transposition events, need to be considered: (i) the *MAL1* locus, or one of its alleles, is ubiquitous in all *Saccharomyces* strains examined; (ii) the five unlinked functional *MAL* loci are functionally and structurally related; (iii) the absence of a functional *MAL* locus is usually correlated with the absence of *MAL*-specific gene-sequences at all loci except *MAL1*; (iv) the *MAL1* through *MAL4* loci map to telomeric regions (*MAL6* has not yet been mapped precisely); (v) the *MAL3* and *MAL6* loci exhibit duplicated sequences of the *MALR* gene. Taken together, these findings suggest that the *MAL1* locus may be the progenitor of all the other *MAL* loci.

The duplication of genes at some of the *MAL* loci may be the result of some instability or exchange of the sequences at the telomeres of their respective host chromosome (*86, 87*). Transposition or duplication may have occurred by the inclusion of the entire *MAL* locus, or a portion thereof, in a sequence-sharing unequal homologous recombination as described in a model proposed to account for the maintenance of tandem telomeric repeats as seen in *Tetrahymena* (*88, 89*). In support of such a model is the observation that the genes encoding the variant antigens in *Trypanosoma* (*90*) are selected for expression by transposition from silent chromosomal locations to transcriptionally active telomeric sites via recombination at sites similar to the Y sequences found in *Saccharomyces* (*91, 92*). The existence of the polymeric *SUC* and *MAL* gene families is consistent with this model, since all but one (*SUC2*) of the *SUC* genes and all of the mapped *MAL* loci are telomeric and are even occasionally closely linked on the same chromosomes. Indeed, most of the *SUC* loci are embedded in telomere-adjacent sequences, the telomere being 3′ to the *SUC* sequence (*28*).

Meiotic gene conversion studies (*93*) show that nonreciprocal recombination (gene conversion) between repeated sequences on nonhomologous chromosomes in *S. cerevisiae* can occur at a frequency as high as 0.5%. This finding raises the possibility that the loci in a multigene family could have been generated by meiotic gene conversion of repeated sequences on nonhomologous chromosomes. The presence of repetitive DNA sequences near most telomeres enhances the likelihood that genes embedded in telomere-specific repetitive sequences could have been transposed to telomeric regions of nonhomologous chromosomes by a meiotic gene conversion mechanism. Both meiotic and less frequent mitotic telomeric exchange events have been observed in *Plasmodium falciparum* (*94*).

It is possible that the high frequency of meiotic gene conversion in *S. cerevisiae* would act to make members of a repeated gene family more homogeneous, thereby maintaining a high degree of homology within a multigene family. However, the different *MAL* and *mal1*^{0} loci may not have resided within a single strain long enough to allow their DNA sequences to become homogeneous. Also, it is unknown whether the maintenance and propagation of *Saccharomyces* strains in nature and in the laboratory have allowed sufficient opportunity for meiotic gene conversion to occur. Most of the functional *MAL* loci are highly homologous, especially within their respective *MALS* genes, suggesting that a gene conversion mechanism could have made the components of the gene family homogeneous. In contrast, the data obtained with regard to the *mal1*^{0} nonfunctional allelic loci show that substantial sequence divergences occurred in the regulatory gene regions of these loci. It is also possible that the conservation of specific domains of the regulatory protein (i.e., the "zinc-finger" and "helix-turn-helix" domains) may be sufficient for sequence-specific DNA binding and transcriptional activation of the *MAL* structural genes. This possibility would tolerate extensive sequence divergence in nonessential regions of the gene.

II. Regulation of *MAL* Gene Expression

A. Maltose Induction and Carbon-Catabolite Repression of Maltase and Maltose Permease, and Catabolite Inactivation of Maltose Permease

In most Mal$^+$ *Saccharomyces* strains, synthesis of maltase and maltose permease are coordinately induced by maltose and exhibit carbon-catabolite repression by glucose (*24, 95–99*). In addition, maltose permease levels are also controlled by catabolite inactivation (*4, 99, 100*). It is currently not known whether maltose is the actual inducer or whether the sugar has to be modified, as in the case of the *E. coli lac* operon (for which allolactose, rather than lactose, is a true inducer), to exert its effect. It is also not known whether galactose is the actual inducer of the *GAL*/*MEL* regulon in *S. cerevisiae*: the available genetic evidence suggests a higher rate of induction in strains carrying the *GAL3* gene (*30, 101*).

Expression of the *MAL* genes has been studied under both steady-state growth and transient inducing conditions. Under inducing conditions (typically 2% maltose), yeast cells synthesize fairly large amounts of maltase (1–2% of total cellular proteins), while the enzyme is essentially undetectable under repressing (2 or 5% glucose) (*69*) and nonrepressing (2% raffinose; 3% glycerol plus 2% ethanol) growth conditions (*98*). Likewise, the *MALS* transcript is fairly abundant in maltose-grown cells (about 26 copies per cell as determined by quantitative R-loop analysis), but is undetectable under repressing conditions (*43*). When the carbon source in the medium is switched from glucose to maltose, the synthesis of higher levels of maltase is preceded by *de novo* synthesis of *MALS*-specific mRNA sequences. This was monitored by DNA–RNA hybridization using a maltase structural gene probe, as well as by assaying functional maltase mRNA by *in vitro* RNA-directed synthesis of immunologically reactive maltase. Once maltase has accumulated, late in induction, further synthesis of the enzyme is inhibited, due to reduced *MALS* mRNA levels (*102*). Maltase synthesis is reduced significantly within 30 minutes after glucose addition to induced cells, due to the disappearance of *MALS*-specific mRNA sequences. This results from a reduction in both the transcription rate and the mRNA half-life. However, the stability of maltase itself, once induced, is not affected by the addition of glucose (*103*).

The induced level of the maltose uptake system is up to 150-fold higher than that observed under repressing conditions (*24, 43, 99*). The 1.9-kb *MALT* mRNA is present at about ten copies per cell under inducing conditions, but undetectable under repressing conditions (*43*). When the steady-state kinetics of *MAL1S* and *MAL1T* mRNA were examined by Northern blotting, the two mRNAs were found to be induced coordinately; their initial appearance was detectable as early as 50–60 minutes after the addition of maltose (*45*). Glucose has a second effect on maltose uptake, besides carbon-catabolite repression of mRNA synthesis, namely, catabolite inactivation that is dependent upon physiological conditions. In the absence of a nitrogen source, glucose brings about a rapid disappearance of the maltose/H^+ symport, in a process presumably involving proteolysis. This effect is nonspecific, since maltose gives very similar, though less dramatic, results. In growing cells, glucose induces a reversible decrease in the affinity of the H^+ symport that could be recovered by starvation. It has been proposed that the initial loss in affinity is the result of a reversible chemical modification of the permease (*100*). The best-known example of a catabolite-inactivated enzyme in yeast is fructose 1,6-bisphosphatase (EC 3.1.3.11) (*4, 104*). Inactivation of the enzyme is biphasic: the initial step involves a reversible cAMP-dependent phosphorylation (*105–107*). The target residue has been identified as Ser-11, located within a sequence (Arg-Arg-X-Ser) that matches the consensus sequence for cAMP-dependent protein-kinase substrates (*108, 109*). Following phosphorylation, an irreversible (proteolytic) loss of enzyme activity occurs that correlates with disappearance of immunoprecipitable material (*110, 111*). Whether catabolite inactivation of maltose permease is mediated by a similar chain of events remains to be seen. It should be noted that a potential phosphorylation site for cAMP-dependent protein kinases (Arg-Arg-Arg-Thr) is in fact present in the deduced sequence of the MAL6T protein (*48b*).

B. *cis*-Acting Sequences and *trans*-Acting Genes Affecting Expression of the *MAL* Genes

Before examining the promoter elements and the regulatory genes that affect the expression of the maltose-inducible *MAL* structural genes, a brief overview of the structure of yeast promoter elements and regulatory proteins is in order. Inducible yeast promoters contain two essential elements: a TATA box and an UAS that mediates induction by responding to an appropriate stimulus (*46, 47, 112–114*). Since the yeast genome is (A+T)-rich, several putative TATA elements can usually be identified by inspecting the DNA sequences upstream from the ATG translation-initiation codon of any given gene, but deletion analyses show that not all of these putative elements are actually used (*115, 116*). In fact, some genes may be transcribed even in the absence of any functional TATA element (*117*). Generally, the TATA element in yeast directs initiation between 60 and 120 bp downstream from the element. The preferred transcription initiation sites are TCGA and RRYRR. Precise spacing between the UAS and the TATA elements is not required for proper UAS activity, the average distance between them being 150–400 bp (*113*).

The best-studied example of a UAS in yeast is that of *GAL* genes termed UAS_{GAL}, located roughly midway between the divergently transcribed *GAL1* and *GAL10* genes, whose transcription is induced by galactose and repressed by glucose (reviewed in *30*). UAS_{GAL} was originally identified as a 365-bp DNA sequence able to confer

GAL4/galactose-dependent induction to a promoter–*CYC1/lacZ* fusion (*118*).

Deletion analysis has narrowed the boundaries of the UAS_{GAL} to a tandemly repeated sequence of 17 bp that confers galactose-inducible/glucose-repressible expression to a promoter–reporter gene fusion (*119–121*). The level of inducible expression depends on the number of copies of the 17-bp repeat. UAS_{GAL} confers transcriptional activation in both orientations, but is much less effective when moved more than 400 bp upstream from the transcription-initiation site; it is completely inactive when placed downstream from a reporter gene (*121a*).

UAS activity depends upon the presence of specific transcription factors, such as GAL4. GAL4 is a 100,000-Da protein (*122*) composed of several domains that have been mapped by deletion analysis. These include domains required for nuclear transport (*123*), DNA binding, (*124–126*), interaction with GAL80 [a negatively acting regulatory protein (*127, 128*)], and transcriptional activation (*57*). At present, it is unclear how a UAS mediates transcriptional activation at a distance, although experimental evidence favors a model in which DNA binding brings RNA polymerase into contact with transcription factors (reviewed in *30*).

1. Identification of the *cis*-Acting Upstream Regulatory Sequences of the Maltase Gene

The mechanisms by which transcription factors activate transcription seem to be wellconserved throughout evolution (*129*). In fact, GAL4 may activate transcription of a UAS_{GAL}-controlled reporter gene in organisms as diverse as mammals (*130, 131*), *Drosophila* (*132*), and plants (*133*). A complementary approach has shown that the products of the human proto-oncogenes c-*fos*, and c-*myc*, both nuclear proteins that may play roles in the regulation of gene expression, can activate transcription in yeast (*134*). Also, the human estrogen receptor, can induce transcription in yeast from a minimal promoter composed of a TATA box and an estrogen-responsive element (*135*). These results imply that the mechanisms of transcription activation, believed to involve interaction of the acidic domain of the upstream binding protein with some basic components of the transcription machinery, are highly conserved (*114*). The finding that yeast TATA-binding proteins can substitute for their mammalian counterparts in a mammalian *in vitro* transcription system (*136, 136a*) and the homology, including DNA binding specificity, between GCN4 and the product of the chicken oncogene *jun*, and its mammalian homologue AP-1 (*137, 137a, 137b*) also point strongly in this direction.

The *cis*-acting control elements were delimited by means of a series of nested upstream and internal deletions in the IG region flanking the maltase structural gene *MAL6S* (*138*). The *MAL6S* coding region and the accompanying deletions in its 5′-flanking promoter region were then inserted into a yeast centromere-based vector, to control the copy number, and transformed into a *mal1S*-Δ yeast strain. Analyses of maltase activity and of the steady-state, maltase-specific mRNA levels showed that deletion of the region from 334 to 380 bp upstream from the translation-initiation codon was clearly associated with the loss of inducible expression of *MAL6S* mRNA and of maltase activity.

Examination of the pertinent DNA sequences responding to maltose induction revealed a large inverted repeat sequence within and neighboring the region necessary for the induction of the maltase gene. It extends from −327 to −361, with 22 of its 34 bases paired. Moreover, four copies of a 7-bp direct-repeat sequence with the consensus sequence AAANTTT were found within and around this inverted repeat structure. Several yeast UASs contain one or more short inverted repeat sequences; good examples are the *ADH2* gene (*139*) and the *GAL7*, *GAL10* gene cluster (*120, 140*). In *CYC1* (*141, 142*) and *SUC2* (*143*), the UASs contain several copies of short direct repeats. The positions of four direct-repeat sequences in UAS_{MAL} and the fact that the endpoint of one of the upstream deletion mutants fell exactly in the middle of the inverted repeat sequence at position −344, and showed about one-half the normal induced level of the maltase expression, suggest that these structures are involved in the induction of the maltase gene. It still must be determined whether the *MAL6S* UAS_{MAL} plays a role in activating transcription of the *MAL6T* gene as well, or whether a separate *cis*-acting element regulates transcription of the divergently transcribed maltose permease gene.

A T-rich sequence is found 237–254 bp upstream from the maltase translation-initiation codon. The glucose-repressed levels of maltase activity and mRNA increased up to 5-fold in mutant constructions in which the T-rich sequence was deleted. Mutants that contain a long stretch of the homopolymer poly(dT) upstream from the *ADR2* gene exhibit high-level constitutive expression of the gene (*144*). Also, a poly(dT) sequence almost identical to the T-rich sequences found in the *MAL6S* upstream region acts in a bidirectional manner as a component of the upstream promoter elements for the constitutive basal-level transcription of some genes in *S. cerevisiae* (*46, 145*).

The results of the deletion experiments of the *MAL6S* promoter region favor the model that the T-rich stretch may be responsible in part for the carbon-catabolite repression by glucose rather than the basal constitutive-level expression of the *MAL6S* gene. This T-rich sequence is likely to act bidirectionally by participating in catabolite repression of the *MAL6T* gene, since no such other large poly(dT) sequence is present in the IG region. The *cis*-acting control regions are also summarized in Fig. 1.

2. The Maltose Regulatory Gene Encodes a Positive Activator

a. Constitutive Mutants. Constitutive mutants that synthesize maltase in the absence of maltose for all the *MAL* loci have been isolated. Not all of the constitutive mutants displayed identical phenotypes in terms of their dominance to the wild-type allele or their sensitivity to glucose repression. Genetic and molecular analyses (including molecular cloning and gene disruption of the mutant alleles, and gene-dosage experiments with *MAL6* genes subcloned on multicopy plasmids) showed that at least three different mechanisms may give rise to a constitutive phenotype. The first class includes *MAL6*^{c} mutants. These mutants have been obtained by reversion of different *mal6R* (*mal63*) maltose nonfermenter strains, which included a null mutant generated by gene disruption, to maltose fermenters (*41, 85, 145a*). This class of constitutive mutants was still sensitive to carbon-catabolite repression and was recessive to the wild-type allele.

Initially, the constitutive *MAL6R* regulatory mutant revertants were assigned to the *MAL6* regulatory gene (*85, 145a*), but later work has shown that constitutivity may have been caused by a mutation in a previously unidentified gene (*MAL64*) at the *MAL6* locus, mapping 2.3 centiMorgans to the left of the *MAL6R* (*MAL63*) gene (*41*). The molecular cloning of this gene (*80*) showed that it is structurally homologous to the *MAL63* gene, and in fact homologous to the *MAL6RR* sequence identified by the Southern cross experiments described in Section I,C,2. The mutationally activated constitutive *MAL64* gene encodes a positive *trans*-acting protein required for the constitutive expression of the *MAL6S* and *MAL6T* genes, while its unactivated wild-type counterpart (*MAL64*) appears dispensable for maltose-induced expression of the *MAL6S* and *MAL6T* genes. The nature of the constitutive mutation(s) in these strains is unknown. However, it is worth noting that *MAL64*^{c} is epistatic to *MAL6R* (*MAL63*) (*41, 80*) while being dominant to wild-type *MAL64*. These observations suggest that the *MAL64*^{c}-, *MAL64*- and *MAL6R*-encoded proteins either interact physically with each other, or compete for a target site, most likely the UAS_{MAL} located in the *MAL6S–MAL6T* IG region. This type of constitutive expression appears to be limited to the *MAL6*, and possibly to the *MAL3* loci, where one of the identified *MALR* repeats could also be functionally analogous to the *MAL64* gene. Accordingly, no Mal^{+} revertants have been obtained in a *mal1R*-Δ strain (*38*, *40*).

A second class of constitutive mutants includes those isolated at the *MAL1*, *MAL2*, *MAL3*, and *MAL4* loci (*25*, *146*, *147*). Most of these mutants are dominant to the wild-type allele and insensitive, although to different degrees, to carbon-catabolite repression. Molecular cloning of the regulatory gene of the *MAL4* locus (*39*, *76*, *148*) and of *MAL2-8*^{c} has been reported. Transformation of the *MAL4R*^{c} gene into inducible strains caused synthesis of both maltase and maltose permease to be constitutive and insensitive to glucose repression. The cloned *MAL4R*^{c} gene was dominant over the wild-type *MAL6R* gene when a single copy was transformed into *MAL6* strains (*148*). Also, *mal*^{0} strains transformed with a *MAL2-8*^{c}-carrying plasmid showed a phenotype with respect to regulation similar to that of the *MAL2-8*^{c} dominant strain (*149*). When the *MAL6R* was introduced in multiple copies into a *MAL1R*^{c} dominant strain, loss of the constitutive phenotype was observed, presumably because the wild-type MAL6R protein, being present at a higher concentration, could overcome the constitutivity brought about by the MAL1R^{c} protein (*45*). These results indicate that the constitutive regulatory genes encode proteins that can stimulate transcription of the *MALS* and *MALT* genes in the absence of inducer, and in the presence of glucose.

A third type of constitutive mutation has been isolated at the *MAL1* locus by reversion of a *mal1-1* mutant allele. Cloning of this *MAL1R* constitutive allele supported the conclusion that the constitutive mutation arose by a rearrangement between the original *mal1-1* mutant allele and sequences having different locations in the genome (*150*). The internal deletions of the pyrimidine-rich sequence in the I-G region may represent a fourth type of constitutive mutant in which basal-level expression is somewhat elevated constitutively, even in the presence of glucose (*138*) and in the absence of a functional regulatory gene (M. Vanoni, unpublished.

b. Uninducible Mutants. Maltose-nonfermenting mutants have long been sought as derivatives of maltose-fermenting strains. An extensive study conducted on *MAL6* strains generated several Mal^{-}

mutants mapping at the *MAL6* locus. The mutants were all assigned to a single complentation group (*18*), proposed to define the regulatory gene of the *MAL6* locus (*18, 26*). The undetected presence of additional, partially functional *mal* loci precluded the isolation of mutants in the structural genes for maltase and maltose permease. Genetic and molecular analyses of several *mal1R* mutants have been performed. Twenty-four regulatory mutants and three permease mutants were used to construct a meiotic map of the *MAL1R* gene that covers a region expressable by a recombination frequency of 0.33%.

The accumulation of *MAL1S*, *MAL1T*, and *MAL1R* mRNAs under inducing conditions was examined by Northern blot analysis and compared to the pattern of expression observed in *mal1R*-Δ strains (*40*). All of the *mal1R* mutants accumulated normal or nearly normal amounts of *MAL1R* mRNA. This finding suggests that a functional MAL1R protein is not required to control transcription of the gene encoding it, neither in a positive nor in a negative manner. Even though some kind of autogenous regulation is widespread among regulatory genes (*151, 152*), nonautogenous regulation would be in keeping with the enhanced transcription of *MAL6R* on mutlicopy plasmids (*45*). However, other genes could have effects on the regulation of *MAL1R* transcription, as suggested by the finding that a mutant mapping outside the *MAL1* locus apparently abolishes the accumulation of *MAL1R* mRNA (see Section II,B,3). All of the *mal1R* regulatory mutants, regardless of their position on the meiotic map, accumulated no mature-sized *MAL1T* mRNA nor *MAL1S* mRNA, i.e., no uncoupling of *MAL1S* and *MAL1T* expression was apparent, all of the mutants closely resembling the phenotype of the *mal1R*-Δ strain.

c. Gene-Dosage Experiments. One approach to analyzing the regulatory circuits controlling gene expression is to perturb the system by altering the dosage of the regulatory and structural genes. Such an analysis can give insights into how the cell controls gene expression. The first studies on the dosage of *MAL* genes involved the construction of strains carrying different numbers of the *MAL* loci and led to the speculation that the *MAL* loci are regulatory in nature and are the limiting factor in maltase expression (*153, 154*). Later, the steady-state levels and the induction kinetics of both the mRNAs and the functional activities of the maltase and maltose permease genes were studied in Mal^+ and Mal^- strains transformed with subcloned *MAL* genes. These studies, together with the gene disruption experiments previously described, confirmed that *MALR* encodes a *trans*-acting positive regulatory gene product (*45*). Interestingly, the presence of *MAL6R* on a multicopy plasmid does not result in constitutive expression of the structural genes for maltase and maltose permease, i.e., maltose is still necessary for the inducing activity of the regulatory gene.

In the cases of the two best-characterized yeast transcriptional activators, GAL4 and GCN4, that control, respectively, the genes required for galactose utilization (*30*) and the cross-pathway or general amino-acid control genes (reviewed in *155*), no organic cofactors are required for either sequence-specific DNA binding or transcriptional activation. The *GAL4* gene is expressed constitutively and regulation of GAL4 activity is achieved via protein–protein interaction (modulated by galactose) with the product of the *GAL80* gene, while regulation of the GCN4 activity occurs at the level of *GCN4* mRNA translation. In the case of HAP1, which regulates the transcription of the *CYC1* gene by binding to its *UAS1* upstream from the gene, a heme cofactor was shown to be required to activate sequence-specific DNA binding for modulation (*156*). Maltose might be required to modulate the activity of the MALR protein that would be inactive in its absence, but there is no direct evidence for this at present.

Two lines of evidence suggest that maltose is required only for transcriptional activation of MALR and not, as in the case of *HAP1*, for sequence-specific DNA binding. First: when *MAL6R* cloned into a multicopy plasmid is introduced into a strain bearing a *trans*-acting constitutive mutation mapping at the *MAL1* locus, it restores inducibility, suggesting that both the wild-type MAL6R and the constitutive MAL1R protein each bind to the same site, the UAS_{MAL}, even in the absence of maltose. Since the plasmid-encoded MAL6R protein, likely to be present in higher concentration, does not activate transcription in the absence of maltose, a loss of the constitutive phenotype is observed (*45*). Second: more compelling evidence comes from *in vitro* DNA binding studies showing that maltose is not required for binding of the MALR protein to UAS_{MAL} as detected by gel retardation assays (*138*). Possible roles for maltose in activating MALR as a transcription factor might include affecting its nuclear localization, dimerization, posttranslational modification (e.g., phosphorylation of MALR), or interaction with other transcriptional factors.

The regulation of the levels of mature *MALS* and *MALT* transcripts shows differences despite their coordinate induction by maltose. In fact, no constitutive expression of the maltose permease gene was observed when the gene was present on a multicopy plasmid, in contrast to what was seen in *MAL1* strains transformed with a multicopy *MAL6S* gene. Moreover, while the induced maltase levels were increased about 5-fold compared to the untransformed strains, no increases in the induced levels of maltose permease were observed. The increased maltase synthesis observed when *MALS* is present in multicopy is indeed made up of two separate components: an increased basal level that approaches the level found in the untransformed induced wild type, and a small extent of induction. The basal *MALS* transcription is independent of the presence of the regulatory gene, since it is also detectable in *mal1R*-Δ mutants, while induced expression is absolutely dependent on a functional regulatory gene (*45*; M. Vanoni, unpublished). Thus, even though the total level of maltase synthesis is increased in the transformed cells, the induction ratio for each copy of the *MALS* gene is much smaller than when it is present as a single genomic copy. Finally, the levels of maltose transport are rate-limiting for expression of the *MAL* genes, at least under transient inducing conditions, as cells carrying a constitutively expressed *MAL6T* gene show a dramatic reduction in the lag required to synthesize maltase upon maltose induction (*45*).

3. Other Genes Affecting *MAL* Gene Expression

Several genes besides the *MAL* genes themselves can affect maltose utilization. Most of these genes are part of a global regulatory system, carbon-catabolite repression, that affects expression of sugar fermentation genes, as well as of genes encoding enzymes of the tricarboxylic-acid cycle, the glyoxylate shunt, and gluconeogenesis (*29, 31, 32, 156a*). So far, little is known about the biochemistry of catabolite repression in yeast, although it is clear that cAMP, the signal for bacterial catabolite repression (*185–187*), is not involved. Most studies on yeast catabolite-repression have involved the isolation of mutants that are either resistant to glucose repression, or that are unable to derepress sensitive enzymes even in the absence of glucose (and in the presence of the required inducer, when appropriate). A variety of strains with widely different genetic backgrounds has been used, thus making difficult the characterization of the mutants and comparisons of the results obtained in different laboratories. Most of the carbon-catabolite repression mutants vary in their extent of pleiotropy.

Table I lists genes that affect maltose metabolism. Biochemical functions are known only for *SNF1* (*188*), which encodes membrane-associated serine/threonine protein kinase (*189*), *SNF3*, which en-

TABLE I

GENES AFFECTING MALTOSE UTILIZATION[a]

Gene	Growth on			*MAL* specific	Effect on		Reference
	Maltose	Glycerol	Sucrose		Maltase	Maltose Permease	
CCR1 = *SNF1* = *CAT1*	–	–	–	No	D.I.	N.T.	*188, 203, 206*
SNF2 = *MNU1*	N.T. (–)[b]	–(+)[b]	(+)[b]	No	D.I.	D.I.	*40, 188*
SNF3		+	–	No	N.T.	N.T.	*188*
CAT3	–	+	N.T.	No	D.I.	N.T.	*203*
MNU2–MNU6	–	+	+	Yes[c]	D.I.	D.I.	*40*
MNU7[d]	–	+	+	Yes[c]	D.I.	D.I.	*40*
HEX1 = *HXK2*	+	+	+	No	D.R.	N.T.	*203*
HEX2	–	+	+	No	D.R.	D.R.[e]	*212*
CAT80	+	+	+	No	D.R.	N.T.	*203*
FLK1 = *CYC9*	+			No	D.R.	N.T.	*213*
CID1	+	+	+	No	D.R.	N.T.	*211*
SSN6 = *CYC8*	+		+	No	D.R.	N.T.	*198*
PMU1	–[f]			Yes	D.I.[f]	N.T.	*207, 210*

[a] Abbreviations: D.I., defective induction, i.e., no synthesis of maltase or maltose permease in maltose-grown cells; D.R., defective repression, i.e., constitutive synthesis of maltase or maltose permease in glucose-grown cells; N.T., not tried.

[b] The original *snf2* mutant did not grow on glycerol or sucrose as carbon sources, while the *mnu1* mutant isolated in our laboratory grew on both types of media.

[c] As judged by growth on sucrose-, raffinose-, ethanol- and glycerol-containing media.

[d] The mutant is dominant over the wild-type, so allelism tests were not performed.

[e] Addition of maltose results in uncontrolled and excessive maltose uptake, leading to a 60-fold increase in intracellular glucose and eventually to cell death.

[f] *pmu1* strains are Mal^- only in the absence of functional mitochondria.

codes the high-affinity glucose transporter (*190, 191*), the *HEX1*, the structural gene for hexokinase isozyme II (*192–194*). The *SNF1* gene is constitutively transcribed, independently of the presence of a functional *SNF1* product (*195*), and its subcellular localization is not altered in either repressed or derepressed cells (*189*). The *SNF1* gene product is required for *SUC2* gene expression at the RNA level (*196*).

It is currently not known at what level *SNF1* affects expression of the *MAL* genes, nor are the target proteins of the SNF1 kinase known. Possible targets for the SNF1 protein kinase include the MALR and GAL4 proteins. The SNF1 protein probably has separate targets for the regulation of different genes, since some *snf1* extragenic suppressors selected on the basis of sucrose fermentation, such as *ssn6* (*197*), do not relieve the defect in galactose and maltose fermentation in an *snf1* background, although allowing considerable constitutive syntheses of both maltase and of a *GAL1–lacZ* gene fusion (*198*). On the basis of epistasis studies, *HEX1* has been assigned to a very early step in glucose repression, consistent with the finding that hexose phosphorylation appears to be an obligatory step in catabolite repression (*199–201*). Epistasis relationships between *SNF1* (*CAT1*) and *HEX1* are as yet unresolved, as conflicting results have been reported by different laboratories, using different *snf1* (*cat1*) alleles (*202, 203*). These differences may reflect allele specificity [the *cat1-1* allele described by Entian and Zimmerman, (*203*), for instance, showed no defect in invertase derepression, i.e., the phenotype used to isolate the *snf1* alleles], or differences in the genetic background, or both. The observation that some *hex1* (*hexk2*) mutations affect only glucose repression without affecting hexokinase activity has led to the proposal that the protein has a separate catalytic and regulatory domain (*192*). However, *in vitro* mutagenesis studies have failed to reveal such a two-domain structure (quoted in *204*). SNF3 has now been identified as encoding a glucose transporter (*205*).

We have recently isolated, in a *MAL1* genetic background, several maltose nonutilizing (*mnu*) mutants (*40*). All of the mutants but one were recessive. Those mutants grew on glycerol, ethanol, raffinose, and sucrose, but not maltose. Because the parent strain was Gal$^-$, the effect on galactose metabolism could not be tested. All mutant strains produce very low amounts (10% wild-type levels) of both maltase and maltose permease, with the exceptions of *mnu2* and *mnu6* mutants, whose maltose permease levels approached one-third and one-sixth, respectively, of those found in the parent strain (*40*). In determining the requirement for maltose in the regulation of the maltase gene by the MALR protein, it came as no surprise that all mutants made no

MAL1S mRNA (M. Vanoni, unpublished). The more interesting mutants appear to be *mnu1* and *mnu7*. *mnu1* is allelic to *snf2*, a gene required for high-level derepression of *SUC2* as well as of acid phosphatase genes (*206*), but that may play no regulatory role in carbon-catabolite repression. The *mnu7* mutation is dominant and abolishes transcription of the regulatory gene. This indicates that although *MALR* transcription is constitutive, there may exist other gene(s) that may regulate its expression.

PMU1, a gene that affects maltose utilization only in *petite* mutants, has been described (*207*). Such a gene is reminiscent of the *IMP1* gene that affects galactose utilization in *rho*$^-$ cells (*208, 209*). The mode of action of this gene is unclear, but it has been suggested that it may be involved in an oxidative pathway for maltose utilization (*210*).

Based on our current understanding of *MAL* gene expression, several potential sites of action of the regulatory gene (including but not restricted to those involved in catabolite repression) may be indicated and are briefly discussed here. An initial regulatory step might be the transport of maltose itself by the maltose permease. Other gene products might be required, either as subunits of the permease itself or as processing enzymes. The effect of glucose on both *MALT* transcription and maltose permease stability has already been mentioned and is not discussed further. Once inside the cell, glucose might interfere with binding of the inducer with the MALR protein, possibly by directly competing with maltose (if maltose is the true intracellular inducer) for the same allosteric site on the regulatory protein. A possible mechanism would be by either preventing MALR from binding to UAS_{MAL}, or by altering MALR to retain an inactive conformation. In both cases, glucose might act by directly binding to the regulatory protein, or acting via another, yet unidentified protein. A possible candidate could be the product of CID1, since, when grown under repressing conditions, *cid1*-226 mutants produce about 60% versus 2.65% for wild-type *CID1* of the fully induced maltase levels. However, it is not known if this effect is peculiar to *MAL3*.

In the *MAL* system, glucose-insensitivity of maltase and maltose permease expression at the *MAL4* locus is caused by an alteration of the *MAL4R* (*MAL43*) gene, thus suggesting that the maltose regulatory protein may directly participate in catabolite repression (*148*). Glucose-catabolite repression might also be mediated by allowing other repressive proteins to bind to *cis*-acting sites within the promoter regions of the *MAL* genes, as suggested by the deletion experiments reported in Section II,B,1. Additionally, other means of regulating *MAL* gene expression may be operative, as suggested by the *mnu7* strain in which the *MALR* gene is not transcribed.

C. Effect of the *MAL* Genes on the Utilization and Accumulation of Other Sugars

Over the past few years, evidence has been presented that the *MAL* genes affect not only maltose fermentation, but may also influence the utilization and accumulation of other sugars, namely, trehalose and α-methylglucoside. Unfortunately, most data are difficult to evaluate and compare, because the *MAL* genotype of the strains employed in the different studies was often poorly characterized. No attempt is made in this review to cover the metabolism of these sugars in detail; instead, we focus on those aspects that are more pertinent to the *MAL* system.

1. Trehalose Metabolism

Trehalose is an important storage carbohydrate in fungi (*157*). In yeast cells, it is accumulated at levels exceeding 15% of the cell dry weight during periods of reduced growth rate, e.g., during carbon, nitrogen, or phosphorus starvation (*157, 158*). Synthesis of trehalose appears to be a function of the combined activities of trehalose-6-P synthase (EC 2.4.1.15) which catalyzes the formation of trehalose-6-P from UDP-glucose and glucose-6-P, and trehalose-6-P phosphohydrolase (EC 3.1.3.12) (*159*). Stimulation of resting yeast cells to proliferate is associated with the rapid conversion of trehalose to glucose, presumably mediated by the hydrolytic activity of trehalase (EC 3.2.1.28) (reviewed in *157*). Both trehalose-6-P synthase (*160*) and trehalase (*161, 162*) activities are regulated by reversible cAMP-dependent phosphorylation in an antagonistic manner: the synthase complex is essentially inactive in its phosphorylated state (*160*), while trehalase is activated by phosphorylation (*161–163*). Indeed, it has been shown that the cell-cycle-dependent oscillation pattern of trehalase activity is the result of a phosphorylation/dephosphorylation mechanism (*164*).

A relationship between trehalose accumulation and maltose metabolism has been known since the early 1960s. Using radioactive maltose, it was shown that maltose-grown cells accumulated the sugar and converted it into labeled trehalose with concomitant exhaustion of the preexisting pool of trehalose (*165*); the net result was an increased trehalose turnover. Later experiments indicated that trehalose accumulation is subject to glucose repression and that growth on maltose increases the accumulation of trehalose (*166*). These studies have been extended to show a more direct link between trehalose metabolism and the *MAL* genes.

Genetic analysis showed a tight link between the Tac$^+$ phenotype (the ability to accumulate trehalose during growth on glucose-containing medium) and constitutive *MAL* alleles (*167, 168*), either resistant or sensitive to glucose repression. The absence of any correlation between the Tac$^+$ phenotype and trehalase levels indicated that the accumulation of trehalose was not due to decreased activity of trehalase. The inability of *sst*1 strains (defective in trehalose-P synthase) to accumulate trehalose unless grown in a maltose-supplemented medium (provided an inducible *MAL* locus is present) or even in glucose (in a *MAL*^{c} background) suggests that the *MAL* gene regulates components of the trehalose accumulation system that can operate independently of the UDP-glucose-dependent trehalose synthetase (*169*). Since maltose is not an obligatory substrate for trehalose synthesis in *sst*1 *MAL*^{c} strains, it seems likely that maltose itself is not directly used as a glucose donor in the transglycation reaction, but rather that the glucose derived from maltose hydrolysis is used as a substrate for trehalose synthesis (*170*). No biochemical evidence has been presented to characterize this putative second system of trehalose synthesis.

The same authors reported the molecular cloning of the *MALR* gene from a strain carrying a constitutive *MAL4* locus. The cloned DNA transformed a Mal$^-$ strain to Mal$^+$, but surprisingly failed to confer to the transformants both the Mal$^+$ constitutive and the Tac$^+$ phenotypes (*171*). Since later reports showed that cloned *MAL4R* gene can confer constitutive expression to both maltase and maltose permease activities (*148*), the possibility exists that the original cloned DNA is not the *MAL4R* gene, so that the significance of the result mentioned above remains uncertain.

Recent results from our laboratory suggest that active maltose metabolism in a *MAL1* strain decreases trehalose accumulation; upon maltase induction, there is an inverse correlation between maltase activity and trehalose levels. Moreover, strains carrying either point mutations or disruption/replacements in the *MAL1R* gene accumulate up to 5-fold more trehalose than the isogenic wild-type strain (Goldenthal and Vanoni, unpublished). A decrease in trehalose content during exponential growth in maltose-containing media has recently been reported (*172*) and is consistent with our findings, as well as with previous reports of increased trehalose turnover in maltose-grown cells (*165*).

At the moment, it appears premature to draw any conclusion that can incorporate all of the available data, some of which are conflicting. The expressed *MAL* genes, maltose or one of its metabolites, might affect trehalose accumulation by several mechanisms. These include compartmentalization of trehalose or trehalase (*157*) or alterations in the levels of phosphorylation of trehalase via a maltose-mediated transient increase in AMP (*173*). A direct interaction of maltase or maltose permease with some components of the trehalose accumulation pathway or trehalose itself cannot be excluded. In this respect, it should be remembered that there is evidence that maltase can bind trehalose, but not hydrolyze it (*5*), and that trehalose may interact with the maltose transport system (*174*). Finally, *MALR* might directly regulate transcription of genes encoding enzymes involved in trehalose metabolism. Also, at least in the cases of the *MAL3* and *MAL6* loci, the possibility that the *MALR* repeat may effect trehalose metabolism, as already shown to be the case for melezitose and α-methylglucoside metabolism (*175*), cannot presently be ruled out.

2. α-Methylglucoside Metabolism

The genetic analysis of α-methylglucoside metabolism in yeast has revealed a complex network of interacting genes. Any of the following complementary gene pairs is necessary for a yeast strain to utilize α-methylglucoside: *MGL3 MGL2*, *MGL1 MGL2*, *MGL3 MAL4*^{c}, *MGL1 MAL1*, *MGL4 MGL1*, *MGL4 MAL1* (*84, 175–177*). It is currently unknown whether each *MGL* locus encodes a regulatory or a structural function, or both, as there are conflicting reports (*178, 179*). Some features that appear to be shared between the *MGL* and *MAL* systems deserve mention. The hydrolysis of α-methylglucoside is accomplished by an oligo-1,6-glucosidase (EC 3.2.1.10) that is usually referred to as α-methylglucosidase. The enzyme hydrolyzes isomaltose, panose, and the chromogenic substrate *p*-nitrophenyl-α-D-glucopyranoside (*p*NPG), but not maltose. Conversely, maltase cannot hydrolyze α-methylglucoside and isomaltose, though it does hydrolyze *p*NPG (*5, 72*). The difficulty in obtaining mutations in the structural gene for α-methylglucosidase, suggests that, by analogy with the *MAL* system, multiple structural genes encoding the enzyme exist, with some of them possibly being cryptic.

The transport of α-methylglucoside into yeast cells is mediated by an inducible permease distinct from the maltose transport system.

Early findings assigned the transport function to *MGL2*, a gene tightly linked to *MAL3*, located near a chromosome II telomere (*15, 180*). Studies of α-methylglucoside transport have often made use of labeled thioethyl-glucopyranoside (TEG), a nonmetabolizable analogue of α-methylglucoside. Uninduced *MGL2* strains transport TEG by facilitated diffusion, whereas α-methylglucoside-induced cells contain an active transport system (*180*). As in the case of maltose, transport of α-methylglucoside is accompanied by proton influx and K^+ efflux in a stoichiometric ratio of 1 : 1 : 1 (*66*), and is inhibited by uncoupling agents such as sodium azide and dinitrophenol (*181*). Strains with the recessive *mgl2* allele are permeable to maltose but not a α-methylglucoside or TEG. Glucose, trehalose, and maltose are effective competitive inhibitors of TEG accumulation in α-methylglucoside-induced cells. The maltose transport system appears to be more specific for maltose (*182*), even though maltose permease is apparently able to transport TEG as well (*24, 183*).

The hypothesis that distinct transport systems exist for maltose and α-methylglucoside is also supported by the finding that the inducible maltose and α-methylglucoside carriers have different half-lives *in vivo*, although the hypothesis that the two transport systems share a common component(s) has been proposed (*24, 64*). An intriguing possibility is that the α-methylglucoside and maltose transport systems are members of a family of related sugar carriers. According to this view, one might speculate that the *MALT* and *MALR* repeats at the *MAL3* locus represent the α-methylglucoside transport and regulatory genes, respectively; they could possibly be identical to the *MGL2* determinant identified by mutational analysis. Both gene pairs could be the product of divergent evolution after a duplication event of an ancestor gene(s) regulating the transport of α-methylglucoside and maltose.

There is clearly a relationship between *MGL* and *MAL* gene regulation. Maltose, for instance, induces both α-methylglucosidase and TEG transport activities (*85, 184*), mediated most likely through the action of the *MALR* gene. This is suggested by the fact that a number of constitutive *MAL* mutants that synthesize α-methylglucosidase constitutively have been obtained. In particular, *MAL64* has been implicated in the control of α-methylglucosidase expression, and it has been proposed to be a regulator of α-methylglucoside metabolism (*175*).

A precise description of the *MGL* system requires the cloning of one or more of the *MGL* genes, to be able to answer basic questions on the nature of the genes and to assess in more detail the relationship between maltose and α-methylglucoside metabolism.

III. Perspectives

In this essay we have summarized the major regulatory features of the *MAL* system and we have highlighted new directions of research. Maltose utilization is regulated through the interplay of two different, although possibly partially overlapping, systems: carbon-catabolite repression and inactivation, and maltose induction. This implies that maltose utilization requires not only the presence of a functional *MAL* locus (i.e., a maltose regulatory gene, a maltose permease gene, and a maltase gene), but of several other genes as well that are required at several levels to promote or repress the activity of the *MAL* gene products or the expression of the *MAL* genes themselves. Figure 4 shows the components of the *MAL* gene regulatory circuit and a speculative scheme of their interactions; some of these interactions have not yet been proven, and thus have to be considered only as working hypotheses.

Glucose might affect expression of the *MAL* genes by several mechanisms: (i) inhibition of maltose transport by repressing maltose permease transcription [see (2) and (3)] and by inactivation of preexisting maltose permease; (ii) direct inhibition of *MAL* promoters as suggested by the increased constitutive maltose synthesis observed when a T-rich region is deleted from the maltose promoters (*138*); (iii) inhibition of MALR binding to UAS_{MAL} or prevention of the conformational change induced by the inducer that is required before the MALR protein can act as a transcriptional activator.

Several genes are involved in catabolite repression in yeast. Thus, partially independent pathways, dependent or different sets of genes, branch from a common point that is more likely defined by the action of the hexokinase isozyme II (encoded by *HEX1*). SNF1-mediated phosphorylation of as yet unidentified proteins is required as an early step in expression. Possible substrates of the SNF1 protein kinase include the MALR and GAL4 regulatory proteins, as well as proteins whose function is located farther upstream in the regulatory cascade, such as the product of the *SSN6* gene.

In the absence of glucose and in the presence of maltose, transcription of the maltase and maltose permease genes is turned on. Transcriptional activation requires a *cis*-acting DNA sequence and

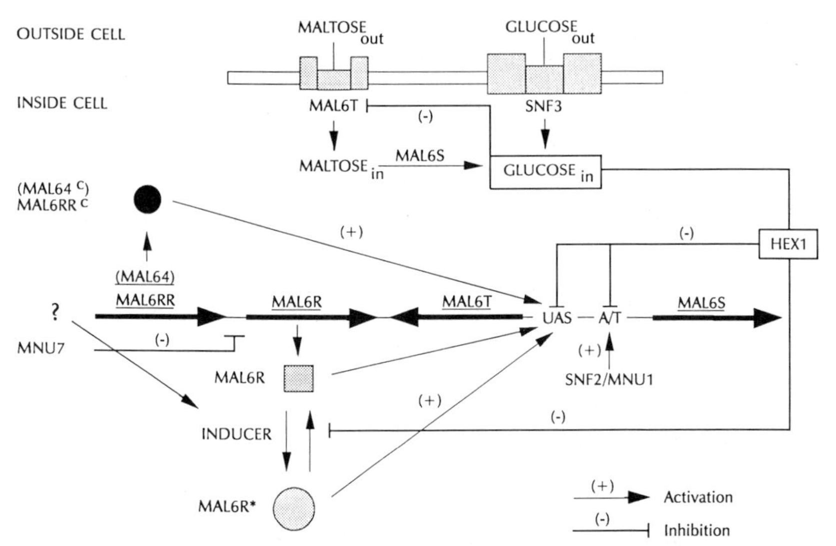

FIG. 4. Model for control of maltose metabolism in *Saccharomyces*. Proposed model for the interactions of known factors involved in maltose metabolism and the regulation of gene expression at the *MAL6* locus. This model is based upon interpretation of existing data and is considered speculative. Lines with arrows and (+) symbols indicate activation or stimulation. Bold lines with bars and (−) symbols indicate repression or antagonism. Lines without symbols reflect metabolic interconversions. Inactive regulatory proteins are represented by stippled squares. Active regulatory proteins are represented by stippled (MAL6R*) or solid (MAL6RR^c = MAL64^c) circles. *Indicates the activated form of MAL6R and a superscript c denotes that the maltose-independent, mutationally activated *MAL64*-encoded gene product is expressed constitutively. Inactive MAL6R protein binds to the UAS$_{MAL}$ *in vitro* in the absence of inducer; however, it is not known if this occurs *in vivo*. Positive-acting regulatory activity only occurs in the presence of inducer.

the product of a functional regulatory gene. The requirement for a functional *MALR* gene may be bypassed in *MAL6* strains by mutational activation of a structurally homologous gene, *MAL64*, resulting in recessive constitutive expression of the *MAL* structural genes. Transcription of the regulatory gene is constitutive, does not require a functional MALR protein, but is dependent on other *trans*-acting genes, such as *MNU7*. Transcriptional activation by MALR, but not sequence-specific binding to UAS$_{MAL}$, requires maltose. Probably, DNA binding is conferred by the MALR cysteine-rich "zinc finger," whereas sequence-specific binding is dependent on other regions, such as the helix-turn-helix motif adjacent to the zinc-finger domain. By analogy with other sequence-specific DNA-binding proteins, the α-helical acidic tail of the regulatory protein may represent the transcriptional activation domain (*214*). Full induction of the *MAL* structural gene requires other gene products, including the SNF2/MNU1 protein.

Although much has been learned about carbohydrate metabolism in *Saccharomyces*, there are many facets of the complex network of controls involved that are not yet well understood. The study of the *MAL* genes offers excellent opportunities for dissection of these control mechanisms. The system is easily manipulated, induction by maltose is readily obtained, repression by glucose is easily observable, and mutants are readily isolated. The genes of the *MAL6* locus have been completely sequenced and some structural and functional features of these genes are known. We are now studying mechanistically the control of gene expression at this locus. *MALR* appears to have global regulatory functions (*175*), and a thorough analysis of its properties should lead to interesting insights of the regulation of α-glucoside metabolism in yeast.

Acknowledgments

We thank S. H. Hong for his comments and Joy Speyer and Grace Sullivan for their excellent preparation of this manuscript. This work was supported by National Science Foundation Grant PCM-8310812, National Institutes of Health Cancer Core Grant P30-CA-133330, and U.S. Public Health Service Grant 2R01-GM28572.

References

1. J. Andrews and R. B. Gilliland, *J. Inst. Brew.* **58,** 189 (1952).
2. N. J. W. Kreger-van Rij, in "The Yeasts: Biology of Yeasts" (A. H. Rose and J. H. Harrison, eds.), Vol. 1, p. 5. Academic Press, London, 1987.
3. B. Magasanik, *CSHSQB* **26,** 249 (1961).
4. H. Holzer, *TIBS* **1,** 178 (1976).
5. J. A. Barnett, *Adv. Carbohydr. Chem. Biochem.* **32,** 125 (1976).
6. C. T. Kelly and W. M. Fogarty, *Process Biochem* **18,** 6 (1983).
7. H. O. Halvorson and L. Ellias, *BBA* **30,** 28 (1958).
8. O. Winge and C. Roberts, *C. R. Lab. Carlsberg, Ser. Physiol.* **24,** 263 (1948).
9. O. Winge and C. Roberts, *C. R. Lab. Carlsberg Ser. Physiol.* **25,** 35 (1950).
10. R. B. Gilliland, *Nature* **173,** 409 (1954).
11. O. Winge and C. Roberts, *C. R. Lab. Carlsberg, Ser. Physiol.* **26,** 331 (1952).
12. H. O. Halvorson, S. Winderman and J. Gorman, *BBA* **67,** 42 (1963).
13. F. Rudert and H. O. Halvorson, *Bull. Res. Counc. Isr., Sect. E* **E11 A4,** 337 (1963).
14. I. S. Pretorius, T. Chow and J. Marmur, *MGG* **203,** 36 (1986).
15. R. K. Mortimer and D. Schild, *Microbiol. Rev* **49,** 181 (1985).
16. P. Sollitti and J. Marmur, *MGG* **213,** 56 (1988).
17. P. Tauro and H. O. Halvorson, *JBact* **92,** 652 (1966).
18. A.M.A. ten Berge, A. Zoutewelle and K. W. van de Poll, *MGG* **123,** 233 (1973).
19. G. I. Naumov, *Genetika* **5,** 142 (1969).
20. G. I. Naumov, *Genetika* **6,** 20 (1970).

21. G. I. Naumov, *Genetika* **6,** 121 (1970).
22. G. I. Naumov, *Genetika* **7,** 141 (1971).
23. G. I. Naumov, *Genetika* **12,** 87 (1976).
24. R. A. de Kroon and V. V. Koningsberger, *BBA* **204,** 590 (1970).
25. M. J. Goldenthal, J. D. Cohen and J. Marmur, *Curr. Genet.* **7,** 195 (1983).
26. H. J. Federoff, J. D. Cohen, T. R. Eccleshall, R. B. Needleman, B. A. Buchferer, J. Giacalone and J. Marmur, *JBact* **116,** 1064 (1982).
27. M. Carlson, D. Osmond and D. Botstein, *CSHSQB* **45,** 799 (1980).
28. M. Carlson, J. L. Celenza and F. Eng, *MCBiol* **5,** 2894 (1985).
28a. M. J. Charron and C. Michels, *Genetics* **120,** 83 (1988).
29. M. Carlson, *J. Bact.* **169,** 4873 (1987).
30. M. Johnston, *Microbiol. Rev.* **51,** 458 (1987).
31. K. D. Entian, *Microbiol. Sci.* **3,** 366 (1986).
32. J. M. Gancedo and C. Gancedo, *FEMS Microbiol. Rev.,* **32,** 179 (1986).
33. K. Lam and J. Marmur, *JBact* **130,** 746 (1977).
34. J. D. Cohen, M. J. Goldenthal, B. Buchferer and J. Marmur, *MGG* **196,** 208 (1984).
35. T. Chow, M. J. Goldenthal, J. D. Cohen, M. Hegde and J. Marmur, *MGG* **191,** 366 (1983).
36. C. A. Michels and R. B. Needleman, *JBact* **157,** 949 (1984).
37. J. D. Cohen, M. J. Goldenthal, T. Chow, B. Buchferer and J. Marmur, *MGG* **200,** 1 (1985).
38. M. J. Charron, R. A. Dubin and C. A. Michels, *MCBiol* **6,** 3891 (1986).
39. T. Chow, Ph.D. thesis. Albert Einstein College of Medicine, New York, 1987.
40. M. Vanoni, S. H. Hong, M. J. Goldenthal and J. Marmur, *in* "Gene Expression and Regulation: The Legacy of Luigi Gorini" (M. Bissel, G. Deho, G. Sironi and A. Torriani, eds.), p. 247. Elsevier, Amsterdam, 1988.
41. R. A. Dubin, E. L. Perkins, R. B. Needleman and C. A. Michels, *MCBiol* **6,** 2757 (1986).
42. R. A. Dubin, R. B. Needleman, D. Gossett and C. A. Michels, *JBact* **164,** 605 (1985).
43. R. B. Needleman, D. B. Kaback, R. A. Dubin, E. L. Perkins, N. G. Rosenberg, K. A. Sutherland, D. B. Forrest and C. A. Michels, *PNAS* **81,** 2811 (1984).
44. S. H. Hong and J. Marmur, *Gene* **41,** 75 (1986).
45. M. J. Goldenthal, M. Varoni, B. Buchferer and J. Marmur, *MGG* **209,** 508 (1987).
46. K. Struhl, W. Chen, D. E. Hill, I. A. Hope and M. A. Oettinger, *CSHSQB* **50,** 489 (1985).
47. K. Struhl, *MCBiol* **6,** 3847 (1986).
48. S. Hohmann and D. Gozalbo, *MGG* **211,** 446 (1988).
48a. S. B. Baim and F. Sherman, *MCBiol* **8,** 1591 (1988).
48b. B. Yao, P. Sollitti and J. Marmur, *Gene,* in press (1989).
49. J. L. Bennentzen and B. D. Hall, *JBC* **257,** 3026 (1982).
50. R. Losson, R. P. P. Fuchs and F. Lacroute, *EMBO J.* **2,** 2179 (1983).
51. A. Laughon and R. F. Gesteland, *PNAS* **79,** 6827 (1982).
52. R. Koren, J. A. LeVitre and K. A. Bostian, *Gene* **41,** 2 (1986).
53. E. Wingender, *NARes* **16,** 1879 (1988).
54. R. M. Evans and S. M. Hollenberg, *Cell* **52,** 1 (1988).
55. C. O. Pabo and R. T. Sauer, *ARB* **53,** 293 (1984).
56. M. Johnston, J. Dover, J. Kim and C. A. Michels, *Proc. Int. Conf. Yeast Genet. Mol. Biol., 14th, Helsinki, August 1988,* S404.
57. J. Ma and M. Ptashne, *Cell* **48,** 847 (1987).

58. G. Gill and M. Ptashne, *Cell* **51,** 121 (1987).
59. I. Hope and K. Struhl, *Cell* **46,** 885 (1986).
60. E. Giniger and M. Ptashne, *Nature* **330,** 670 (1987).
61. I. A. Hope, S. Mahadevan and K. Struhl, *Nature* **333,** 635 (1988).
62. A. Harris and C. C. Thompson, *BBA* **52,** 176 (1961).
63. G. de La Fuente and A. Sols, *BBA* **56,** 49 (1962).
64. M. Siro and T. Lovgren, *Bakers Dig.* **53,** 24 (1979).
65. A. Seaston, C. Inkson and A. Eddy, *BJ* **134,** 1031 (1973).
66. R. Blocklehurst, D. Gardner and A. Eddy, *BJ* **162,** 591 (1977).
67. R. Serrano, *EJB* **80,** 97 (1977).
68. M. C. Loureiro-Dias and J. M. Pemado, *BJ* **222,** 293 (1984).
69. R. B. Needleman, H. J. Federoff, T. R. Eccleshall, B. Buchferer and J. Marmur, *Bchem* **17,** 4657 (1978).
70. N. R. Eaton and F. K. Zimmerman, *MGG* **148,** 199 (1976).
71. S. Tabata, T. Ide, Y. Umemura and K. Torii, *BBA* **797,** 231 (1984).
72. N. A. Khan and N. R. Eaton, *BBA* **146,** 173 (1967).
73. B. Svensson, *FEBS Lett.* **230,** 72 (1988).
74. I. Yamashita, K. Suzuki and S. Fukui, *JBact* **161,** 567 (1985).
75. M. Snyder and N. Davidson, *JMB* **166,** 101 (1983).
76. R. Rodicio and F. K. Zimmermann, *Curr. Genet* **9,** 547 (1985).
76a. T. H. C. Chow, P. Sollitti and J. Marmur, *MGG* in press (1989).
77. J. L. Celenza and M. Carlson, *Genetics* **109,** 661 (1985).
78. C. S. M. Chan and B. K. Tye, *Cell* **33,** 563 (1983).
79. R. W. Walmsley, C. S. M. Chan, B. K. Tye and T. D. Petes, *Nature* **310,** 157 (1984).
80. R. A. Dubin, M. J. Charron, S. H. Haut, R. B. Needleman and C. A. Michels, *MCBiol* **8,** 1027 (1988).
81. M. J. Holland, J. P. Holland, G. P. Thill and K. A. Jackson, *JBC* **256,** 1385 (1981).
82. D. Rogers, J. M. Lemire and K. A. Bostian, *PNAS* **79,** 2157 (1982).
83. U. Lutstorf and R. Megnet, *ABB* **126,** 933 (1968).
84. A. M. A. ten Berge, *MGG* **115,** 80 (1972).
85. A. M. A. ten Berge, G. Zoutwelle and R. B. Needleman, *MGG* **131,** 113 (1974).
86. R. M. Walmsley, *Yeast* **3,** 139 (1987).
87. A. M. Weiner, *Cell* **52,** 155 (1988).
88. G. B. Moria and T. R. Cech, *Cell* **46,** 873 (1986).
89. G. B. Moria and T. R. Cech, *Cell* **52,** 367 (1988).
90. L. H. T. Van der Ploeg, D. C. Schwartz, C. R. Cantor and P. Borst, *Cell* **37,** 77 (1984).
91. H. Horowitz, P. Thorburn and J. E. Haber, *MCBiol* **4,** 2509 (1984).
92. H. Horowitz and J. E. Haber, *MCBiol* **5,** 2369 (1985).
93. S. Jinks-Robertson and T. Petes, *PNAS* **82,** 3350 (1985).
94. L. M. Corcoran, J. K. Thompson, D. Walliker and D. J. Kemp, *Cell* **53,** 807 (1988).
95. R. van Wijk, *Proc. K. Ned. Akad. Wet., Ser. C* **71,** 60 (1968).
96. R. van Wijk, *Proc. K. Ned. Akad. Wet., Ser. C* **71,** 137 (1968).
97. R. van Wijk, *Proc. K. Ned. Akad. Wet., Ser. C* **71,** 293 (1968).
98. R. van Wijk, H. Ouwehand, T. van den Bos and V. V. Koninsberger, *BBA* **186,** 178 (1969).
99. C. P. M. Gorts, *BBA* **184,** 299 (1969).
100. J. M. Peinado and M. C. Loureiro-Dias, *BBA* **856,** 189 (1986).
101. W. Bajwa, T. E. Torchia and J. E. Hopper, *MCBiol* **8,** 3439 (1988).
102. H. J. Federoff, T. R. Eccleshall and J. Marmur, *JBact* **154,** 1301 (1983).

103. H. J. Federoff, T. R. Eccleshall and J. Marmur, *JBact* **156,** 301 (1983).
104. C. Gancedo, *JBact* **107,** 401 (1971).
105. A. G. Lenz and H. Holzer, *FEBS Lett.* **109,** 272 (1980).
106. M. J. Mazon, J. M. Gancedo and C. Gancedo, *JBC* **257,** 1128 (1982).
107. G. Pohlig and H. Holzer, *JBC* **260,** 13818 (1985).
108. J. Rittenhouse, P. B. Harrsch, J. N. Kim and F. Marcus, *JBC* **261,** 3939 (1986).
109. P. Cohen, *EJB* **151,** 439 (1985).
110. S. Funayama, J. M. Gancedo and C. Gancedo, *EJB* **109,** 61 (1980).
111. P. Tortora, M. Bictel, A. G. Lenz and H. Holzer, *BBRC* **100,** 688 (1981).
112. K. Struhl, *Cell* **49,** 295 (1987).
113. L. Guarente, *ARGen* **21,** 425 (1987).
114. L. Guarente, *Cell* **52,** 303 (1988).
115. F. Nagawa and G. R. Fink, *PNAS* **82,** 8557 (1985).
116. K. T. Arndt, C. Styles and G. R. Fink, *Science* **237,** 874 (1987).
117. J. E. Ogden, C. Stanway, S. Kim, J. Mellor, A. J. Kingsman and S. M. Kingsman, *MCBiol* **6,** 4335 (1986).
118. L. Guarente, R. R. Yocum and P. Gifford, *PNAS* **79,** 7410 (1982).
119. R. R. Yocum, S. Hanley, R. W. West, Jr., and M. Ptashne, *MCBiol* **4,** 1985 (1984).
120. R. W. West, Jr., R. R. Yocum and M. Ptashne, *MCBiol* **4,** 2467 (1984).
121. E. Giniger, S. Varnum and M. Ptashne, *Cell* **40,** 767 (1985).
121a. K. Struhl, *PNAS* **81,** 7865 (1984).
122. A. Laughon and R. F. Gesteland, *MCBiol* **4,** 260 (1984).
123. P. A. Silver, L. P. Keegan and M. Ptashne, *PNAS* **81,** 5951 (1984).
124. L. Keegan, G. Gill and M. Ptashne, *Science* **231,** 699 (1986).
125. M. Johnston, *Nature* **328,** 353 (1987).
126. M. Johnston and J. Dover, *PNAS* **84,** 2401 (1987).
127. S. A. Johnston, J. M. Salmeron, Jr., and S. S. Dincher, *Cell* **50,** 143 (1987).
128. J. Ma and M. Ptashne, *Cell* **50,** 137 (1987).
129. M. Ptashne, *Nature* **332,** 697 (1986).
129a. M. Ptashne, *Nature* **335,** 683 (1988).
130. H. Kakidani and M. Ptashne, *Cell* **52,** 161 (1988).
131. M. Webster, J. R. Jin, S. Green, M. Hollis and P. Chambon, *Cell* **52,** 169 (1988).
132. J. A. Fischer, E. Giniger, T. Maniatis and M. Ptashne, *Nature* **332,** 853 (1988).
133. J. Ma, E. Przibilla, J. Hu, L. Bogarad and M. Ptashne, *Nature* **334,** 631 (1988).
134. K. Lech, K. Anderson and R. Brent, *Cell* **52,** 179 (1988).
135. D. Metzeger, J. White and P. Chambon, *Nature* **334,** 31 (1988).
136. S. Buratowski, S. Hahn, P. A. Sharpand and L. Guarente, *Nature* **334,** 37 (1988).
136a. B. Cavallini, J. Huet, J. L. Plassat, A. Sentenac, J. M. Egly and P. Chambon, *Nature* **334,** 77 (1988).
137. P. K. Vogt, T. J. Bos and R. F. Doolittle, *PNAS* **84,** 3316 (1987).
137a. K. Struhl, *Cell* **50,** 841 (1987).
137b. D. Bohmann, T. J. Bos, A. Admon, T. Nishimura, P. K. Vogt and R. Tjian, *Science* **238,** 1386 (1987).
138. S. H. Hong and J. Marmur, *MCBiol* **7,** 2477 (1987).
139. J. Shuster, J. Yu, D. Cox, R. L. V. Chan, M. Smith and E. Young, *MCBiol* **6,** 1894 (1986).
140. M. Johnston and R. W. Davis, *MCBiol* **4,** 1440 (1984).
141. L. Guarente and T. Mason, *Cell* **32,** 1279 (1983).
142. L. Guarente, B. Lalonde, P. Gifford and A. Alani, *Cell* **36,** 503 (1984).
143. L. Sarokin and M. Carlson, *MCBiol* **6,** 2324 (1986).

144. D. W. Russell, M. Smith, D. Cox, V. M. Williamson and E. T. Young, *Nature* **304,** 652 (1983).
145. K. Struhl, *PNAS* **82,** 8419 (1985).
145a. A. M. A. ten Berge, G. Zoutewelle, K. W. van de Poll and H. P. J. Bloemers, *MGG* **125,** 139 (1973).
146. N. A. Khan and N. R. Eaton, *MGG* **112,** 317 (1971).
147. F. K. Zimmermann and N. R. Eaton, *MGG* **134,** 261 (1974).
148. M. J. Charron and C. A. Michels, *Genetics* **116,** 23 (1987).
149. R. Rodicio and F. K. Zimmermann, *Curr. Genet.* **9,** 539 (1985).
150. R. Rodicio, *Curr. Genet.* **11,** 235 (1986).
151. O. Raibaud and M. Schwartz, *ARGen* **18,** 173 (1984).
152. M. Igarashi, T. Segawa, Y. Nogi, Y. Suzuki and T. Fukosawa, *MGG* **207,** 273 (1987).
153. D. B. Mowshowitz, *JBact* **137,** 1200 (1979).
154. D. B. Mowshowitz, *Genetics* **98,** 713 (1981).
155. A. G. Hinnebush, *Microbiol. Rev.* **52,** 248 (1988).
156. K. Pfeifer, B. Arcangoli and L. Guarente, *Cell* **49,** 9 (1987).
156a. A. Manhart and H. Holzer, *Yeast* **4,** 227 (1988).
157. J. M. Thevellin, *Microbiol. Rev.* **48,** 42 (1984).
158. S. H. Lillie and J. R. Pringle, *JBact* **143,** 1384 (1980).
159. D. Manners, *in* "The Yeasts" (A. H. Rose and J. S. Harrison, eds.), Vol. 2, p. 418. Academic Press, London, 1987.
160. A. C. Panek, P. S. deAraujo, V. Moura Neto and A. D. Panek, *Curr. Genet.* **11,** 459 (1987).
161. C. H. D. Ortiz, J. C. C. Maia, M. N. Tenan, G. R. Braz-Padrao, J. R. Mattoon and A. D. Panek, *JBact* **153,** 644 (1983).
162. I. Uno, K. Matsumoto, K. Adachi and T. Ishikawa, *JBC* **258,** 10867 (1983).
163. G. M. della Mora-Ortiz, C. H. D. Ortiz, J. C. C. Maia and A. D. Panek, *ABB* **251,** 205 (1986).
164. J. Van Doorn, M. E. Scholte, P. W. Postma, R. V. Driel and K. van Dam, *J. Gen. Microbiol.* **134,** 785 (1988).
165. G. Avigad, *BBA* **40,** 124 (1960).
166. A. D. Panek and J. R. Mattoon, *ABB* **183,** 306 (1976).
167. A. D. Panek, A. L. Sampaio, G. C. Braz, S. J. Baker and J. R. Mattoon, *Cell. Mol. Biol.* **25,** 334 (1979).
168. D. E. de Oliveira, E. G. C. Rodriguez, J. R. Mattoon and A. D. Panek, *Curr. Genet.* **3,** 235 (1981).
169. M. S. Operti, D. E. de Oliveira, A. B. Freitas-Valle, E. G. Oestreicher, J. R. Mattoon and A. D. Panek, *Curr. Genet.* **5,** 69 (1982).
170. V. M. F. Paschoalin, V. L. A. Costa-Carvalho and A. D. Panek, *Curr. Genet.* **10,** 725 (1986).
171. D. E. de Oliveira, M. Arrese, G. Kidane, A. D. Panek and J. R. Mattoon, *Curr. Genet.* **11,** 97 (1986).
172. K. F. MacKenzie, K. K. Singh and A. D. Brown, *J. Gen. Microbiol.* **134,** 1661 (1988).
173. M. Beullens, K. Mbonyi, L. Geerts, D. Gladines, K. Detremerie, A. W. H. Jans and J. M. Thevelein, *EJB* **172,** 227 (1988).
174. A. Kotyk and D. Michaljanicova, *J. Gen. Microbiol.* **110,** 323 (1979).
175. E. L. Perkins and R. B. Needleman, *Curr. Genet.* **13,** 369 (1988).
176. D. Hawthorne, *Heredity* **12,** 273 (1958).
177. G. Naumov and T. Bashkirowa, *Genetika* **279,** 1496 (1984).
178. N. A. Khan and R. H. Haynes, *MGG* **118,** 279 (1972).

179. A. M. A. ten Berge, Ph.D. thesis. University of Utrecht, Utrecht, The Netherlands, 1973.
180. H. Okada and H. O. Halvorson, *BBA* **82,** 538 (1964).
181. A. Alonso and A. Kotyk, *Folia Microbiol.* **23,** 118 (1978).
182. H. Okada and H. O. Halvorson, *BBA* **82,** 547 (1964).
183. P. Houtera and Lovgren, *J. Inst. Brew.* **81,** 309 (1975).
184. J. Ouwehand and R. van Wijk, *MGG* **117,** 30 (1972).
185. K. Matsumoto, I. Uno, A. Toh-e, T. Ishikawa and Y. Oshima, *JBact* **150,** 277 (1982).
186. K. Matsumoto, I. Uno, T. Ishikawa and Y. Oshima, *JBact* **156,** 898 (1983).
187. P. Eraso and J. M. Gancedo, *EJB* **141,** 195 (1984).
188. M. Carlson, B. C. Osmond and D. Botstein, *Genetics* **98,** 25 (1981).
189. J. L. Celenza and M. Carlson, *Science* **233,** 1175 (1986).
190. L. F. Bisson, L. Neigeborn, M. Carlson and D. G. Fraenkel, *JBact* **169,** 1656 (1987).
191. J. L. Celenza, L. Marshall-Carlson and M. Carlson, *PNAS* **85,** 2130 (1988).
192. K. D. Entian, *MGG* **178,** 633 (1980).
193. K. D. Entian and K. V. Froheich, *JBact* **158,** 29 (1984).
194. K. D. Entian, F. Hieberg, H. Opitz and D. Mecke, *MCBiol* **5,** 3035 (1985).
195. J. L. Celenza and M. Carlson, *MCBiol* **4,** 49 (1984).
196. M. Carlson and D. Botstein, *Cell* **28,** 145 (1982).
197. M. Carlson, B. C. Osmond, L. Neigeborn and D. Botstein, *Genetics* **107,** 19 (1984).
198. J. Schultz and M. Carlson, *MCBiol* **7,** 3637 (1987).
199. I. Witt, R. Kronau and H. Holzer, *BBA* **118,** 522 (1966).
200. C. Gancedo and J. M. Gancedo, *EJB* **148,** 593 (1985).
201. J. M. Siverio, M. D. Valdes-Hevia and G. Gancedo, *FEBS Lett.* **194,** 39 (1986).
202. L. Neigeborn and M. Carlson, *Genetics* **108,** 845 (1984).
203. K. D. Entian and F. K. Zimmermann, *JBact* **151,** 1123 (1982).
204. H. Ma and D. Botstein, *MCBiol* **6,** 4046 (1986).
205. J. L. Celenza, L. Marshall-Carlson and M. Carlson, *PNAS* **85,** 2130 (1988).
206. E. Abrams, L. Neigeborn and M. Carlson, *MCBiol* **6,** 3643 (1986).
207. N. A. Khan, *MGG* **186,** 40 (1982).
208. A. A. Algeri, L. Bianchi, A. M. Viola, P. P. Puglisi and N. Marmiroli, *Genetics* **97,** 27 (1981).
209. I. Ferrero, R. Rambaldelli, A. M. Genga, C. Donnini and P. P. Puglisi, *Curr. Genet.* **8,** 407 (1984).
210. N. A. Khan, *Curr. Genet.* **10,** 111 (1985).
211. L. Neigeborn and M. Carlson, *Genetics* **115,** 247 (1987).
212. K. D. Entian, *MGG* **179,** 169 (1980).
213. D. H. J. Schamhart, A. M. A. ten Berge and K. W. van de Poll, *JBact* **121,** 747 (1975).
214. J. Ma and M. Ptashne, *Cell* **51,** 113 (1987).

Index

A

B

C

D

E

F

N

O

P